Industrial powder coating

Judith Pietschmann

Industrial powder coating

Basics, Methods, Practical Application

Springer

Judith Pietschmann
Solothurn, Switzerland

ISBN 978-3-658-37591-1 ISBN 978-3-658-37592-8 (eBook)
https://doi.org/10.1007/978-3-658-37592-8

The translation was done with the help of artificial intelligence (machine translation by the service DeepL.com). A subsequent human revision was done primarily in terms of content.

This Springer imprint is published by the registered company Springer Fachmedien Wiesbaden GmbH, part of Springer Nature.
The registered company address is: Abraham-Lincoln-Str. 46, 65189 Wiesbaden, Germany

Preface

The impetus for the development and application of powder coatings was, as so often, society's ecological and economic requirements. In the USA, it was Rules 66 that for the first time required environmental aspects to be taken into account in the coating process. Later, similar regulations were introduced in many industrialised countries.

The first developments of powder coatings were made in the 1950s. After initial restraint in industrial application, a true triumphal procession followed a short time later. Today, powder coating is established in many industries. When it comes to opening up new markets and, above all, establishing itself in the market, powder coating often has a hard time. The reasons for this include the curing conditions and curing times.

The most important application areas of powder coating technology are the automotive supply industry, construction, mechanical engineering, the furniture industry, and the huge market of the household appliance industry. Driving factors are both environmental regulations such as the EU VOC Directive as well as the industry's demand for cost reduction, waste reduction, and the improvement of computer-aided process automation. Globally, powder coatings account for approximately 7% of all coating systems produced.

Developments in powder coatings are focused on significantly reducing curing temperatures and fast curing times, as well as developing systems for new base materials and special applications.

The topics of this technical book cover the complete powder coating process. Starting with the coating material, its production, and application, with practical tips in case of system malfunctions; continuing on to hanger and conveyor technology, baking or curing of coatings, and cleaning and pretreatment of metals; and concluding with measuring and testing technology, as well as detailed information, descriptions, and possible solutions for defects in the powder coating film.

In the fifth edition, the chapters have been revised editorially and in terms of content and adapted to the current state of the art. Chap. 2 has been revised in the area of systems engineering. Chap. 6 is new in terms of content, with examples of developments and areas of application. The new Chap. 10 focuses on environmental and energy aspects in the manufacture and application of powder coatings.

The book is aimed at people who enjoy powder coating and want to deepen their knowledge in the field of surface technology.

I would like to thank all those who have contributed to the success of this book in various ways. My special thanks go to Sabine Molitor, Thomas Gohmann, Marco Capizzi Hendrik Fliedner, Lars Sebralla, Michael Topp, Roman Mlakar, Nils Reinke, Anton Ritter, Frank Berg, and Michael Seeger. I would also like to thank the mechanical engineering editorial staff at Springer Verlag, Ellen-Susanne Klabunde and Thomas Zipsner, for their constructive cooperation and editing support. I would also like to thank Jochen Kecht for his suggestions for the fifth edition.

I welcome your feedback, because despite careful research, improvements and more up-to-date information will be found. Suggestions are always welcome and will be considered in the sixth edition.

Solothurn, Switzerland
March 2019

Judith Pietschmann

Contents

History of Powder Coating

The driving factors behind the development and introduction of powder coating can be traced back to the ecological and petroleum resource constraints of the late 1960s and early 1970s. In the USA, it was "Rule 66" in 1966 which for the first time demanded that environmental aspects be taken into account in painting, before this was followed in Germany by the Federal Immission Control Act (Bundes-Immissionsschutz-Gesetz – BimschG) in 1974 and the Technical Instructions on Air Quality Control (TA Luft) in 1986.

Coatings with powdery, fusible coating compounds were already being applied in 1940 with thermoplastic powders. The process itself would be called "flame spraying" today. In 1952, the so-called whirl sintering process was developed by E. Gemmer (Knapsack AG, Frankfurt/Main), through which larger quantities of powder were used for the first time in the coating sector [1]. The applications were initially limited to electrical insulation and pipe coating, with layer thicknesses in the range of 200 to 300 μm. Melting of the protective film onto sufficiently preheated workpieces occurs during immersion in the fluidised bed of a plastic powder. The powder material was initially polyethylene, later followed by other thermoplastics such as polyamide and PVC. It was not until the early 1960s that a thermosetting material came onto the market in the form of epoxy resins. It was the Bosch company that developed the basic type of epoxy resin powder in its search for a suitable electrical insulating material [2]. These were so-called "long-life powders", which required 20 to 30 min at temperatures of 200 °C to cross-link. The suitability of this completely solvent-free coating material also for decorative purposes was so obvious that only a suitable application method and finer grinding for thinner coating thicknesses were missing. With the electrostatic spray guns from the SAMES company, which proposed the term "samesing" for electrostatic application, this hurdle was also cleared in the mid-1960s. The spray gun and high-voltage generator were closely based on the electrostatic spray guns for liquid paint, and the system was complete with a powder container in the form of a paint pressure vessel [3]. The powder coating adhered to cold workpieces with the same electrostatic forces for a sufficiently long time, that is, at least until subsequent melting and curing. Initially, people were disappointed with this application method because the application efficiency, which was expected to be as high as with liquid paint, was much lower due to pneumatic powder conveying. It was not until the possibility of recovering the overspray directly, which was completely new in the field of painting, that powder painting was able to get off to a successful start, as the overall efficiency was now available for evaluation, taking into account the powder cycle. This was at the end of 1966. Precursors of an electrostatic application of fine particles had already existed in electrostatic smoking and the dusting of cigars [4].

The next development step was to supplement the epoxy powders with non-yellowing epoxy-polyester hybrid powders. Thus, the first breakthrough was made in Europe in 1968, after these powders with shorter curing times and constant qualities became available in larger quantities and, at the same time, the electrostatic application equipment was further improved and appeared increasingly suitable for large-scale plants [5]. However, the development of polyester resins containing carboxyl groups with significantly reduced resistance to yellowing was not yet a solution with regard to weather resistance [6]. The number of powder coating plants in Germany alone increased from 4 in 1966 to 51 in 1970 [7]. At the beginning of the 1970s, a hardener was found in 1,3,5-triglycidyl isocyanurate, or TGIC for short, which made it possible to produce a powder coating system with a wide variety of qualities for the widest range of applications over the next two decades. Step by step, reactivity, flow, flexibility, and excellent weather resistance were achieved. Alternatives were polyurethane systems, initially with caprolactam as a blocking agent. Acrylic powders were also available, but were unable to gain a foothold in Europe at the time due to incompatibility in manufacture and processing with other powder materials. The need for separate plants hindered the use of this powder system in Europe. It was only used in Japan and the USA.

The epoxy systems lost importance but remained indispensable for the interior and especially because of their chemical resistance to alkalis and acids.

The largest volume increase in powder consumption was achieved in 1978/1979 [6]. The following binder systems were available at that time: epoxy, polyester/epoxy, polyester/TGIC, polyurethane, and acrylic. A distinction was made between powder coatings for interior use and qualities for weather-resistant and chalking-resistant exterior use.

In 1972, patents for triboelectric charging of the powder demonstrated a further possibility for electrostatic powder application in addition to charging by corona in the high-voltage field. Patents were granted for the material polytetrafluoroethylene (PTFE polytetrafluoroethylene) as a friction partner in [7] and the design solution of the charging unit with annular gap in [8] and later, in 1980, for charging in hoses and spraying from finger nozzles in [9]. However, triboelectric charging requires a special chemical equipment of the powder, which was given from the outset in the case of epoxy powders, but had to be created first in the case of polyester and hybrid powders with easily electron-donating molecular groups during powder formulation.

At the beginning of the 1970s, a flood of ideas began to propose completely different solutions for powder spraying equipment in place of spray guns. They were referred to as "second generation powder sprayers" [10]. These included, for example, the ribbon atomiser introduced by AEG in 1970. A perforated plastic belt, which takes up the powder from a fluidised powder supply, carries it upwards and air nozzles blow the powder through the holes in the perforation, through a large number of parallel corona wires and onto the workpieces to be coated. Another solution was offered by Mueller with the powder jet, a system of individual nozzles coupled to one another, each of which is fed directly, that is, without a hose line, from its own fluid container as a powder supply. A similar solution without a powder hose was the powder ramp introduced in France by the Somip company in 1971. Several manufacturers propagated the rotating disc atomiser and, in a variation, the powder centrifugal wheel, in order to support uniform powder distribution with centrifugal force. Only the powder disc survived until a few years ago, in an example for coating washing machines at Bosch in Giengen. At the same time, GEMA in Switzerland achieved a leap forward in the development of corona guns by succeeding in accommodating the high-voltage transformer and the DC cascade in the gun itself, thus dispensing with the "stiff" high-voltage cable [11]. This high-voltage generation was made possible by progress in the miniaturisation of the components – capacitors and rectifiers for television technology – and the dielectric strength of cast-resin insulation.

The use of high voltage involves hazards, which on the one hand lie in the approach or contact of live or highly charged system parts and on the other hand in the possible explosion and fire hazard of ignitable powder-air mixtures due to spark discharges. Consequently, it was necessary to prevent these dangers with appropriate safety regulations. Thus, the first guideline entitled "Powder coating" was published by the Employer's Liability Insurance Association as ZH 1/444 with an issue date of 10.1971 and the stipulation that the requirements of the accident prevention regulation "Paint spraying, dipping and painting work" (VGB 23) should be applied mutatis mutandis due to the similarity of the work process to paint spraying. One requirement that could not be fulfilled was that for grounding control, which required the conveyor to be stopped if a ground leakage resistance greater than 10 kOhm was measured when a workpiece or hanger entered the conveyor. It took sound arguments before this value was corrected to 1 MOhm years later. Meanwhile the cabin or spray stand with forced ventilation became a necessary prerequisite. In addition, the German Engineering Federation (Verband Deutscher Maschinen- und Anlagenbau – VDMA) imposed its own guidelines for compliance in March 1974 with the VDMA standard sheet 24 371, which was based on the results of experimental tests with regard to explosion protection. Since October 1977, electrostatic manual testing equipment has had to undergo a type test in accordance with DIN 57 745 (VDE 0745) to ensure that this equipment is safe to touch and that it does not represent an ignition source for powder-air mixtures. To counter possible damage to health, the European Committee for Coating Powders of the CEPE (European Federation of Paint, Printing Ink and Artists' Colours Manufacturers' Associations) arranged for toxicological tests to be carried out with various coating powders, based on several years of checks on employees in powder production. The results are summarised in a brochure with recommendations for hazard prevention [12]. The brochure was first published in 1985 and is now available in several reprints.

Today, powder coatings have the reputation of being an environmentally friendly alternative to solvent-based coatings. Above all, legislation and requirements for VOC reduction have contributed to the switch to solvent-free painting with powder coatings in new planning, conversion, or due to internal environmental guidelines. In recent years, developments in the field of plant technology have concentrated on increasing efficiency, optimising colour changes, and the time required. A major topic is automation, and here the focus is primarily on control and operation.

Today, powder coatings have been able to win over market segments that were considered classic wet paint segments, such as agricultural and construction machinery weighing up to several tons.

Literature on the History of Powder Coating

The history of powder coating was kindly compiled by Prof. Dr.-Ing. habil. W. Kleber.

1. Gemmer, E.: Plastics Review 7 (1960)
2. Sträter, F.-J.: Stand der Pulverentwicklung. In: Reports of the IV. Tag. "Elektrostatisches und Elektrophoretisches Beschichten", HfV Dresden, pp. 70–71 (1969)
3. Auerbach, D.: Development of applicators for electrostatic plastic coating. Plaste Kautsch. **14**, 34–38 (1967)
4. Neubert, U.: Elektrostatik in der Technik, pp. 150–151. Oldenbourg, Munich (1954).
5. Gunia, G.: Der Trend zur Automatik, EPS-Bilanz 72. documentation of the journal JOT. mi-Publikationsgesellschaft, Munich (1972)
6. Chemism, Production and Technological Properties of Powder Coatings, Pulverlack 73rd Conference Proceedings Powder Coatings Conference, Hamburg (1973)
7. Noll, G.: 20 Jahre Pulverlacke. Fachtagung Pulverlack Hamburg, Conference Proceedings (1985)
8. Kleber, W.: Triboelectric powder charging. Metal Surface. **50**(7), 564–567 (1996)
9. Ruud, J.: Powder spraying device. Patent specification DE 3 100 002 Swedish priority 04.01.80
10. Gebhardt, O.: Powder coating plants of the second generation. In: Reports of the V. Conference "Electrostatic and Electrophoretic Coating", HfV Dresden, pp. 144–149 (1972)
11. Braun, F.: Electrostatic powder coating gun with built-in high-voltage generator; Electrostatic charges. In: DECHEMA Monographs, Vol. 72, pp. 281–292. Verlag Chemie, Weinheim (1974)
12. anonymous: Results of experimental toxicology study on thermosetting powder coatings. European Committee of Paint, Printing and Artists' Colours Manufacturers' Associations, Brussels (1985)

1 Powder Coatings

Powder coatings are coating materials which, after application to the substrates, are melted or chemically cross-linked by the action of heat, resulting in closed, well-adherent coatings. Like most surface treatments, they have two main functions: a decorative and/or a functional one (Table 1.1).

1.1 Different Types of Powder Coating

In general, powder coatings are composed of:

- Binders (resins, hardeners, accelerators)
- Pigments and dyes
- Fillers (extenders)
- Additives

Based on their physical and chemical behaviour, a distinction is made between thermoplastics and thermosets in the case of binders/film formers.

1.1.1 Film Former/Binder

Table 1.2 gives an overview of the most important film former systems for powder coatings today. For chemically curing film formers, the ratios of glass transition temperature Tg, average molar mass Mn, average functionality fn and reactivity must be precisely balanced. It must be possible to melt the premixed powder coating in the extruder without any noticeable crosslinking. In addition, the powder coating material must not sinter too much

J. Pietschmann, *Industrial powder coating*,
https://doi.org/10.1007/978-3-658-37592-8_1

Table 1.1 Functions of a powder coating layer

Decoration	Protection
Color	Mechanical loads
Gloss	Corrosion resistance
Course/structure	Weathering
	Chemical resistance
	Electrical insulation

Table 1.2 Binder systems for powder coatings [1]

Common abbreviation	Film former		Area of application
Thermoplastic systems			
PE	Polyethylene (LDPE, LLDPE, HDPE)		Inside
PA	Polyamide 11 or 12		Inside
SP	Polyester		Inside
EVOH	Ethylene-vinyl alcohol copolymer		Inside and outside
PVC	Polyvinyl chloride (and copolymers)		Inside and outside
PVDF	Polyvinylidene fluoride (and copolymers)		Inside and outside
Duromer systems			
	Resin	*Hardener*	
EP	Epoxy resin	Phenolic hardener Imidazoline derivatives Anhydride adducts	Inside
EP-DCD	Epoxy resin	Modified dicyandiamide	Inside
EP-SP	COOH polyester resin	Epoxy resin	Inside
SP-TGIC	COOH polyester resin	TGIC	Inside and outside
SP-HAA	COOH polyester resin	Hydroxyalkylamide hardener	Inside and outside
SP-GE	COOH polyester resin	Aromatic glycidyl esters	Inside and outside
SP-PUR	OH polyester resin	Isocyanate adduct Blocked uretdiones	Inside and outside
AC-DDA	Glycidyl acrylate resin	Dodecanedicarboxylic acid	Inside and outside
AC-PUR	OH acrylic resin	Isocyanate adduct	Inside and outside

during storage, but should melt during baking in such a way that it runs to the desired film before the expected film properties are then achieved through crosslinking. The resins are therefore usually amorphous polymers with a sufficiently high glass transition temperature (of at least 40 to 50 °C) and with a molar mass of several thousand g/mol, so that sintering during storage is suppressed.

1.1.2 Thermoplastic Binders

The reversibly melting thermoplastics form a film on the substrate by heating above the melting point, which solidifies into a non-porous coating after cooling. Today, the following requirements are placed on thermoplastic powder coating materials:

- solid at room temperature (T = 25 °C)
- storage stable at room temperature (25 °C) and elevated temperatures up to at least 40 °C without lump formation
- meltable without decomposition
- melting temperature must not be too high
- melt viscosity must allow film formation
- good adhesion to various substrates
- coatings must be colourable

Since thermoplastic binders do not crosslink during melting, the macromolecule and thus the molecular weight must already be formed before melting. As a result, thermoplastic binders have relatively high melt viscosities at the application temperatures, which then lead to high coating thicknesses, e.g. >100 µm. Further disadvantages of thermoplastic powder coatings are, the limited possibility of exposure to heat (softening) due to the softening range of the binder, as well as more complex grinding processes, since due to the ductility of thermoplastics, grinding must take place at lower temperatures compared to thermosets. In addition to the more complex grinding processes, the higher layer thicknesses compared to thermosets also lead to higher costs. Another disadvantage is the use of adhesion promoters, which is necessary in most cases. The systems are not solvent-resistant and have low temperature stability.

1.1.3 Thermoset Binder

Thermosets, also commonly referred to as thermosetting powder coatings, are chemically crosslinked by the action of heat after sintering onto the object to be coated, whereby they lose their initial thermoplastic properties and are not remelted by subsequent heat action. Chemically, two types of crosslinking reactions are possible for thermosetting powder coatings: polyaddition and polycondensation.

Polyaddition is the formation of polymers or networks by repeated addition of di- or polyfunctional monomers or low molecular weight building blocks without the elimination of volatile substances. In polycondensation, polymers are formed by condensation reactions between di- or higher-functional monomers or low-molecular-weight building blocks with the elimination of volatile substances, such as water or alcohol. Another type of cross-linking known in polymer chemistry, the so-called polymerisation reaction, only plays a role in the powder coatings industry with UV-curing powder coatings.

The following requirements are placed on thermoset powder coating materials today:

- solid at room temperature (melting point > approx. 65 °C)
- grinds well at room temperature
- melting without separation
- melting temperature must not be too high

- low melt viscosity in the temperature range usual for curing
- physically and chemically stable during storage up to at least 40 °C without clumping, chemical cross-linking or deterioration of the flow properties, normally stable during storage for 28 days, exception: low temperature powder coatings
- sufficient functionality to crosslink to thermosets in combination with appropriate crosslinkers
- good adhesion to various materials without adhesion promoter
- possible for pigmentation
- Volume resistivity 10^{10}–10^{16} Ωcm
- relative dielectric constant approx. 2–6.

The advantages of thermosetting powder coatings over thermoplastic ones are the relatively low curing temperatures (approx. 120–200 °C), a low melt viscosity in the temperature range usual for curing, and the resulting good substrate wetting and good surface properties. Furthermore, the brittle properties in the non-crosslinked state allow simple and economical grinding processes. There have already been many efforts to lower the curing temperature for thermoset powder coatings. However, a major obstacle to these efforts is the melting point required to maintain sufficient storage stability and the low melt viscosity required for substrate wetting and to achieve good flow properties. Table 1.3 shows an overview of the thermosetting powder coating systems used today which are based on an addition reaction. The systems based on a condensation reaction are summarized in Table 1.4.

Bisphenol A has been included in the REACH-SVHC candidate list since January 2018. Therefore, there is an information obligation within the supply chain. It has to be disclosed if the articles exceed a concentration of 0.1% [3].

Table 1.3 Powder coating systems with a polyaddition reaction during crosslinking [2]

Resin component	Hardener component
Epoxy resin (bisphenol A type)	Dicycandiamide: Accelerated or modified; subst. imidazoles; BF3 complexes; polycarboxylic acid anhydrides; acid polyesters; polyphenols
Polyester resins (containing carboxyl groups)	Epoxy resins (bisphenol A type); triglycidyl isocyanurate (TGIC); PES containing hydroxyl groups; oxazolines; aromatic glycidyl esters
Polyester resins (containing hydroxyl groups)	Capped polyisocyanates; carboxylic acid anhydrides; modified melamine or urea resins
Polyacrylates (containing epoxy groups)	Carboxylic acid anhydrides; dicarboxylic acids; acid polyesters; acid acrylates; normal EP crosslinkers
Polyacrylates (containing hydroxyl groups)	Carboxylic anhydrides; capped polyisocyanates; acid acrylates or polyesters; modified melamines or urea resins
Polyacrylates (containing carboxyl groups)	Triglycidyl isocyanurate (TGIC); acrylates containing epoxy groups; acrylates or polyesters containing hydroxyl groups; oxazolines

Table 1.4 Powder coating systems with a polycondensation reaction during crosslinking [2]

Resin component	Hardener component
Polyester resins (containing carboxyl groups)	Hydroxyalkylamide (Primid XL 552), cleavage product: Water
Polyester resins (containing hydroxyl groups)	Tetramethoxymethyl glycoluril (Powderlink 1174), elimination product: methanol

1.1.4 Epoxides

The epoxy resins used in the manufacture of powder coatings are mainly solid types with a melting range of between 60 and approx. 90 °C according to Kofler. The most important type of epoxy resins are bisphenol A (2,2-bis(4-hydroxyphenylpropane) and epichlorohydrin (1-chloro-2,3-epoxypropane) condensed in the presence of sodium hydroxide solution. Figure 1.1 shows the essential molecular structure.

The properties of epoxy resin powder coatings are decisively influenced by the hardener component used. Modified or substituted dicyandiamides, polyphenols or also low molecular esters of polycarboxylic acids are preferably used. The curing of epoxy resins with dicyandiamide, especially also with basic accelerated or modified types, is primarily a polyaddition reaction with the formation of N-alkylcyanoguanidines (Fig. 1.2).

In addition to this polyaddition reaction, especially at higher temperatures, an anionic catalyzed polymerization reaction can take place between epoxide groups still present with the formation of ether bridges as well as an addition of OH groups to the nitrile triple bond. Crosslinking with substituted imidazolines proceeds in a similar manner, i.e. via an addition and anionic catalyzed polymerization reaction. Crosslinking with carboxylic acid anhydrides proceeds in two stages. In the first stage, semi esters are formed by addition of an anhydride group to a hydroxyl group, and the resulting carboxyl group then adds to an epoxide group to form an ester and hydroxyl group. Due to the carboxyl groups present during the cross-linking reaction and also relatively high temperatures, a cationic polymerization of epoxide groups as well as the addition of hydroxyl to epoxide groups also occur as side reactions.

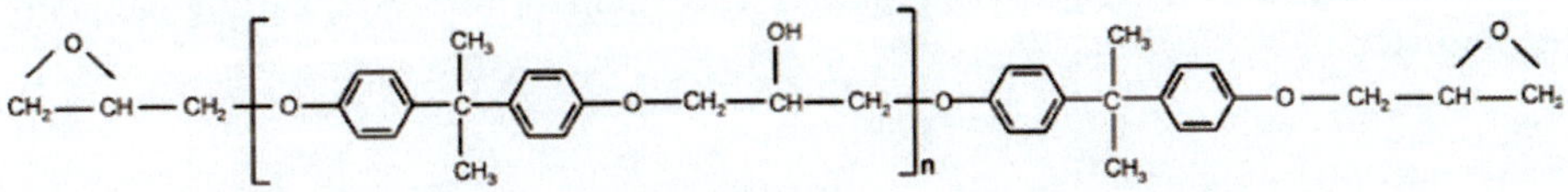

Fig. 1.1 Idealized structure of a Bis-A resin [1]

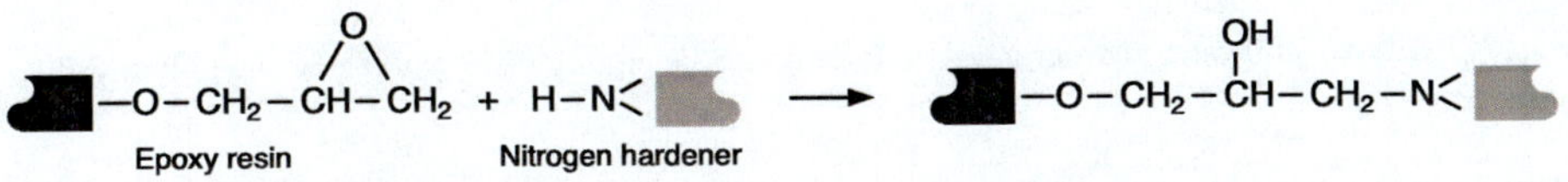

Fig. 1.2 Epoxy coating powder; reaction of an epoxy resin with a nitrogen hardener [4]

The crosslinking with polyphenols essentially proceeds via a polyaddition reaction. Especially the so-called "low temperature hardeners"are based on polyphenols.

In addition to very good adhesion to a wide variety of substrates and due to their very low melt viscosity, epoxy powder coatings have excellent flow and good technical coating properties. The curing conditions range from 120 °C/20 min to 200 °C/5 min.

Other positive properties include good resistance to solvents, acids and alkalis. Negative aspects are the poor overbaking (yellowing) due to the aromatic character and the chalking under UV exposure.

Epoxy powder coatings are therefore nowadays used almost exclusively in the functional sector, e.g. for automotive parts, in the electrical and electronics industry, for fittings and reinforcing iron, and for coating pipes, pipelines, etc.

1.1.5 Hybrid

In the production of epoxy resin/polyester mixed powder coatings, so-called hybrids, suitable polyester resins are used which contain terminal, free carboxyl groups in the molecule which cause spatial crosslinking via addition to epoxy groups. In the SP-EP system (hybrid system), COOH-functional polyester resins with a molar mass of several thousand g/mol are used. Since their molar mass is greater than that of the epoxy resins used, it is referred to here as the parent resin. The mixing ratio varies from 60: 40 to 10: 90 of epoxy resin to polyester. The exact mixing ratio is determined by the specific customer requirements and areas of application. The mixing ratios predominantly used today are 50: 50 EP: PES up to 30: 70 EP: PES. The background to this is the rising prices for epoxy, so that the trend today is towards higher polyester proportions (Fig. 1.3).

Hybrids have similar properties to epoxy powder coatings, but better yellowing stability during baking and less tendency to chalking under UV exposure. A disadvantage compared to epoxy powder coatings is the poorer solvent resistance. The curing range is between 130 °C/15 min and 200 °C/5 min. The areas of application are particularly in the decorative sector, shop and shelf construction, metal office furniture, household appliances, refrigeration, garden and camping furniture, ceiling elements and radiators.

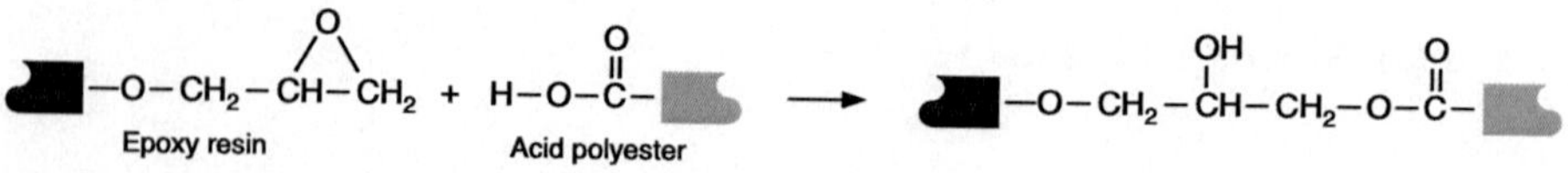

Fig. 1.3 Epoxy-polyester powder coating. Reaction of an epoxy resin with an acidic polyester [4]

1.1.6 Polyester/TGIC

In the past, polyester resins containing free carboxyl groups in combination with triglycidyl isocyanurate (TGIC) have proved successful in the manufacture of weather-resistant powder coatings. TGIC reacts with the carboxyl groups of the polyester resin via its three reactive epoxy groups, forming a three-dimensional network. Polyester/TGIC powder coatings are characterized by very good weathering and chalking resistance. They have very good over-bake stability and technical paint properties. The curing conditions are between 160 °C/15 min and 200 °C/5 min. However, the solvent resistance is lower compared to epoxies and hybrids.

Today, TGIC polyester systems are no longer used in Europe. The reasons for this were the labelling of TGIC as "toxic" and the obligation to also label powder coatings with a content of ≥0.1% TGIC in the formulation. TGIC is mildly irritating to the skin and highly irritating to the eye. Like many other glycidyl compounds, TGIC also shows a skin sensitizing effect.

1.1.7 Polyester/Hydroxyalkylamide

As an alternative to TGIC, the class of β-hydroxyalkylamides has been available since 1990, see Table 1.5. In contrast to the other crosslinking reactions, no epoxide or isocyanate groups are involved in the reaction here. The linkage takes place via particularly reactive hydroxyl groups. Hydroxyalkylamides are toxicologically completely harmless and the OH groups in the ß-position to the amide group have a high reactivity from approx. 160 °C onwards. Below this temperature, polyester/hydroxyalkylamide powder coatings are characterized by very good chemical storage stability. The crosslinking between carboxyl and ß-hydroxyalkyl groups is a condensation reaction with elimination of <1% water (Figs. 1.4 and 1.5).

By adding suitable degassing additives to remove released reaction water, sufficiently high film thicknesses can also be achieved with this powder coating system. In addition to very good coating properties (curing conditions 160 °C/20 min to 200 °C/5 min), polyester/hydroxyalkylamide powder coatings also have excellent weathering and chalking resistance. They can be used as an alternative to polyester and polyurethane powder coatings in exterior applications.

1.1.8 Aromatic Glycidyl Esters

The crosslinking reaction of the glycidyl esters with the carboxyl groups of the polyester resins is comparable to that of the TGIC. A polyaddition reaction takes place without cleavage of by-products. Glycidyl esters are comparable to TGIC in many properties, such

Table 1.5 Overview of TGIC alternatives [6]

Labelling in the EU	Typical curing conditions	hardening as of	Typical mixing ratios with polyester, acid value 33	Spin-off products	Reaction principle	Reaction partner	Reactive group	Manufacturer	Product name	Chemical class
–	10 min at 180 °C	170 °C	95:5 and 94.5:5.5 respectively.	<1% water	Polycondensation	Carboxyl polyester	Hydroxyl-	EMS-Primid	Primid XL-552; Primid QM-1260	β-Hydroxyalkylamide
Depending on quantity, hardness content ≥3.0%: H317, H318, H360 F, H412 Pictogram: GHS08 + GHS07 + GHS05 Hardener content ≥1%: H317, H318, H360F, H412 Pictogram: GHS08 + GHS07	15 min 180 °C	160 °C	91:9	–	Polyaddition	Carboxyl polyester	Epoxy	Huntsman	Araldite PT 910/912	Aromatic glycidyl esters
–[c]	10–15 min 200 °C	180 °C	88:12 to 80:20[b]	<5% Caprolactam[a]	Polyaddition	Hydroxil polyester	Isocyanate-	Bayer, Degussa-Hüls	Crelan Vestagon	Polyurethane
–[c]	10–15 min 160 °C	130 °C	70–80% epoxy acrylate 20–30% dicarboxylic acid anhydride	–	Polyaddition	Dicarboxylic acids or anhydrides	Epoxy	Solutia, Anderson	Synthacryl	Epoxyacrylate
Attention H315; 319	15 min 180 °C	160 °C	90:10	–	Polyaddition	Carboxyl polyester	Epoxy	Nissan	MT-239	Methyl-substituted TGIC
Danger H301; 302; 311; 317;318; 331; 340; 372	10 min 180 °C	150 °C	93:7	–	Polyaddition	Carboxyl polyester	Epoxy	Vantico, Nissan	Araldite PT 810 Tepic	Triglycidyl isocyanurate (TGIC)

[a]Internblocked types are free of cracking products
[b]With hydroxyl polyesters, hydroxyl value 42
[c]Polymers do not need to be notified

Fig. 1.4 Hydroxyalkylamide, Primid® XL 552 [5]

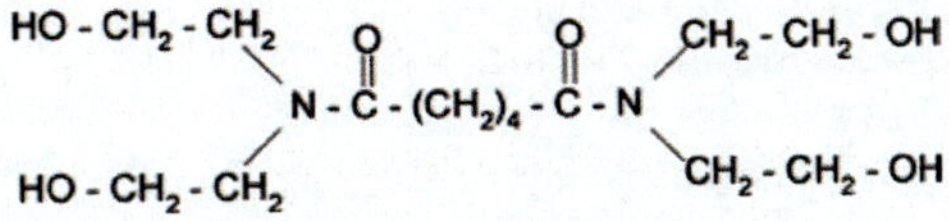

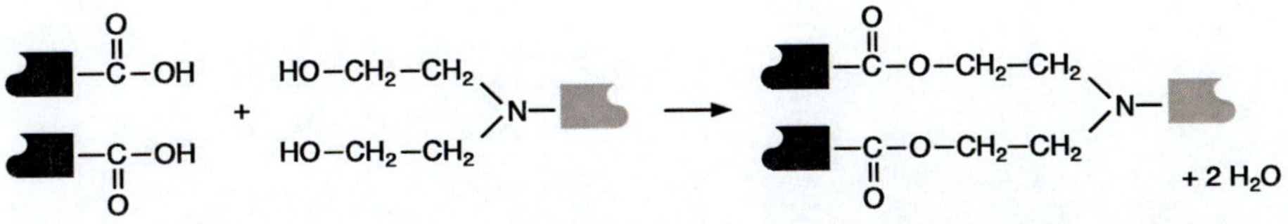

Fig. 1.5 Polyester containing carboxyl groups cured with ß-hydroxyalkylamide [4]

as thermostability, weather resistance and gas oven resistance. The reactivity of Araldite PT 910® is slightly below that of TGIC [6].

The limited storage stability of the coatings and the poorer chemical resistance should be mentioned as disadvantages. Powder coatings with a content of ≥0.3% PT 910® are subject to labelling, see Table 1.5.

Coating systems with this type of hardener require a higher extrusion effort, since the polyester resins specially developed for this type of crosslinker have a higher melt viscosity.

To reduce the disadvantages of single TGIC alternatives, mixtures of β-hydroxyalkylamide and e.g. PT 910® are also used, see Fig. 1.6. Compared to pure β-hydroxyalkylamide systems, the maximum layer thickness can thus be increased without pinholes [6].

An advantage of aromatic glycidyl esters is that they can be catalyzed and the curing conditions can be decreased (Figs. 1.6 and 1.7).

1.1.9 Polyurethane

PUR powder coatings are based on free polyester resins containing hydroxyl groups, which are crosslinked with polyisocyanates by an addition reaction. Two types of isocyanate curing agents are currently available. Isophorone diamine (IPDI) capped with caprolactam and capped IPDI with urethdione structure without caprolactam. Aromatic isocyanates, e.g. based on methylenedianiline (MDI), have a high yellowing potential and are therefore usually not used. Relatively high temperatures are required to split off the caprolactam or to break up the urethdione structure. The isocyanate groups then react with the hydroxyl groups of the polyester to form urethane bonds.

Therefore, baking temperatures of at least 180 °C are required. A disadvantage of blocked systems is that the blocking reagent is split off during baking. Thus, the VOC content in powder coatings blocked with ε-caprolactam may well be 2 to 4%. Today, however, isocyanate prepolymers that have been dimerized to uretdiones are available, which naturally no longer release any cracking products during baking. In order to reduce the

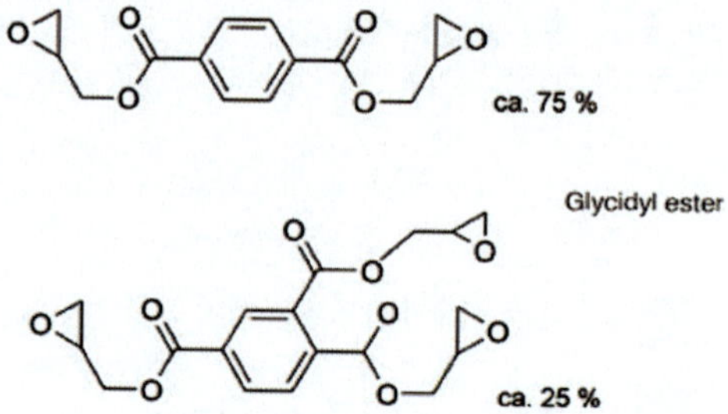

Fig. 1.6 Aromatic glycidyl esters, Araldite PT 910®. PT 910® is a mixture of di- and trifunctional diclycidyl esters in the ratio 75:25 [7]

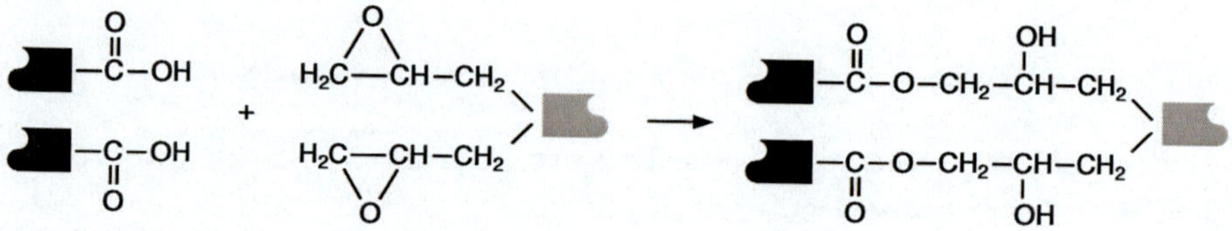

Fig. 1.7 Polyester containing carboxyl groups cured with glycidyl ester [4]

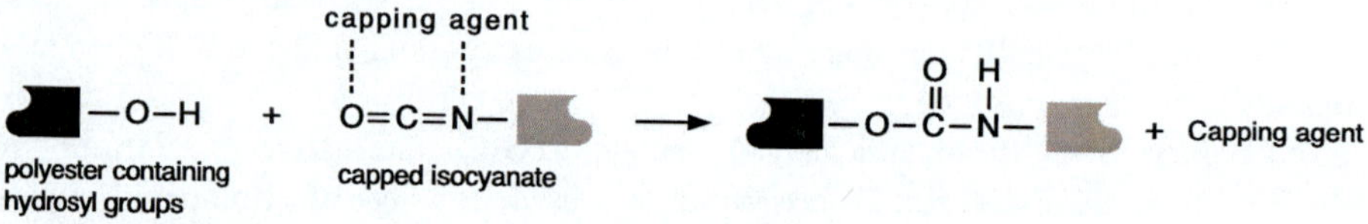

Fig. 1.8 Polyurethane powder coating. Reaction of a hydroxyl group-containing polyester with a capped isocyanate [4]

curing temperatures, these systems are catalyzed by dibutyltin dilaurate (DBTDL) or tertiary amines (Fig. 1.8).

Systems formulated with OH-functional polyesters flow better than those with COOH-functional ones, the SP-PUR systems usually have better flow than other powder coatings. Finally, AC-PUR powder coatings are characterized by their excellent weather resistance.

A disadvantage of caprolactam-capped IPDI, apart from the separation of the capping agent, is the insufficient edge coverage. Otherwise, PUR powder coatings have very good weathering and chalking resistance, combined with very good flow and coating properties. The field of application of PUR powder coatings is therefore congruent with that of polyester powder coatings.

1.1.10 Acrylates

The product group of weather-resistant acrylic powder coatings is based on acrylic resins that can be crosslinked with different hardeners. Dicarboxylic acids, anhydrides of dicarboxylic acids or isocyanates are used today. The other recipe components, as well as the manufacturing processes, differ little from conventional powder coatings.

For three reasons, people were rather hesitant about acrylic powder coatings at the beginning: The raw materials have a very high price, the finished powder coatings were not storage stable under common conditions and the compatibility with other powder coatings was not given. Improvements could be achieved here. The storage stability at max. 20 °C is adequate, the compatibility could be improved. Today, acrylates are less sensitive to contamination than, for example, hybrids. In general, the compatibility of the systems must be checked. The properties of the finished coating have a flow similar to solvent-based liquid coatings, very high brilliance and significantly greater weather stability.

The search for low-emission or emission-free coatings has drawn the attention of the automotive industry to acrylic powder coatings. With polyester or polyurethane powders, the properties and requirements of the standard coatings could not be achieved.

Advantages of acrylic systems:

- emission-free and low-waste generating coating
- leveling comparable to liquid coatings in the automotive industry
- very clear coating. High brilliance of the clear coats on metallic base coats.
- very good weather stability (5 years Florida)
- crack-free clear coatings
- stoving temperatures of 140 °C and below are possible

Disadvantages:

- price
- storage stability at room temperature limited, cooling necessary
- air-conditioned and purified supply air necessary
- compatibility: on other powder coatings effect like silicone. Separation from equipment with conventional powder coatings necessary.
- suitable plant technology necessary

The first user in the automotive industry was Harley-Davidson in the USA. In Germany, powder coatings were first used on the car body at BMW in Dingolfing, for serial painting on water- soluble basecoats. At Mercedes Benz in Rastatt, acrylic powder was used as a "powder slurry", i.e. dispersed in water and applied conventionally, and at Smart the steel body was coated with acrylic powder in metallic anthracite and in silver. Today, BMW and Smart use liquid clearcoats on the body again. Reasons for the "step back" were mainly the costs, but also the insufficient reparability and car wash resistance. Today, the most important areas of application for acrylics are bicycle frame painting and the aluminum wheel industry.

Acrylic resin is prepared by a polymerization of glycidyl methacrylate (GMA) and other monomer building blocks. The resulting glycidyl polyacrylate (GPA) is cured with

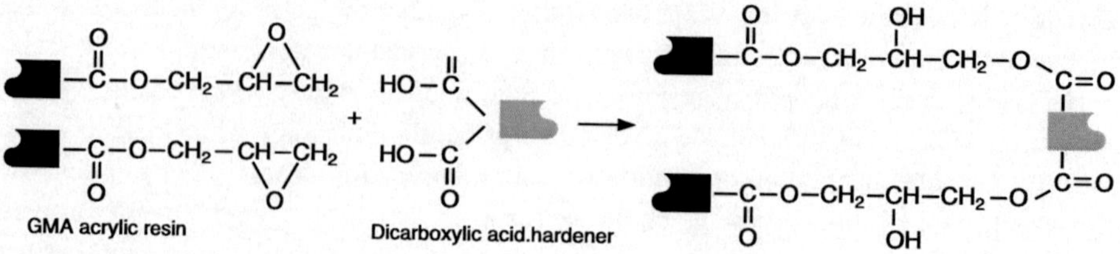

Fig. 1.9 Simplified illustration of the crosslinking of powder coatings based on glycidyl methacrylate (GMA) [4]

dicarboxylic acid anhydride in a polyaddition reaction (Fig. 1.9). These systems are characterized by excellent weathering resistance, high abrasion resistance, very good flow and excellent brilliance. However, they are so highly reactive (curing possible from T = 130 °C) that they are not stable in storage under common conditions. Cooling during transport and storage is necessary, as well as air conditioning of the coating plant is advisable.

If the crosslinking is realized via hydroxy/isocyanate reactions, the baking temperatures are at object temperatures of T > 180 °C due to the capping of the isocyanate. These systems are also more stable in storage. Another area of application is in the field of matt coatings. Polyester acrylates, COOH polyesters and GMA matt hardeners are frequently used here. These are relatively well compatible with glossy polyester variants.

1.1.11 Methyl-Substituted TGIC

An alternative in the field of crosslinkers is a product from Nissan, which is very similar to the structure of TGIC. The only difference lies in the methyl groups of the glycidyl residues. The technical properties therefore largely correspond to those of TGIC. The disadvantage is the low reactivity, but above all the manufacturer's labelling with danger and the corresponding H-phrases, see Table 1.5.

1.1.12 Additives

Additives for powder coatings include leveling agents, degassing agents, matting additives, waxes to influence surface hardness, structure and texture additives, catalysts and anti-caking agents.

Especially the anti-caking agents, aluminium oxides or pyrogenic silicas, are important for a smooth processability of the powder coatings at the customer (fluidization). Powder coatings have a relatively high surface tension in the molten state due to the absence of solvents. On insufficiently cleaned or pre-treated metal surfaces, this can lead to poor wetting, edge alignment and poor adhesion. For this reason, powder coatings often contain 1 to 2% leveling agents, usually polyacrylates, which at the same time reduce the tendency to crater formation. However, a too high dosage can lead to an orange peel effect. Many powder coatings also contain 0.1 to 1% benzoin. It serves as a degassing agent and also counteracts cratering. Heat stabilizers against yellowing, thermo yellowing caused by nitrous gases in directly heated ovens and stabilizers to improve adhesion to subsequent coats are also used. In clear coats, light protection combinations of HALS and UV absorbers are always used for exterior applications. Generally, additives are added in low concentrations, usually 0.1 to 1%, because their effect can turn into the opposite if overdosed and have negative effects on application or film properties. For example, an overdose of degassing additives leads to greater condensate formation in the oven and poorer printability and glueability of the coating surface.

1.1.13 Pigments

Pigments provide an opaque, colored impression of the coating. High demands are placed on them in powder coatings:

- thermal stability at the baking temperature
- increase in melt viscosity as low as possible
- no reaction with other powder coating components and
- stability against the shear forces occurring during extrusion and grinding

These criteria are best met by inorganic pigments. In the case of organic pigments, high-quality types must be used.

Although the pigment range for powder coatings is limited due to the high baking temperatures, almost all color shades can be produced. However, in the area of intense chromatic colors (e.g. strong reds and yellows), it is necessary to use high color and pure organic pigments, which are considerably more expensive, which also has an effect on the price of the finished powder coating. In the past, heavy metal-containing (inorganic) pigments based on lead and cadmium compounds were used for this purpose. However, they are no longer used today because of their toxicological properties. Since organic pigments are much more finely divided than inorganic pigments, greater effort is required during dispersion in order to break down the pigment agglomerates into their primary particle size. Inorganic pigments, mainly from the group of metal oxides or oxidic mixed-phase pigments of the rutile and spinel type, are easier to handle. This group of pigments is characterized by a coarser particle size and thus easier dispersibility. The color strength of this group is usually lower than that of organic pigments, but they have a greater scattering power, hiding power and a predominantly greater temperature resistance.

A special group are the functional pigments, which consists mainly of corrosion protection pigments. Here, it is not so much the achievable color impression that is the goal, but rather the protective effect for the substrate that counts. Zinc phosphates and other, actively acting types are used here.

Unlike pigments, dyes are soluble in the coating material. This results in a glazing, i.e. non-covering color impression. Coatings tinted with dyes can create an interesting effect, but this requires a flawless, as visible, substrate. However, the color impression or the intensity is strongly influenced by the layer thickness. The disadvantage of using dyes is the lower light and weather resistance.

The incorporation of gloss pigments is difficult with an extruder. The platelets are bent or broken in the process. Thus, effect pigments are either only physically mixed in or, in order to avoid segregation, fixed to the powder coating particles in a bonding process.

1.1.14 Fillers

The fillers used are mainly natural minerals: barite, feldspar, chalk or quartz sand [8].

Each filler has specific properties and thus also gives the powder coating film its characteristic properties. The fillers influence many macroscopic phenomena of the coating:

- uniform paint distribution
- Edge coverage
- Edge run-off behaviour
- Flexibility (cupping, bending)
- Impact resistance
- Gloss
- Chemical resistance
- Density, yield

1.2 Low Bake and Radiation Curing Systems

Radiation-curing coatings are coating systems that are activated and crosslinked by high-energy radiation. The formulation of the systems, the type and intensity of the radiation, and the arrangement of the radiation sources must be matched to the substrate to be coated. Most formulations for radiation-curable systems consist of a balanced mixture of [9]:

- oligomeric binders (resin)
- monomeric reaction partners (cross-linking component)
- photoinitiators (only for UV curing)
- Auxiliaries (these are not involved in networking)

Proven binders are unsaturated polyesters, epoxy resins or polyurethanes. One or more acrylic or methacrylic groups are attached to the ends of these base molecules. At such chain ends with a stable chemical double bond $CH_2 = CH$- the energy-rich rays attack. They generate free radicals and initiate a chemical reaction.

1.2.1 NT Powder Coatings

The LB (low bake) curing process corresponds to the well-known thermoreactive cross-linking process based on epoxy resin powder coating systems, which are characterized by a one-step melting and curing process. Melting and curing cannot be separated from each other. The energy input into the system takes place via convection heat and/or medium-wave IR radiation. The heat conduction always takes place via the coating material into the substrate. As a result of the relatively long curing time, it cannot be prevented that the mass of the substrate, even with MDF and plastics, is completely heated through (see Fig. 1.10).

Crosslinking of polyester systems at temperatures of 100–120 °C is not possible. Thus, only epoxy resin powder coatings are available for the NT range, with the known limited properties.

With specially selected epoxy resin and hardener components, a chemical crosslinking reaction is already possible at temperatures below 100 °C. The practical implementation in powder coating formulations is critical with regard to storage stability (avoidance of sticking of the powder particles at normal room temperature), which does not allow a further reduction of the molecular weight.

The NT powder coatings available today still require curing conditions of at least 120 to 130 °C for 10–20 min. The reasons for this are complex, the most important being the achievement of better adhesion, better flow and a more resistant surface. The application of this special NT powder coating on thermally sensitive substrates is a sensible option due to the comparatively low temperature balance for the entire melting and curing process.

Due to the well-known cross-linking process, almost all colours can be provided.

The chemistry for matting epoxy resin systems only works >160 °C cross-linking temperature, so that the provision of matt products (gloss <60) must currently still be

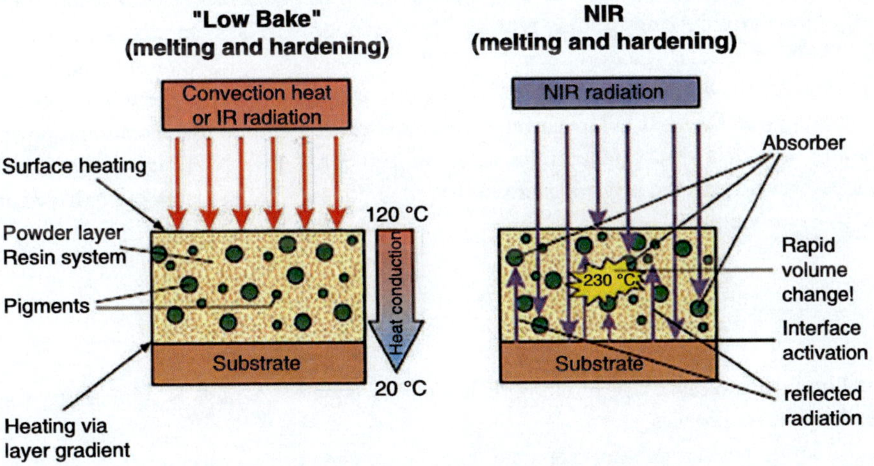

Fig. 1.10 Comparison of the curing principles "Low Bake (NT)" and NIR [10]

dispensed with. Epoxy resin systems are not light- or UV-resistant, which does not allow the outdoor use of these systems under direct sun and weather influence. Despite these limitations, NT powder coatings are already widely used for the very economical coating of MDF and plastics and meet the predominant decorative requirements of the previously used liquid coating systems.

With hybrid systems, temperatures of 140–160 °C are possible with a holding time of approx. 10–15 min. With acrylates, temperatures from 120 °C are possible today. Formulations that cure in the temperature range of approx. 150 °C are predominantly used. For polyester systems, an object temperature and holding time of 170 °C/15 min is the lowest curing limit today. Here, new developments in the resin sector are on the way which allow a further reduction of the temperature.

1.2.2 NIR Powder Coatings

Powder coatings for this curing can theoretically be formulated with all known thermoreactive binders based on epoxy and polyester resins or acrylate resins. However, since all these binders show transparency to NIR radiation (wavelengths between 0.75 and 1.2 µm), conversion of radiation to heat is not possible. To ensure this conversion, the powder coatings must have an appropriate absorption characteristic. Commercially available powder coatings are unsuitable for NIR curing because they do not automatically possess these properties. NIR powder coatings, on the other hand, can also be cured with the conventional curing ovens like commercial powder coatings.

NIR technology has not yet been able to establish itself on the market.

NIR-curable powder coatings are specially optimized for these applications and ensure that the maximum possible amount of energy can be utilized for the crosslinking process. The electromagnetic activation and the radiation-induced heat mechanism complete the melting and crosslinking process within a few seconds. The system works particularly effectively on metal surfaces. The portion of the radiation transmitted through the powder layer is reflected by the metal surface and leads to the use of increased energy density in the coating (see Fig. 1.10). The curing process, compared to that of conventional powder coatings, can thus be shortened by a factor of 100–1000 [10]. NIR radiation offers the shortest wavelength and highest energy infrared radiation with an extremely narrow band energy distribution.

1.2.3 UV Powder Coatings

The film formation and cross-linking of UV-curing powder coatings is subject to a completely different process.

UV-curing powder coatings separate the melting and curing processes (see Fig. 1.11). On the one hand, this requires specially coordinated process management, but on the other

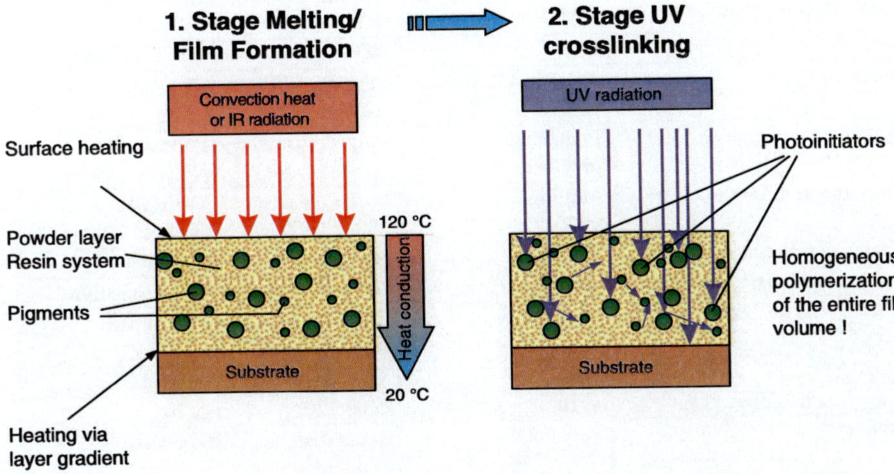

Fig. 1.11 Principle of UV curing of powder coatings [10]

hand it opens up major technological advantages when coating porous substrates, for example, which tend to release low-molecular and other volatile substances from the pores when subjected to higher thermal stress. Particularly important are MDF (medium density fibreboard), many plastics and also zinc, aluminium or magnesium die castings.

If the crosslinking process starts in thermoreactive powder coatings, this leads to molecular growth and inevitably to an increase in melt viscosity. Under these conditions, escaping gaseous substances from the substrate surface leave tiny channels (craters) in the coating film that can no longer close by themselves (pinholes, bubbles, craters by outgasing).

The situation is different when a UV powder coating is melted separately from the curing process. Here the viscosity of the melt does not change at a certain temperature as long as there is no UV light. This means that degassing can be influenced in a controlled manner.

It is also advantageous that a melting temperature of only approx. 100–120 °C for 0.5–2 min is required to achieve good flow. This also means that the substrate is only exposed to low thermal stress. Subsequently, the coated parts are irradiated with UV light for approx. 5–20 s and the coating is completely crosslinked.

UV powder coatings can be formulated on the basis of polyester, epoxy or polyurethane acrylates. Restrictions exist in the variety of colours and gloss ranges. For physical reasons of UV light absorption, yellow and yellow-containing inks cannot be crosslinked, especially not at the typical film thickness level of 70 μm for powder coatings.

Matt products are also not available, as UV powder coatings in particular are subject to different film formation mechanisms than liquid coatings. Matt products are a key requirement for the use of UV powder coatings for coating MDF in the furniture industry. At the moment, finely structured powder coatings are used to achieve matt surfaces.

Table 1.6 Properties of the systems [10]

	UV	NIR	NT (Low bake)
Binder systems	AC, PE-AC, U-AC, EP-AC	PE, EP-PE, EP, AC	EP, (AC)
Hardening principle	Photochemical polymerization	Polyaddition	Polyaddition
Temperature/time	Melting <100 °C/t = 40 s	t = 1 s, peak 230–250 °C	T = 100/140 °C, t = 10 to 20 min
Chemical resistance	Very good	Standard	Standard
UV light resistance	Medium to high	Medium to high	Low to medium
Hot steam (100 °C/20 min)	Constantly	PE, AC = resistant	Not durable
Pencil hardness	>2 H	Max. H	Max. H
Pendulum hardness	>170	<100/120	<120
Course	Very good	Better than standard	Standard
Layer thickness	20–80 μm	50–100 μm	80–300 μm
Energy/m^2 and part (related to conventional powder system)	60–80%	40–80%	90–95%
Plant size (related to conventional powder plant)	20–40%	5–10%	95–100%

Today, UV powder coating technology is still a niche application. The hoped-for market penetration could not be realized. Raw material and powder coating manufacturers are also showing reluctance to push this technology.

Table 1.6 shows a comparison of the properties of the different systems.

1.3 Effect Coatings

Metal and pearlescent effect pigments have a platelet-like structure and, in the case of metal effect pigments, consist of metallic elements or alloys such as aluminium, copper, brass or zinc. Pearlescent pigments consist of synthetic mica coated with high optical density metal oxides (e.g. TiO_2) [11]. In general, metallic and pearlescent pigments in powder coatings have the following advantages:

- metallic and pearlescent appearance
- polychromatic colour scheme
- functional protection

The formation of the effect is essentially determined by the pigment properties, particle size and pigment orientation (leafing or non-leafing behavior). Leafing pigments prefer to orient themselves plane-parallel at the upper interface of the paint film, while non-leafing pigments distribute themselves evenly throughout the paint film. In powder coatings, the non-leafing pigments are by far the most important. However, formulations with leafing pigments based on aluminum are increasingly being used for interior applications.

A distinction is made between metal effect powder coatings based on the method of manufacture or the effects to be achieved [12]:

Method of Manufacture:

- single-extruded metallic pigments (are an exception)
- dry blended metallic pigments (dry blend)
- bonded metal pigments

Metal effects:

- Hammer effect: Here, layer thicknesses of >100 μm are necessary to obtain a uniform appearance and corresponding effect. The recovery of the powders is limited due to segregation of added metallic pigment. Effect variations can occur due to different layer thicknesses and curing conditions (differences in the material thickness of the workpiece).
- Metallised effect mixed in: If the metallised effect is only slightly pronounced, colour tones are only conditionally stable in circulation.
- Metallise effect bonded: These powders are circulation stable.
- Full metallic mixed in: Light, silver/chrome shades (e.g. similar to RAL 9006). These powders are not circulation stable and are more or less sensitive to scratches (black markings). There is a safety risk with a proportion of more than 5% aluminium pigment (deflagration).
- Full metallic bonded: Light, silver/chrome shades (e.g. similar to RAL 9006). Circulation-stable powder, whereby the paint film is more or less scratch-sensitive. Therefore, an overcoating with transparent lacquer systems is recommended to protect the metallic effect or in case of very high demands on appearance and durability.
- Special effects: pearl effect, sparkling effect, mica, etc. Here, the formulation and processing must be checked in each individual case.

The bonded powder coatings are characterized by good cycle stability with high color and effect consistency and good processability. A disadvantage is the higher price due to the additional finishing process. Usually, the proportion of effect pigments for bonded powder coatings is in the range of 1 to 6% [11].

Non-bonded powder coatings are cheaper. However, difficulties arise in processing, especially with regard to cycle stability and color consistency. However, with the use of automatic or regular metering of fresh powder and the use of suitable spraying equipment, a satisfactory result can also be achieved here. With the use of a newly developed process, it should be possible to reliably recirculate dry blend systems [13]. The mixing ratio is monitored by means of the frictional electric charge of the batch to be sprayed. The process is used for coating aluminum extrusion material with brazing powder.

1.3.1 Extrusion

In this case, the effect pigments are added and extruded during premixing, just like all the other components of the coating. This has the advantage of less effort and the segregation stability of the end product. However, the sensitive pigments are destroyed during the subsequent grinding process. Therefore, only a few effects can be adjusted.

1.3.2 Dry Blend

After sieving the powder, the metal effect pigments are added and mixed again. Here the metal effect pigments are present next to the powder particles. As already mentioned, the product is not segregation-stable, which already forces compliance with narrow tolerances during application when checking the color shade.

1.3.3 Bonding Procedure

The effect pigments are mixed with the powder coating in an additional production process. The end product is largely segregation-stable due to the binding of the pigments to the powder coating grains. The range of effects is greater than with dry mixing, since a higher pigmentation is possible [14].

The bonding process can be described in 4 process steps. In the first step, the uni-powder coating is placed in the mixer. The process is carried out in an inert atmosphere and the oxygen content in the mixer is controlled. In step 2, energy is added to the mix by friction until an optimum approach to the Tg (glass transition temperature) of the resin is achieved, then the effect pigments are added. In step 3, the temperature is held constant for a period of time until the effect pigment is bound to the resin. To avoid the formation of excessive agglomerates, the mixed material is cooled gently in step 4. This can be done in the process mixer (with intensive cooling) or in a downstream cooling mixer [15] (Figs. 1.12 and 1.13).

1.3.4 Safety When Handling Aluminium Pigment Powders

Information on the safe handling of aluminium pigment powder can be found in the brochure “Recommendations for storage and handling of aluminium powders and paste” [16] or in the information brochure “Safety instructions for handling and processing aluminium

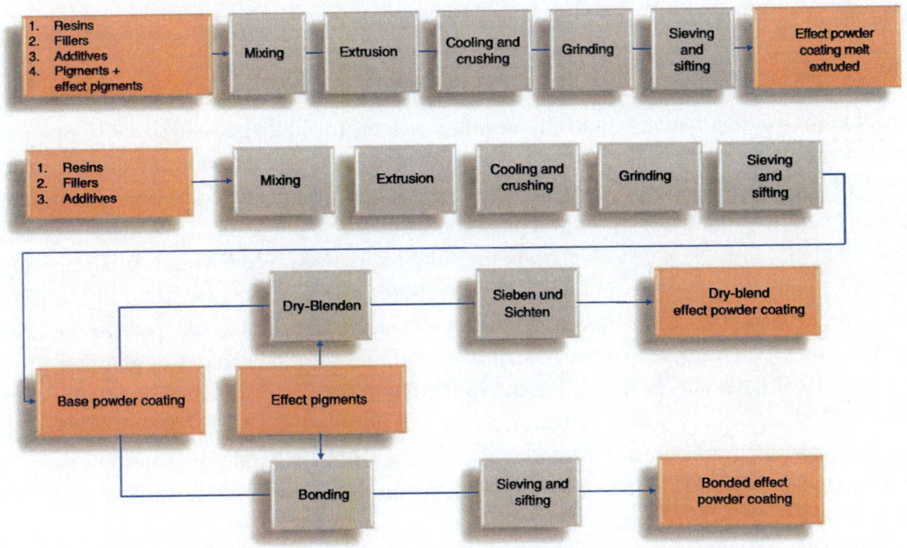

Fig. 1.12 Possibilities for the production of metallic and pearlescent effect powder coatings [11]

Fig. 1.13 SEM image of a bonded metal effect powder coating. The metal pigments "stick" to the powder grain

powder [17]. Information on the subject of "Static electricity - ignition hazards and protective measures" can be found in [18].

The minimum ignition energy is partly determined by the size of the surface of the pigment particles. For example, fine aluminium pigments (average particle size <10 μm) can have a minimum ignition limit of less than 1 mJ. It is imperative to observe the necessary safety measures during processing. Segregation, accumulation and build-up of aluminium pigments must be avoided during application throughout the plant. Spraying equipment must comply with the applicable guidelines and regulations, see Chap. 2. It is important that all parts and personnel are adequately earthed. The dust concentration in the room must be kept as low as possible and the extraction system must be adapted to the local conditions. Open flames, sparks or overheated surfaces must be avoided at all costs.

Detailed studies on the determination of the explosion characteristics of powder coatings and aluminium pigmented metal effect powder coatings were carried out by the Federal Institute for Materials Research and Testing (Bundesanstalt für Materialforschung und –prüfung – BAM) in Berlin and the Physikalisch-Technische Bundanstalt (PTB) in Braunschweig [19, 20] (Tables 1.7 and 1.8).

An epoxy powder was used as the base powder coating. The tests showed that the addition of 6% aluminium pigment does not change the minimum ignition energy. Likewise, there is no change whether it is a bonded system or a dry blend. However, the maximum explosion pressure and the explosion severity (KST value) increase by about 10%.

Table 1.7 Explosion characteristics of base powder coatings with effect pigments [11]

Powder mixture	Lower explosion limit [g/m3]	Ignition temperature [°C]	Minimum ignition energy [mJ]
Basic powder varnish	30	450	<10
Aluminium pigment powder[a]	30	650	<1
Gold bronze pigment powder	750	390	–
Pearlescent pigment powder	–	–	–

[a]values depend on the particle sizes the minimum ignition energies can fall below 1 mJ for very fine-particle aluminium pigments (D50 < 10 μm), while coarse pigments (D50 > 50 μm) have minimum ignition energies of over 200 mJ in some cases

Table 1.8 Minimum ignition energy of a bonded and a dry-blend aluminium powder coating [11, 20]

Powder mixture	Mean particle size [μm]	Minimum ignition energy [mJ]
Basic powder varnish	43	2,9–4,8
Metallic powder coating bonded	43	2,9–4,8
Metallic powder coating dry-blend	43	2,9–4,8

Composition of the powder coating: base powder 94%, aluminium pigment 6%

1.4 Powder Slurry

In powder slurry, a secondary dispersion is produced using binders such as those used in the powder coating sector. In principle, it is a powder coating dispersed in water. The aqueous dispersion has a proportion of non-volatile components of 34 to 37%, the pH value is 6, the volatile organic components have a proportion < 1%.

Since the material is formulated on a water base, it can be processed with the usual wet painting equipment. After an intermediate drying step, the material is cured at 150 °C for 20 min [21].

One advantage of the powder slurry is the average particle diameter of about 1 to 3 μm. By using fine-grained powder, good flow properties can be achieved with a lower dry film density [22]. However, these systems have not been able to establish themselves on the market.

1.5 Film Formation with Powder Coatings

Chemical crosslinking is characterized by an additional increase in the molecular size of the film former during film formation. In general, the longer the molecular chains resulting from crosslinking and the more branched the network, the more resistant the coating is to chemical and mechanical stresses.

The control and monitoring of the curing process is a key aspect of successful powder coating.

To be noted:

- timely achievement of the curing temperature, which means optimised temperature control in the baking oven
- the suitable cross-linking speed and density by appropriate formulation of the paint manufacturer

The temperature and crosslinking rate is crucial for powder coatings for the following reasons [23]:

- the temperature ranges and the "safety margins" in between, starting with the production and ending with the baking, allow sufficient storage stability and also the lowest possible baking temperatures.
- the viscosity curve during baking is the result of the factors temperature increase in the furnace, melt viscosity and viscosity increase due to incipient cross-linking.

The ideal sequence during baking is: melting of the individual powder particles, flow and wetting of the surface and subsequent crosslinking. A very good flow naturally involves

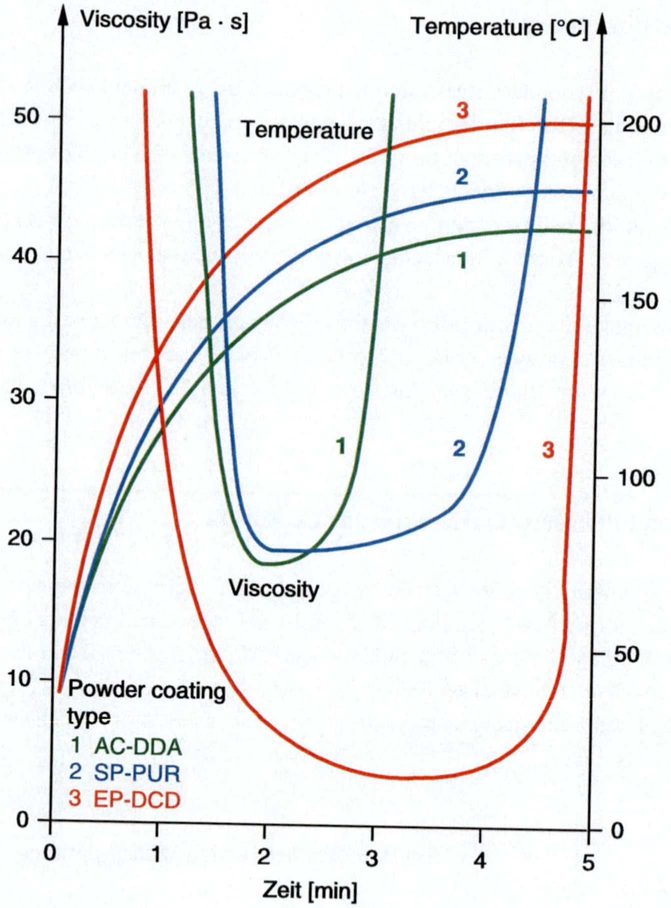

Fig. 1.14 Time course of object temperature and viscosity of a powder coating during baking [1, 25]

the risk of runner formation (sagging of the coating film, drop or nose formation) and insufficient edge coverage due to retraction of the powder melt from the edge [24].

1.5.1 Melt Viscosity and Surface Tension

The viscosity curve as a function of the baking time is shown in Fig. 1.14. After melting, the viscosity initially decreases with increasing temperature. At the same time, with increasing temperature, the crosslinking reaction increases more and more rapidly and finally becomes so fast that the decrease in viscosity with increasing temperature is compensated by the influence of the increasing film former molar mass. The viscosity here passes through a minimum at less than 5–20 Pas and then increases again with increasing

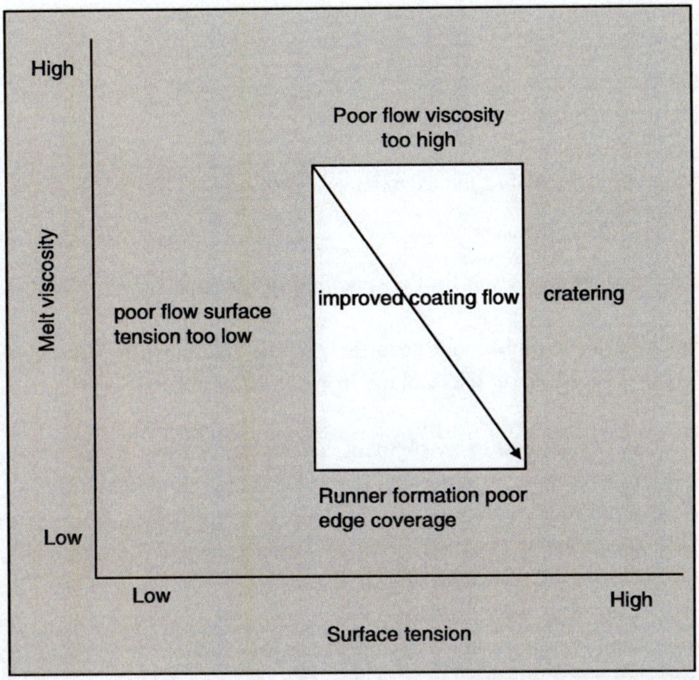

Fig. 1.15 Dependence of the surface quality of a powder coating on the surface tension and the melt viscosity [24]

crosslinking. The crosslinking reaction and thus the increase in viscosity come to a stopp when either a glass transition temperature of about 50 °C above the baking temperature is reached or the film former has completely reacted [1] (Fig. 1.14).

The faster the heating of the powder coating, the deeper and narrower the minimum. The flow and wetting of the substrate are determined by a window of the lowest viscosity and the relevant time period. Here, too, a low viscosity minimum is favourable.

The driving force for the flowing together of the "coating droplets" (coalescence) is the surface tension. Compared to the huge surface area of the powder, the surface area of the coating after coalescence is vanishingly small. Thus, a high surface tension should aid coalescence and smooth the molten coating film. The influences of melt viscosity and surface tension on the appearance of a powder coating are shown in Fig. 1.15.

1.6 Powder Coating Production

The production of powder coatings can be divided into the following steps:

1. Raw material check
2. Weighing of the premix
3. Mixing in the premixer
4. Homogenizing and dispersing in the extruder
5. Cooling of the extrudate to RT
6. Intermediate test (control of the extrudate)
7. rough grinding
8. Fine grinding
9. Testing of the powder coating for the specific properties

The raw materials are weighed out according to the specifications of a recipe. Almost without exception, production takes place in batches, depending on the capacity of the batch tanks.

The mixture of resins, additives, pigments and fillers is then extruded, a process that ensures intensive homogenization (dispersion) of the raw materials. This is achieved by melting the resin particles, combined with intensive shear forces, in order to optimally distribute all substances in the coating, especially the pigments. The pigments in particular are present in the form of so-called "agglomerates", a kind of clumping. These must be broken up and reduced to their primary particle size; this is the only way to achieve the exact and reproducible adjustment of a color tone. In practice, an extruder consists of a heated housing and one or more rotating shafts. The raw material mixture is now conveyed through the extruder via the feed area of the shaft in the direction of the outlet opening. During this process, the resins melt, causing the mass to change its consistency into a highly viscous liquid. In the further course, this melt is kneaded powerful, depending on the extruder design by special configuration of the shaft or by barrier (structure) elements in the housing. At the end, the melt (extrudate) emerges, is cooled, rolled out and broken into small pieces by a shredder.

These chips are then ground. Rotor classifier mills are used for this purpose, which allow an exact particle size distribution to be set. Carried by an air stream, the chips enter the grinding chamber through a channel. They are caught by a fast rotating rotor and thrown against the wall of the grinding chamber (impact chuck). The air stream pulls the particles towards the outlet opening, where they have to pass the rotating classifier. Only the sufficiently ground particles pass unhindered through the so-called "classifier window", the coarser ones are caught by the classifier and thrown back into the grinding chamber where they are ground again. At this point it is also possible to influence the fluidisation and charging behaviour of the powder by adding special additives. Figure 1.16 shows a schematic diagram of the powder production process. Figures 1.17, 1.18, 1.19, 1.20, 1.21 and 1.22 show individual steps in powder coating production.

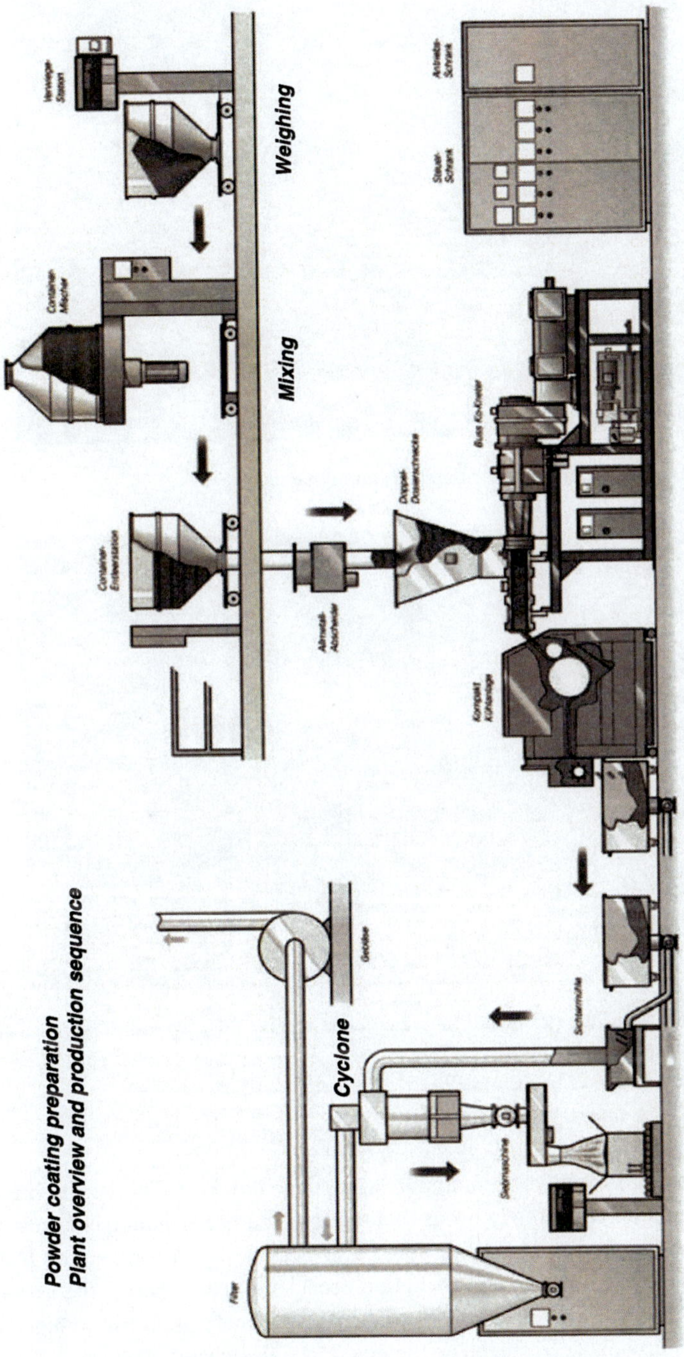

Fig. 1.16 Schematic diagram of powder coating production [26]

Fig. 1.17 Weighing and mixing the additives and pigments

Fig. 1.18 Extrusion of the premixed individual components - homogenization and dispersion of the mixture

1.7 Storage of Powder Coatings

Powder coatings are reactive one-component systems that crosslink during baking. This crosslinking reaction can already occur to a small extent at room temperatures, but especially at higher temperatures, such as may occur near the oven or in sunny places. This cross-linking reaction is called a pre-reaction because it occurs before the actual desired curing on the object. This pre-reaction causes powder coatings to lose a high degree of reactivity. This circumstance can be documented by a decreasing gel time.

Fig. 1.19 Quasi finished powder coating at the end of the extruder

Fig. 1.20 Powder coating after rolling on the cooling belt

The decrease in reactivity of the powder leads to restrictions in the flow with increasing ageing of the coatings, with extremely old materials gel particles and specks appear.

In addition to flow disturbances and the formation of gel particles, improperly stored powder can cause further problems. If, for example, an already aged, pre-reacted powder is mixed with fresh powder, matting of the coating film can occur due to the different reactivities.

Fig. 1.21 Powder coating is broken into chips

Fig. 1.22 Powder coating chips are ground and packaged

Powder coatings stored too hot also tend to physically block. Due to the increased temperature, the powder coating particles stick together and form agglomerates. This has a negative effect on the coating result (flow disturbances, speckling).

The powder coatings should be stored in a cool and dry place. The specifications and recommendations of the powder manufacturer must be observed.

1.8 Measuring and Testing Technology for Powder Coatings

In addition to testing the coating properties on the cured coating film, see Chap. 7, there are a number of tests which can be carried out directly on the powder and which are important for trouble-free processing during coating.

Powder coatings are characterised by the following parameters, the determination of which is standardised in DIN EN ISO 8130, Parts 1 to 14:

- Grain size and grain size distribution
- Density
- curing conditions and curing loss or baking loss
- lower ignition limit
- Block resistance
- Storage stability

In addition to these parameters, properties such as flowability, whirlability, tribo capability and charging behaviour are of particular importance to the processor. Table 1.9 gives an overview of the typical powder coating parameters for processing.

The measurement of the specific resistance is unnecessary for the typical organic powder materials such as epoxy, polyester, polyurethane, polyethylene and acrylates, as they have a sufficiently high resistance which prevents the discharge of the applied powder layer for a sufficiently long time. For lower resistivities, adhesion promoters are necessary to prevent the powder layer from falling off before and during the melting process.

The level of charge must exceed a threshold value so that the electrical adhesive forces hold the powder layer firmly to the workpiece before the adhesive forces in the liquid phase take effect. As a specific charge, for example, at a powder density of 1.6 g/cm^3 already 0.1 µC/g is sufficient for adhesion to vertical surfaces. For application with hand guns, at least 0.5 µC/g and for automatic guns >1 µC/g are recommended [28].

Table 1.9 Influencing parameters of powder coatings during coating [27]

Typical powder coating processing parameters		
Mechanical	Electrical	Chemical
Particle size (shape)	Electrical resistor	Storage stability
Particle size distribution	Tribo capability	Gel time
Bulk density		Burn-in loss
Whirlability Flowability		
Stability against agglomeration		

1.8.1 Pourability (Flow Behaviour)

The flowability is characterized by the discharge time of a certain fluidized powder quantity from a hopper (similar to the viscosity determination of liquid coatings with the DIN cup). The hopper is double-walled with a porous inner wall (e.g. sintered bronze) and has a defined outlet diameter [27].

A previously weighed-in (or fixed in volume) quantity of powder is fluidized via the porous inner hopper with a certain lateral pressure, so that it behaves like a liquid and flows out in a compact jet with a diameter corresponding to the discharge nozzle when the discharge opening is released. In addition to the powder properties (particle shape, particle spectrum), the discharge time depends on the pressure of the fluidizing air, the outlet diameter, the powder volume and, to a lesser extent, climatic factors. A side pressure of $p = 1$ bar, an outlet nozzle of 4 mm diameter, the internal angle of the hopper of 45° and a test volume of $V = 100\ cm^3$ or 60 to 70 g have proved to be favourable device parameters.

The first measured value should be discarded for fresh powder, since the powder has compacted due to transport and storage and thus, depending on the "previous history", usually provides a value that is too high. Impurities of the powder, formation of lumps etc. cause a partial or complete blockage of the powder outlet.

1.8.2 Fluidisation

Fluidization or vorticity of a powder is understood to be its property of changing into a powder-air mixture with a larger volume when air flows through it against the force of gravity and thus assuming the mechanical behavior of a liquid. With the expansion of a powder spill by fluidization or by determining the so-called dancing gas or boundary gas quantity which is required for the use of fluidization, a further quantity for the characterization of a powder coating can be determined metrologically.

The expansion shows itself as a typical parameter for a particular powder coating. It remains independent of the amount of air flowing through it in a relatively wide range until a floating powder cloud is formed, and only sets in when a certain flow rate is exceeded. This quantity of air which leads to sudden expansion has been given the name of boundary or dancing gas quantity.

The maximum height h_{max} reached by fluidization in relation to the height of the resting filling h_0 in a model cylinder constructed as a fluidized bed can thus form the measure of the vorticity. However, it is necessary to assume the same resting filling h_0, since this ratio decreases with increasing filling height [27].

1.8.3 Tribological Capability

The tribo-ability of a powder coating is the property of being unipolar charged in a contact and separation process with another material. The tribo capability is a prerequisite for the processability with tribo guns.

The so-called specific powder charge Q* related to the powder mass or powder volume can serve as a "measurand" for the tribological capability, if it is recorded under uniform measuring conditions (air pressure, air volume, powder throughput and PTFE as the tube material) [29].

1.8.4 Grain Size Distribution

With the effort to reduce the coating thickness, the grain spectrum has shifted to ever smaller average grain diameters. Grinding the powder coating produces a coarse fraction outside the usable range, which can be easily screened out with mesh sizes of 100 to 120 µm, but also a fine fraction. This fine fraction has a negative effect during coating, especially with tribocoating, as it is difficult to charge, forms dust clouds and thus leads to system soiling even outside the booth. The respirable nature of the particles in the size range below 6 µm is also problematic. Particle sizes smaller than 10 µm also impair the flowability of the paint and lead to sintering in the injectors and on the impact cones. In the recovery process, the separation efficiency of the cyclones decreases and the end-filters have to be cleaned more frequently.

A remedy is the use of sifted powder, where the fines are already separated at the powder manufacturer by a complex sifting process. These sifted powders have a higher price, but contribute to a better painting result and there is less cleaning effort and less maintenance of guns and injectors.

1.8.5 Loss on Stoving of Powder Coatings

Today, powder coating ovens are often operated with a high proportion of circulating air, in individual plants with 100% circulating air, for reasons of energy saving. This only works if no substances are released during baking. Powder coatings usually have a loss of 1–2% by weight. If, for example, 1 t of powder is processed, 10 kg of volatile components are produced. These must be discharged via the exhaust air or by other means, otherwise they will settle in cooler areas of the curing oven. If these residues enter the coating system and settle on the workpieces, they cause wetting problems such as craters or fish eyes.

Volatile components in powder are mainly flow additives (polydimethylsiloxane) or waxes, see Chap. 10. The volatile components of powder coatings are determined according to DIN EN ISO 8130-0-7 [30, 31].

1.9 Economic Importance of Powder Coatings

At the beginning of powder coating technology, the functionality of the coating system was primarily in demand. In the 1979s, powder coatings were applied exclusively to metal substrates. The variety of colors was limited and the optical properties of the coating film were initially of secondary importance. The mechanical-technological coating properties, however, had to meet high requirements even then.

However, this has changed fundamentally. Not only the variety of substrates has increased, but also the requirements for color variety, decorative and technological properties. Today, powder coatings are applied to metal, cast materials, glass, wood and plastics. The most important areas of application are the automotive industry and its suppliers, the construction industry, household appliances, the machine market and the furniture industry. Here, a wide range of colours, effect coatings and multi-layer systems are required. The optical properties and the coating quality must meet high standards, but the mechanical-technological coating properties are also still of great importance.

Electrostatic powder coating has currently reached a share of approximately 14% of industrial surface coating in Germany and is today rather in a stagnation than in an upward trend.

In the 1980s and 1990s, powder coating established itself as an economically advantageous and at the same time environmentally friendly process with high growth rates. Today, the conversion of painting plants from liquid to powder coating has been completed for those plants that were eligible for it. The build-up of industrial manufacturing capacity in Eastern Europe has had an increasing impact on Western Europe in recent years. Industrial preliminary products or industrial goods were increasingly imported from Eastern Europe. As a result, there is now significant overcapacity in paint production in Europe. This leads to intense price competition, especially for standard products that can be easily substituted. Rising raw material prices are also contributing to this.

In order to be able to achieve satisfactory growth rates, new areas of application must be specifically developed in powder coating technology. In particular, the area of non-metallic materials still offers potential in this respect.

New developments by powder manufacturers continue to focus on radiation-curing systems for use with temperature-sensitive substrates, powders for thin paint films and the application of powders in the automotive industry.

In order to be able to remove liquid paint coatings on a broad front, much thinner paint films must be achieved in powder coating. As a rule, coating thicknesses today are between 50 and 120 μm. Thin-film powders with film thicknesses of 30 to 50 μm are now available, but for critical color shades, reliable hiding power can only be achieved with higher film thicknesses.

Due to rising energy costs, systems with low baking temperatures in particular have been further developed in recent years. In hybrid systems, temperatures ≤150 °C are possible and already in use in certain areas. In the case of acrylates, it is possible to reduce the baking temperature to 120 °C, and even new developments in polyester resins allow significantly lower baking temperatures.

In addition to functional coatings such as anti-graffiti, better cleanability, lower dirt adhesion and antibacterial paints, highly weather-resistant paints have been able to establish themselves in the construction industry and architecture. Warranties of >10 up to 20 years on gloss and colour, depending on the system used, have enabled powder coatings to gain market share over liquid coatings.

The coating of large and heavy components is a market segment in which powder coating is increasingly being used. The possibility of applying primer and colored paint in one coating pass without the previously necessary baking of the individual layers also opens up new applications (Chap. 6).

1.9.1 Potentials and Future Developments

Opportunities for powder coatings exist in the development of better and greener products, because environmentally compatible solutions are already increasingly being sought in coating. The drivers of this development are customers and legal requirements. Future challenges for powder coatings include rising raw material prices and stagnating powder coatings prices caused by overcapacities, especially in the European market. Since 2010, the average powder coating price has increased by 0.7%. Another issue is the curing temperatures and the limitations in the substrate as well as the size of the workpieces to be coated. Today, a variety of MDF and some hardwoods can already be successfully powder coated. However, this is only a beginning. The market for wood finishes is large and diverse.

Future development trends are:

- Lower and faster curing and cross-linking technologies - Energy savings
- Thinner powder layers, <25 µm with optimum covering capacity even with white
- Super durable binder with good mechanical properties
- Corrosion resistant binder systems to be applied directly to the metal substrate without the need for pretreatment
- Improved performance of powder coatings on non-metallic substrates

References

1. Brock, T., Groteklaes, M., Mischke, P.: Lehrbuch der Lacktechnologie. Vincentz, Hannover (1998)
2. Kaplan, A.: Pulverlacksysteme unter dem Gesichtspunkt des dekorativen und/oder funktionellen Einsatzes, Praxis-Forum Schulung. Technik + Kommunikations, Berlin

3. Kandidatenliste – ECHA – Europa EU. https://echa.europa.eu/de/candidate-list-table, www.reach-info.de/kandidatenliste.htm
4. Thometzek, P.: Pulverlacke, GDCh, Gesellschaft Deutscher Chemiker, die aktuelle Wochenshow. www.aktuelle-wochenshow.de
5. Kaplan, A.: Die Natur der Primid-Bindung, Internationale Primid Tagung, März 1997. Flims, Schweiz
6. Kaube, O.: TGIC-freie Pulversysteme – Härtungsmechanismen und Eigenschaften, GSB-Schulungskurs: Qualitätsprüfung von organisch beschichteten Bauteilen. Bereich Lacke und Applikation, Schwäbisch Gmünd (2000)
7. Alfort, H.J., Hohmann, G., Zimmermann, F.: TGIC-freie Vernetzungsysteme bei Pulverlacken, DFO-Tagung Aluminium Anwendungen. Ther. Ber. **36** (1998)
8. Rademacher J. G.: Pulverlacktypen, ihre Eigenschaften und Einsatzgebiete, DFO-Weiterbildungslehrgang: Pulverlackiertechnik-Grundlegende Zusammenhänge für die Praxis. DFO-Druck, Paderborn (1993)
9. Zorll, U. (Hrsg.): Römpp Lexikon Lacke und Druckfarben. Georg Thieme, Stuttgart (1998)
10. Zimmermann, F.: Thermisch sensible Substrate mit NIR-, UV- oder Low-Bake-Pulverlack beschichten? EPS-Tagung Bad Nauheim, November 2000
11. Metallic- und Perlglanzpigmente für Pulverlacke, Eckart Effect Pigments, Informationsbroschüre (2014)
12. Neubauer, E.: Optimierungs-Potential beim Einsatz von Metallic-Pulverlacken. JOT. **11**, 28–31 (2000)
13. Cudazzo, M., Domnick, J.: Sichere Kreislaufführung von Dry-Blend-Materialien. JOT. **2**, 28–30 (2001)
14. Horvath, G.: Rätsel um die Vorteile des Boning Verfahrens, DFO Berichtsband: 6. DFO Tagung „Pulverlack-Praxis", September 2000, Dresden
15. Zeppelin System GmbH, Infobroschüre. www.zeppelin-systems.de
16. „Recommendations for storage and handling of aluminium powders and paste", Aluminium Association, 900 19th Street, N.W. Washington D.C. USA
17. „Sicherheitshinweise zur Handhabung und Verarbeitung von Aluminiumpulver, Gesamtverband der Aluminiumindustrie e. V., GDA, Am Bonneshof 5, Düsseldorf. www.aluinfo.de
18. „Statische Elektrizität – Zündgefahren und Schutzmaßnahmen", Broschüre der IVSS, ISBN 92-843-7091-4, Herausgeber: Internationale Sektion für die Verhütung von Arbeitsunfällen und Berufskrankheiten in der chemischen Industrie, Heidelberg
19. BAM (Bundesanstalt für Materialforschung und -prüfung) in Berlin
20. von Pidol und Krämer: Vermeidung der Entzündung pulverförmiger Sprühstoffe (Pulverlacke) beim elektrostatischen Pulverbeschichten, Bericht W-50, PTB (Physikalisch-Technische Bundanstalt) in Braunschweig, ISBN 3-89429-225-3 (1992)
21. Hilt, M., Kraus, R.: Integriertes Lacksystem und Pulver-Slurry. Meta. **53**(7), 43–45 (1999)
22. Kraus, R.: Neuer Lackierprozess für die A-Klasse. Meta. **52**(04), 299–301 (1998)
23. Brock, T.: Rund um TGIC-freie Pulverlacke – Filmbildung, Härtung und Eigenschaften im Vergleich. Farbe Lack. **106**(2), 38–45 (2000)
24. Pietschmann, J., Pfeifer, H., Raub, C.J.: Einfluss der Zusammensetzung des Pulverlackes auf die Kantendeckung. Meta. **44** (1990)
25. Wicks, Z.W., Jones, F.N., Pappas, S.P.: Organic Coatings, Science and Technology. Wiley, New York (1992)
26. Völkel, T.: Pulverlackforum. http://www.pulverlackforum.de
27. Kleber, W.: Pulverlackeigenschaften und ihre Auswirkungen auf den Beschichtungsvorgang, Praxis-Forum Schulung. Technik + Kommunikations Verlag, Berlin
28. Kleber, W.: Elektrostatische Oberflächenbeschichtung, Teil 5. Meta. **54**(10), 48–51 (2000)

29. Kleber, W., Bauch, H., Steinmann, F.: Überwachung und Steuerung elektrokinetischer Pulversprühgeräte. Meta. **42**(5), 253–258 (1988)
30. Pudewills Dr., Jens: Einbrennverlust bei Pulverlacken, besser lackieren 2018, 1
31. DIN EN ISO 8130--7, Pulverlacke, Teil7 Bestimmung des Einbrennverlustes von Pulverlacken, Beuth

Further Literature on Powder Coatings

Garratt, P.G.: Strahlenhärtung. Vincentz, Hannover (1996)

Bate, D.: The Science of Powder Coatings. SITA Technology, London (1990)

Bate, D., Copland, J., Floyd, R., Tweddle, E.: The Science of Powder Coatings. Application. SITA Technology, London (1990)

Stoye, D., Freitag, W.: Lackharze, Chemie, Eigenschaften und Anwendungen. Hanser, München (1996)

Howell, M.D.: The Technology, Formulation and Application of Powder Coatings, Powder Coating, vol. Bd. I. Wiley, New York (2000)

Crowley, M.: A Practical Guide to Equipment, Process and Productivity at a Profit, Powder Coating, vol. Bd. II. Wiley, New York (1999)

Deane, R.: Controlling Coating Properties through Management of Powder Particles, PCE Europe, 19–21 Januar in Amsterdam (2000)

Woltering, J.: Powder Slurries for Automotive OEM Finishing, PCE-Powder Coating Europe, in Amsterdam (1998)

Kaplan, A.: Primid-Pulverlacke Kennzeichnungsfreie Pulverlacke auf Basis von ß-Hydroxilalkylamiden, DFO-Tagung Pulverlack-Praxis. Ther. Ber. **33** (1997)

Talbert, R.: Powder Coater's Manual. Vincentz, Hannover (1998)

Images used in Chap. 1

IGP Pulvertechnik AG, CH Wil Abb. 1.17 bis 1.22

Karl Bubenhofer AG, CH Gossau, Abb. 1.13

Application

2

2.1 Introduction

Electrostatics is the study of electric charges at rest and the phenomena associated with them [1].

Modern electrostatic working methods are based on knowledge that goes back to antiquity. Especially in the experiments of the physicists of the seventeenth and eighteenth century the physical basics of these methods can be recognized [2–4].

The attractive effect of amber on light particles has been known for more than 2000 years. Thus, written records are found far before our era by Thaies of Miletus and later by Plato in the work "Timaeus" [5]. A student of Aristotle also mentioned the attractive effect of grated amber [6].

It was not until the middle of the sixteenth century that people began to study this physical phenomenon in greater detail and discovered that other materials, e.g. diamond, also exhibit this "amber effect". At the same time, however, the fundamental difference to the attractive forces of a magnet was recognized. The personal physician of the English queen proved this effect with numerous experiments, whereby he used a needle electroscope invented by him and examined the most diverse substances for their ability to exert a force effect on the moving needle. According to the result, he calls the substances that attract the needle "Electrica"in reference to the Greek word "electron", which means nothing else than "amber". At the same time, he already finds that moisture extinguishes the force effect, while oil does not [7].

Only a little later, it was recognized at the University of Parma that particles are not only attracted and later fall off by their own weight, but are sometimes thrown back to a distance of "three or four finger-widths" [8].

Otto von Guericke, the mayor of Magdeburg, came up with a fundamental and remarkable interpretation of the effect [9].

J. Pietschmann, *Industrial powder coating*,
https://doi.org/10.1007/978-3-658-37592-8_2

Guericke also discovered that the propagation of the effect of force is possible and that the rubbed sphere has the ability of sounding, which manifests itself in hissing, crackling and simultaneous glowing. This luminous phenomenon was correctly interpreted by Leibniz as small sparks.

At the beginning of the eighteenth century, experiments were no longer carried out with a sulphur sphere but with an evacuated glass sphere. Now, for the first time, electricity was available in "larger quantities". Naturalists everywhere rushed with great zeal to the object of investigation "electricity", especially as the means for experiments were easy to obtain. As a result, some discoveries were made simultaneously or shortly after each other in different places by different scholars. Hauksbee was the first to describe the radiating alignment of short strings on the surface of a sphere [10, 11].

As an experimental result, DuFay clearly stated the existence of two types of electricity and named them "glass electricity" and "resin electricity" [12]. A few years later, Benjamin Franklin introduced the terms "plus" and "minus" electricity, which are still in use today. At the same time, experimentation with humans began and initially caused great consternation. Professor Bose from Wittenberg wrote the following lines about the experiments of the Englishman Gray [13]:

> *Brave Briton Gray, do you know enough the power of the incredible and new property? And may you, as the daring one, prevent the electricity from connecting with the people.*

In the course of his experiments, Gray discovered that impact-transmitting sparks could be extracted from human and animal bodies, but also from all metallic bodies [14, 15]. With these sparks it was possible to ignite an explosive mixture, as was demonstrated during an academy session by the field physician Ludolf with a warmed bowl of spirit [16].

A Benedictine monk named Andreas Gordon, a professor of physics in Erfurt, Germany, built the first ion wheel, with which he was able to demonstrate the peak discharge and recoil effect of ions shooting out of sharp points or edges [17].

On closer examination of such peak discharges, Franklin concluded that it must be possible to discharge an electrified body much more easily with a pointed metal rod. To this end, he made the bold and original attempt to feed the thunderstorm electricity in the clouds via a pointed metal rod to an isolated person in a sign house and then to extract electrical sparks from him. This was successful in 1752, and the experiment proved that lightning was an "electrical phenomenon." Soon after, Franklin's first lightning rods were built, and their usefulness has been proven to this day.

The beginning of industrial use can be attributed to Hohlfeld's discovery in 1824, who observed the precipitation of tobacco smoke in the electric field of a corona discharge. However, it took another 75 years before the first industrial application in the form of electrostatic gas cleaning was technically realized by Cottrell in the USA and Müller in Europe at the beginning of the twentieth century.

Now, as a result of new discoveries, such as the electrostatic atomization of liquids by Zeleny and the experience gained in the use of electrostatic precipitators, patent

Table 2.1 Milestones in the development of electrostatic coating

Ancient	Thales of Miletus	Amber effect Discovery of frictional electricity
1599	W. Gilbert	Interpretation of electricity as a natural phenomenon in its own right
1663	O. von Guericke	Attraction of small leaflets and downy feathers by rotating sulphur sphere
1709	F. Hauksbee	Discovery of influence. Observation of the alignment of twine
1731	S. Gay	Discovery of the conductivity of electric force effects
1733	C. DuFay	Interpretation of attraction and repulsion as two types of electricity (resin and glass electricity) recognition of good and bad conductors
1744	C. Ludolf	Proof of the igniting effect of an electric spark
1785	C. A. Coulomb	Formulation of the basic law of electrostatics (Coulomb's law)
1824	H. Hohlfeld	Precipitation of tobacco smoke in the field of a corona discharge
1908	F. Cottrell K. Möller	Industrial use of electrostatic precipitators for gas purification
1911	J. Zeleny	Detection of electrostatic atomization of liquids
1927	E. Pugh	Patent application for the uniform coating of surfaces with electrostatic forces

applications followed for the uniform coating of surfaces by means of electric field forces, which formed the basis for the diverse developments and areas of application of electrostatic coating processes. In 1928 the first coating machine from E. Pugh appeared on the market. It works on the principle of electrostatic precipitators (spraying through corona grids). This was followed by Ransburg's equipment with rotating horizontal disc and bell atomizer. And in 1962 the electrostatic plastic powder coating is introduced.

The milestones in the development of electrostatic coating are briefly presented in chronological order in Table 2.1 [1, 2, 18].

2.2 Electrostatic Surface Coating

The term electrostatic surface coating is a collective term for the application of one or more layers with the aid of electrical field forces. The object to be coated is charged in the opposite direction to the particles of the coating material. In this way, for example, paint particles can be transported to the workpiece surface without an intermediate carrier and deposited there as a coating.

With the help of electrostatic forces based on the attraction of unlike charges and the repulsion of like charges, particles can be uniformly aligned, powder particles can be made to adhere in a thickness-limiting layer after deposition, and liquids of a certain conductivity range can be atomized into small electrically charged droplets via the corrugation mechanism [19].

Since the charges are positioned only at the surface of the particles due to their mutual repulsion and the electrical forces act on the charges, these act as surface forces. Their influence becomes decisive when they predominate over the mass forces, such as gravity and centrifugal force. On the one hand, this requires a large surface-to-mass ratio, which is only possible for small particles in the range of 10 μm to 2 mm and for filaments and films. Also required are both large charge densities on the particle surfaces and high field strengths in the immediate coating zone, in this case between the charged particle cloud (sprayer) and the workpiece surface. All coating processes are based on the following theoretical laws of particle transport, repulsion and attraction.

Particle Transport The electric force F is proportional to the field strength E and is stronger the larger the charge Q is. This is a directional quantity that acts in the direction of the largest voltage field.

$$\overline{F} = Q \cdot \overline{E}$$

$$F = \frac{Q^2}{4\pi\varepsilon_0 a^2} \quad \text{Coulomb's Law}$$

F Electric power
Q Particle charge
E Field strength
A Particle distance
ε_0 Dielectric constant

Rejection:

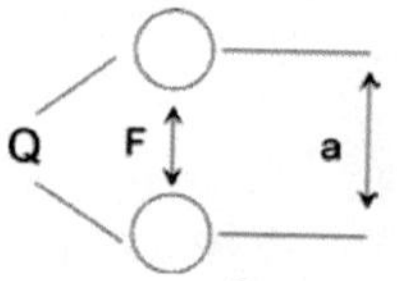

Attraction:

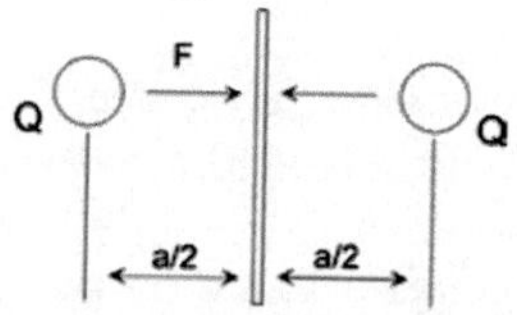

2.3 Physical Principles of Coating Processes

The specific advantages of electrostatic processes result on the one hand from the direct use of electrical energy without energy conversion, and on the other hand from the effect of electrical forces as surface forces. The advantage is that they can act from a

Table 2.2 Electrostatic force effects and their prerequisites

Forces act without Intermediate beam	Attraction Rejection Diversion
Area forces, prerequisite	$\frac{Surface}{Mass} \rightarrow Large$
Requirements are met for:	Small particles 10 μm to 2 mm foils and threads

considerable distance, their lines of action can be specifically influenced and they hardly disturb the medium in which the substances to be treated are located.

The electric field forces act on electric charges, which are distributed on the surfaces of the electrically charged particles due to their mutual repulsion. Their effect as surface forces in relation to mass forces (gravitation, inertia, etc.) is decisive when there is a large ratio of surface to mass.

Electric forces are practically only effective on small particles up to a maximum of 2 mm, rarely larger, or on thin films using very high field strengths. However, they are superior to all other forces.

With the help of electrostatic working methods, the tasks that are necessary in the handling of small particles of an attractive or repulsive or deflecting force effect are solved. The fact that the electrical forces act without an intermediate carrier is of great importance for their technological use. In Table 2.2 the prerequisites for the electrostatic force effect are shown again [20].

In electrostatic coating, the following subtasks are to be fulfilled:

- Formation of an electrically charged powder cloud
- Transport of the charged powder particles to the workpiece surface
- Precipitation and electrical adhesion of powder particles
- Formation of the powder layer and electrical self-limitation of the layer thickness.

The physical basis of electrostatic coating processes is the attraction or repulsion of electrically charged bodies of different or same charge, Table 2.3. These electrostatic phenomena act on the powder particles in such a way that they are transported specifically to the workpiece surface and deposited there as the desired coating. Furthermore, both high charge densities at the particle surfaces and high field strengths in the immediate coating zone, i.e. between the sprayer and the workpiece or the charged particle cloud and the workpiece surface, are required.

2.3.1 Charging Mechanisms

For the electrical charging of solid and liquid bodies, the mechanisms of action compared in Table 2.4 [1] can be used. Electrostatic powder coating requires a high electrical

Table 2.3 Electrostatic solved technological tasks

Phenomenon	Technological task
Attraction Repulsion	Targeted transport of small particles
Rejection	Homogeneous particle cloud
Attraction	Layering
Attraction	Adhesion to the coating surface
Rejection	Self-limiting of the layer thickness
Rejection Attraction	Backspraying

Table 2.4 Mode of action of different charging mechanisms [1]

Charging mechanism		Physical principle of action	
Conduction charging	Contact charging	If a conductive body in the electric field is in contact with a live or grounded electrode, charge carriers can pass from the electrode to the body, or vice versa, and charge it.	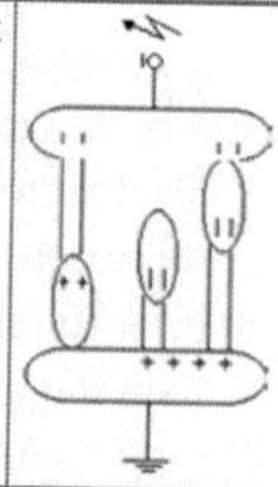
	Influenza Charge	The electric field also penetrates a dielectric body located between the electrodes and leads to a separation of the free charge carriers. When the body is separated by electrical or mechanical forces, the parts are then charged	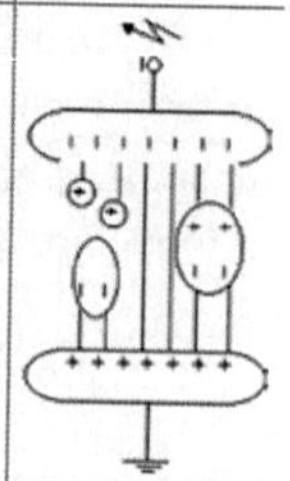
Ionization charging	Ion diffusion	Accumulation of free ions on conductive or dielectric particles due to the disordered thermal motion of the ions in the gas, independent of the external field (predominant for particles with d < 1 mm)	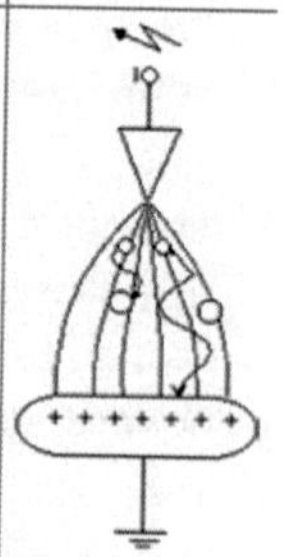
	Ion bom-bardement	Attachment of free ions to conductive or dielectric particles due to their directional motion in the electric field, e.g. a corona discharge (predominantly for particles with d > 1 μm).	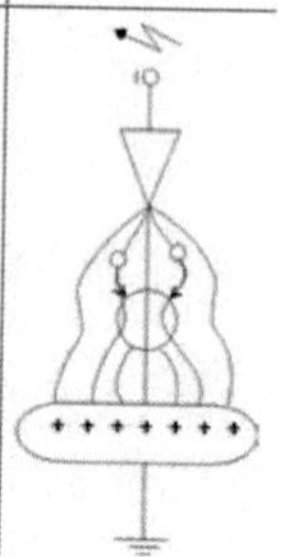

(continued)

(continued)

Charging mechanism		Physical principle of action	
Triboelectric charging	Touch	Due to the different molecular forces of attraction, electrons or ions can transfer when two surfaces are in close contact, and when they separate, both bodies are then charged with different polarities	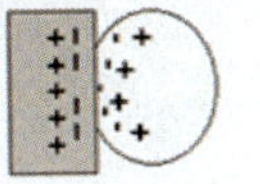 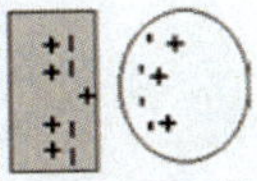
	Friction effects	Charging of similar bodies as a result of special friction effects. The asymmetry of the friction process can, for example, result in different surface temperatures or destroy double layers.	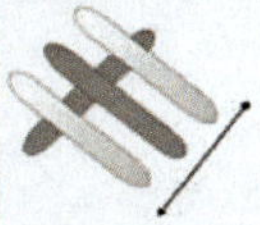
	Electrolytic processes	Under normal atmospheric conditions, there is a very thin layer of water on the surface of most solids which reacts with the solid. If this is destroyed (e.g. by contact), the charge equilibrium is disturbed.	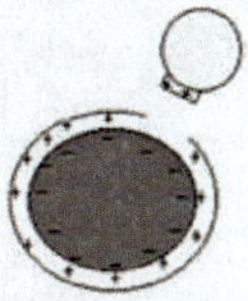

resistance of the coating materials so that the charges cannot flow off and the powder particles adhere until baking due to the Coulomb forces between the powder charge and the charge of the workpiece surface. Thus, only electrically insulating substances are used as powder coatings. Of the possible charging mechanisms, only the triboelectric process and ionization are of importance for powder coating.

Triboelectric Charging Transfer of charge carriers in the contact surface based on interface-specific processes (independent of an external electric field) and subsequent mechanical separation (Table 2.4).

Ionization (Corona Charging) Accumulation of free air ions in the electric field of one or more corona electrodes connected to a high DC voltage (Table 2.4).

2.3.2 Triboelectric Charging

The causes of charging during contact and subsequent separation of solid bodies (also referred to as frictional charging, contact charging or electrification) can be of a very diverse nature (e.g. destruction of adsorbed electrolyte layers, temperature difference, piezoelectric effects, etc.). In most cases, however, an intimate contact between two substances is sufficient; the rubbing itself only increases the intensity and the number of contact points.

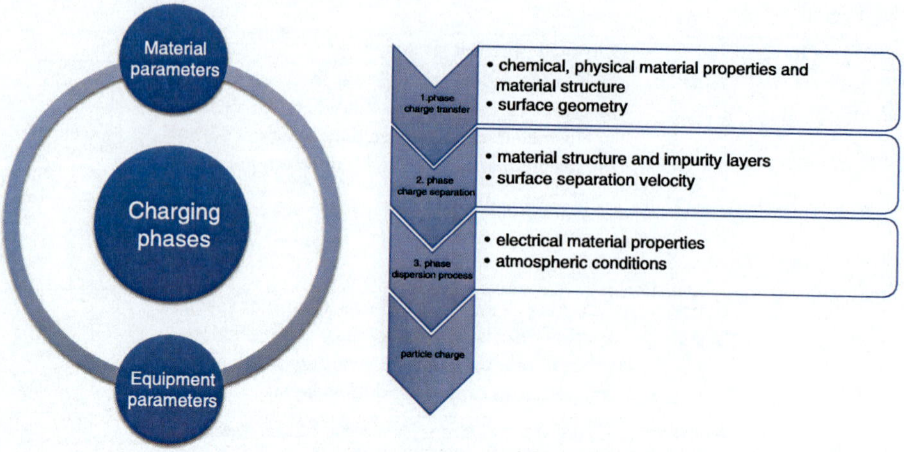

Fig. 2.1 Three-stage process of a tribocharging and its influencing factors [20]

In the actual contact area, the atoms of the two substances come so close to each other that the atomic force fields disturbed at the interfaces overlap and charge carriers cross over, which cannot flow back fast enough during the subsequent mechanical separation of the surfaces. The separated surfaces are then electrically charged with opposite polarity. The triboelectric charge is always bipolar, i.e. the negative charge on the surface of one body is opposed by an equally large positive charge on the other.

Triboelectric charging (frictional charging) takes place in a three-step process (Fig. 2.1). The first step is the intimate contact between the powder particles and the wall of the gun, during which a transfer of electrons from one contact partner occurs, depending on which of the two shows the greater affinity of the surface molecules for electrons. The intimate contact is followed by separation, which must be so rapid that restoration of the original charge distribution is not possible. Thus, the disturbed equilibrium ratio remains after the separation process and the powder particles are positively charged by electron withdrawal in contact, e.g. with PTFE.

In the tribo sprayers, the powders flow through an insulating channel, which is usually annular-gap or spiral-shaped, directly in front of the atomizer head for electrical charging (Fig. 2.2). The material combination powder/channel wall is decisive for the polarity and the level of the charge. Both materials should be as far apart as possible in the electrification series. With regard to a predetermination of the level and polarity of the tribocharge of high-polymer materials, the electron affinity of the elements or functional groups bonded to the carbon chain and their stereometric arrangement in the macromolecule are decisive. The tendency to negative charging increases almost without exception with increasing electron work function (PMMA, PE, PVC, PTFE), while the materials with the lowest work function (PA, PE oxides) more easily donate an electron and become positively charged [21].

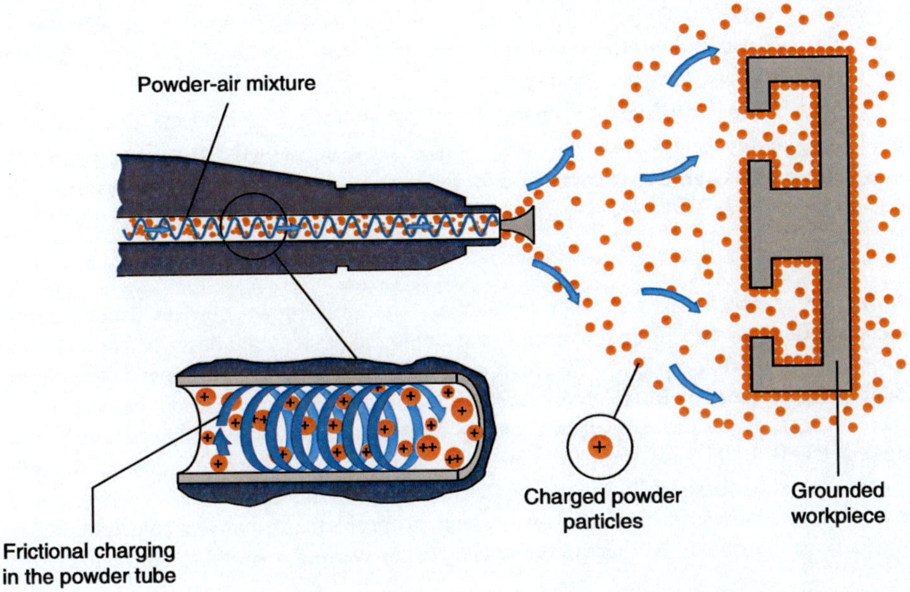

Fig. 2.2 Principle of tribocharging

PTFE is used as a contact material because it is at the negative end of the electrification series. It leads to a high positive charge in epoxy, polyamide, PVC and PE powders. In addition, the material is characterised by very good non-stick properties, which suppress deposits on the wall. To intensify the contact of the powder particles with the wall surfaces of the flow channel, various design measures were provided, such as increasing the turbulence by surface profiling, the installation of flow bodies, by blowing in additional air, the use of extremely long loading tubes and/or the parallel arrangement of many channels with small cross-sections, the generation of a spiral particle path along the tube wall by tangential blowing and the use of curved loading tubes. These variations of the device-technical realization are mostly subject of different, partly patented, solutions.

Since triboelectric charging involves interfacial processes, minor "impurities" have a significant influence. This is a disadvantage that limits the process-engineering application of tribocharging, but it also offers the possibility of making resins that are not prone to tribocharging "tribo-able" by minor admixture of additives or to compensate for differences in charging behavior caused by different pigmentation. The "cold mixing" of alumina with a grain size in the nanometer range, which practically envelops the powder particles, ensures sufficient, possibly even very high tribocharging, but brings with it other process-related problems, such as clogging of the filters, change in chargeability and grain size distribution in the powder circuit, and lack of stability of the mixture when the overspray is recovered. Technically, the problem of using polyesters that are difficult to charge triboelectrically has been solved by the development of so-called "internally modified" grades in which, for example, amine or other salt-like compounds are incorporated into the resin. Epoxy resin, polyurethane and polyamide powders are generally always suitable for the triboelectric charging process [1].

Tribocharging is a mechanical charge separation. It always leads to an equal proportion of positive charges on one contact partner as negative charges on the other. This means that as much positive charge as is carried out of the sprayer by the charged powder remains on the contact surface in the sprayer as negative charge and can be selectively discharged as leakage current [22].

Influence of Mechanical Powder Properties The well-known climate dependence of electrostatic charges is often transferred to tribological devices and described as a decisive disadvantage. Only powders with a high water absorption (e.g. PVC) show a clear dependence of the charge on the relative humidity. Due to the moisture film on the particle surface, an increase in the triboelectric charge even occurs with these materials in contact against PTFE. In the case of the usual thermoreactive powder coatings, however, the influence in terms of chargeability is practically negligible in the range < 95% relative humidity. However, increased humidity always leads to problems in powder conveying during powder coating due to the increased tendency of the coating material to clump together.

Influence of Mechanical Powder Properties The charging behavior and the flow of the powder layer to the coating film during baking, is favored by the spherical shape of the powder particles. At least, convexly curved surfaces and, as far as possible, equal dimensions in the three spatial axes should be aimed for.

With regard to the influence of the particle size on the triboelectric powder charging, two opposing tendencies must be taken into account. With increasing particle diameter d, the chargeable surface area increases, but at the same time the possible surface charge density decreases due to the decreasing maximum field strength at the particle surface. In addition, with larger particles the actual contact area becomes relatively small compared to the total surface area due to the limited number of contacts with the friction partner.

For very small particles, the conditions for charge carrier transfer are also poor, because the impact energy and thus the elastically deformed particle surface remain low. In the case of the usual triboelectric spray devices, a maximum of the specific powder charge Q* therefore occurs as a function of the particle size [22, 23].

$$Q^* = \frac{\text{current} * \text{spraying time}}{\text{sprayed powder quantity}}$$

This value of the specific charge of the powder provides information on the suitability of the powder for tribocharging. Q* must not fall below 0.3–0.5 μC/g for manual guns and 0.5–1 μC/g for automatic guns [22].

2.3.2.1 Influence of the Plant Parameters

The flow velocity v, the particle concentration K in the powder-air dispersion and the pneumatic dimensions of the charging channel (gap width, changes in direction of the powder flow) are decisive for the triboelectric charging in the spraying device from the plant engineering side. For triboelectric charging to be usable in process engineering, at least one velocity of the conveying medium air is required so that the flow is turbulent (Reynolds number > 2320). While the particle charge increases with increasing flow velocity, it decreases with increasing particle concentration. Small channel cross-sections favour powder charging, but at the same time reduce the throughput [24].

Practically, the tribological devices work with a slightly higher air flow rate like high-voltage guns. This can be between 10% and 30% more depending on the conveyor system.

2.3.3 Ionisation Charging (Corona Charging)

In the case of sprayers fed with high voltage, the powder charging takes place practically exclusively by ion bombardment. In this process, the powder particles are guided through the electric field of a corona discharge by appropriate design of the spray head, Fig. 2.3. As a result of the polarization of the powder particles, a part of the field lines ends on their surface and the much faster air ions following the field lines are guided directly there and accumulate until the field of these charges compensates the field of the corona electrode. The maximum attainable charge of a spherical particle of diameter d in the electric field of the corona electrode with the field strength $_{EA}$ in the charging region is, following the Eq. (2.1) developed by Rohmann [25]:

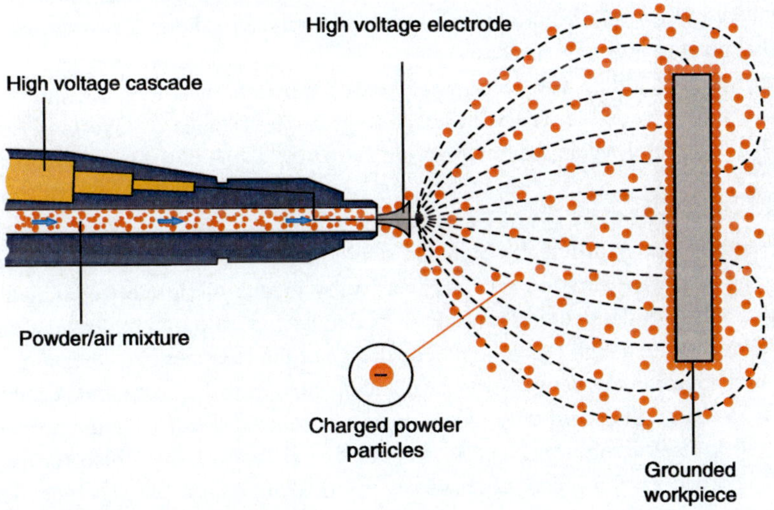

Fig. 2.3 Principle of corona charging

$$Q_{oI} = 3\pi \in_o \frac{\varepsilon_r}{\varepsilon_r + 2} d^2 E_A \tag{2.1}$$

The time course of the charge increase on the particles is described by the Pauthenier Eq. (2.2) [26].

$$Q1 = Q_{oI} \frac{1}{1 + \frac{4\varepsilon_0 E_A}{S_A t}} \tag{2.2}$$

Q_{oI} max. charge of a spherical particle
ε_r Dielectric constant
ε_0 Electric field constant
d Particle diameter
E_A Field strength in the charging area
S_A Current density in the charging region

2.3.3.1 Influence of Powder Properties

Of decisive importance for the charge of the powder particles is their grain size. Since the charge of spherical particles is proportional to the surface area, the absolute value of the charge increases quadratically with the diameter, while the charge Q* related to the particle mass m falls hyperbolically. Although the particle shape of powder coatings deviates more or less from the spherical shape, the formulas derived for spheres can be adopted as good approximate equations as long as the particle dimensions do not differ too much along the three spatial axes. The cross-sectional area perpendicular to the field direction is decisive, i.e. a particle of any shape is to be replaced by a sphere with the same cross-sectional area. However, this principle is also used by most methods of particle size analysis, so that no conversions are necessary.

As with tribocharging, powder particles with a spherical or at least outwardly curved surface are most favourable. The SEM image (Fig. 2.4) of a standard polyester-based powder, with a common particle size distribution, shows the shape of the powder particles.

2.3.3.2 Influence of the Plant Parameters

The prerequisite for a sufficiently high and stable powder charge is first of all a unipolar ion flow. This can be realized relatively easily by means of tip electrodes, which are arranged near the outlet opening (outside or pulled back a little into the powder channel) and are connected to a high DC voltage. The design of the atomizer from the point of view of optimum coordination between ion flow and air flow must be taken into account here. Attempts to charge the powder by means of an electrode system inside the sprayer have not yet led to useful solutions. The reason for this is that a corona also forms on the counter-electrode, even if it does not have any tips or sharp edges, due to powder deposits, so that the space between the electrodes is bipolar ionized.

The powder throughput M has a very large influence on the powder charging. Due to the space charge of the powder cloud, the corona current is suppressed, and the greater the

Fig. 2.4 SEM image of powder particles

powder output, the more strongly, so that the powder charge drops steeply. This effect is particularly evident in sprayers with electrodes retracted into the powder channel. To a certain extent, however, the current-weakening effect of an increase in throughput or workpiece spacing can be compensated for by an increase in voltage. The minimum voltage at the gun is determined by the corona insert voltage and is 30–40 kV. With increasing voltage, the field strength $_{EA}$ and the current density S_A in the charging area in front of the powder outlet increase, but at the same time the undesired portion of the current of free air ions also increases. The maximum applied voltage is about 100 kV [27, 28]. The voltage polarity has little influence on the charging process itself. As a rule, a negative high voltage is selected because the corona has a higher current and is more stable, and the disturbing back-spray effects on the workpiece surface occur to a lesser extent.

2.3.4 Flight Behaviour of Electrically Charged Particles

Three forces act on the charged paint or powder particle in the spray jet (Fig. 2.5):

- the electric force: $F_E = Q \cdot E_T$
- the force of gravity: $F_s = m \cdot g$
- the air resistance: $F_w = 3\pi \cdot \eta_L \cdot d \cdot v$

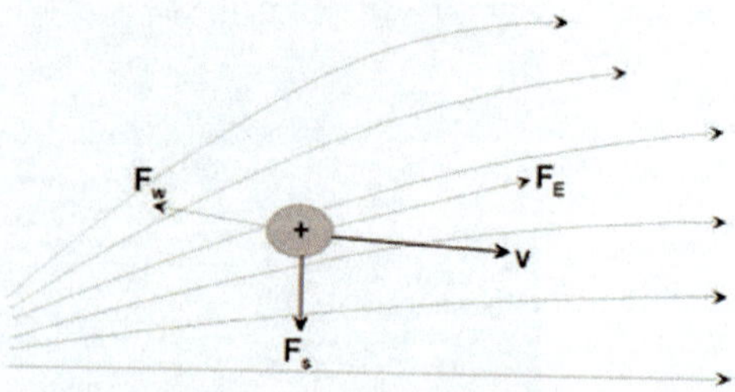

Fig. 2.5 Forces acting on the charged powder particle in the spray jet

The electrical force component is composed of the field force and the Coulomb force of the mirror charge. The mirror charge is only effective in the immediate area of the workpiece surface (approx. 1 mm) and must be taken into account in the adhesive forces. Due to the reciprocal relationships between field strength, particle charge and particle velocity, an extensive homogenization of the field occurs. The field strength E_T, which is decisive for the transport of the charged powder particles, is therefore almost constant in the entire space between the atomizer and the workpiece.

In the case of high-voltage spray equipment, it is only slightly below the mean value E = U/a resulting from voltage U and distance a; it only rises sharply in the immediate vicinity of the corona electrode. In tribo devices it is determined by the space charge of the powder cloud and is of the same order of magnitude.

2.3.5 Relationship Between Field and Gravity Forces

In order for the particles to follow the field lines to a large extent, it is necessary that the electric field force F_E exceeds the gravity force F_S by a multiple. Taking into account Eqs. (2.3) and (2.4), one obtains for the achievable ratio of the two forces at

Tribo charging:

$$\frac{F_E}{F_S} = E_{o\ max} \times \frac{6 \cdot \varepsilon_0}{g \cdot \gamma \cdot d} E_T \tag{2.3}$$

Corona charge:

$$\frac{F_E}{F_S} = \frac{3 \cdot \varepsilon_r}{\varepsilon_T + 2} \cdot E_{A\ max} \cdot \frac{6 \cdot \varepsilon_0}{g \cdot \gamma \cdot d} E_T \tag{2.4}$$

E_A Field strength in the charging area during ionization
E_0 Field strength at the particle surface
E_T Field strength in the transport area
Γ Density of the particle material
ε_r Relative dielectric constant of the particle material
ε_0 Dielectric constant

2.3.6 Separation Behaviour

The charge Q of the surface A is proportional to the square of the diameter d, but since this is only linearly included in the air resistance, the relative velocity with respect to the surrounding air decreases with decreasing particle size. For small particles of 10 µm grain size, for example, it is only 0.75 m/s at $E_A = 5$ kV/cm, and $E = 3$ kV/cm, and is thus not significantly higher than the air movement in the cabin. Even smaller particles are therefore preferentially carried away with the exhaust air and do not reach the workpiece. The recovered powder consequently has a different particle size distribution than the fresh powder. Both fresh and recovered powder should therefore be fed into the supply hopper continuously and in a constant ratio in order to achieve consistent coating results.

In the medium diameter range, the atomized particles largely follow the field lines leading from the corona electrode to the workpiece. Due to their large surface area, the coarser particles have a high charge Q, so that the field force is also very strong and the particles follow the field lines even against the air flow. Their proportion is therefore particularly high in the edge zones and on the back of the workpiece. Only in the case of very coarse powder grains ($d > 60$ µm) does the influence of gravity predominate, so that the application efficiency for these grain fractions decreases again.

2.3.7 Formation of the Powder Layer

2.3.7.1 Electrical Adhesive Force

In order for the powder particles to adhere to the non-preheated workpiece surface outside the electrical field of the spray equipment, it is necessary that

- the Coulomb force between the particle charge and the countercharge (in the form of the mirror charge in the workpiece) is greater than the force of gravity, and
- the specific electrical resistance of the applied powder is so high that the charges cannot flow off before reaching the melting phase in the furnace.

The charge on the powder particle influences an equal area charge of opposite polarity on the grounded workpiece. This charge distribution can be replaced by a point charge at the

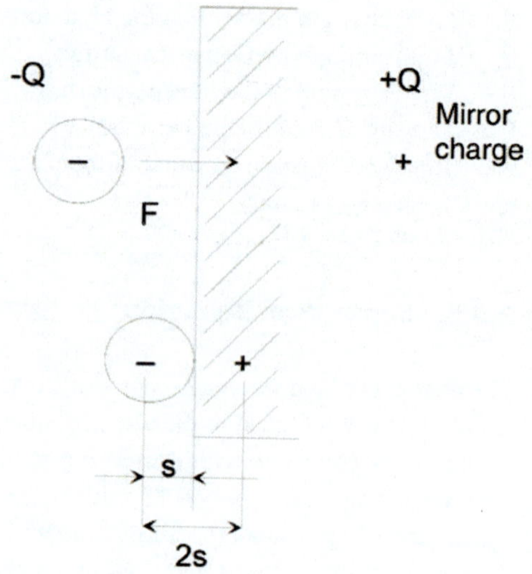

Fig. 2.6 Adhesive force of the powder particles on the workpiece surface due to mirror charging

particle centre and a corresponding mirror charge at the same distance on the other side of the workpiece surface, see Fig. 2.6.

The attractive force between two equal charges of opposite polarity is:

$$F = Q \cdot E = \frac{Q^2}{16 \cdot \pi \cdot \varepsilon_0 \cdot s^2} \tag{2.5}$$

The Coulomb force of the mirror charge falls rapidly with distance and therefore only acts at very short distances (<1 mm). For a powder particle with a diameter d = 50 μm, a density = 1.5 g/cm³ and a specific charge Q* = 1 μC/g, for example, it is already smaller than the force of gravity F_s at a distance s = 150 μm.

S = distance of the charge centre to the electrode.

For the evaluation of the adhesive force of the applied powder, one cannot disregard the effect of the neighboring particle charges. If one considers the powder layer as a homogeneous medium consisting of several layers with the thickness of the average grain diameter with a constant space charge density ρ_s, then the adhesive force of the uppermost layer is no longer dependent on the distance (as in the case of two capacitor plates with constant charge). As a minimum charge for adhesion by electrical forces, one obtains for powder coatings with the usual parameters $Q^*_{min} \approx 10^{-7} \cdot C/g$ [27].

2.3.7.2 Layer Limitation Effect

The process of forming the powder layer on the workpiece surface consists of three characteristic phases that follow one another in time: Layer formation phase, confinement phase and saturation phase.

2.3.7.3 Layer Formation Phase

The applied powder mass initially grows almost linearly with the coating time t. The coating structure of the powder deposited in this phase is relatively homogeneous and dense. The layer structure of the powder deposited in this phase is relatively homogeneous and dense. The spray current I at the corona electrodes is only a fraction of the value in pure air (without powder output); it depends on the set system parameters and the constructive form of the diffuser and is usually in the range I = 5–20 μA.

2.3.7.4 Limitation Phase

As the layer thickness s increases, the field strength (E) in the powder layer also increases. It is greatest directly at the workpiece surface and amounts to:

$$E = \frac{1}{\varepsilon_{rs} \cdot \varepsilon_o} \int_o^s \rho_s(x)\,dx.$$

ε_o Dielectric constant
ε_{Rs} rel. Dielectric constant of the powder layer
ρ_S Space charge density
d Average grain diameter
S Powder layer thickness
x Distance of the charge centre to the electrode

At a certain coating thickness, it has reached such a high value that the electrical strength of the air is exceeded in the cavities caused by the grain shape and ionization channels form in the powder layer. The reionization, which initially starts at individual points, spreads over the entire coated surface within a transition period, causing the gap to the sprayer to become bipolar ionized, which results in a steep increase in the spray current. In contrast, however, the charge of the particles and the proportion of powder reaching the workpiece surface decrease (Fig. 2.7).

2.3.7.5 Saturation Phase and Back-Spray Effects

After reaching a current I which is approximately equal to or greater than the initial value I_0 (without powder output), no more powder is applied. In the now bipolar ionized field in front of the counter-electrode, the subsequent particles are discharged by the accumulation of air ions of opposite polarity to such an extent that they no longer adhere or are repeatedly recharged ("fogged") until they fall out of the field weakened by the re-ionization under the influence of gravity or are carried out by the air flow (Fig. 2.7).

Fig. 2.7 Boundary phase; with increasing layer thickness the field strength in the powder layer increases

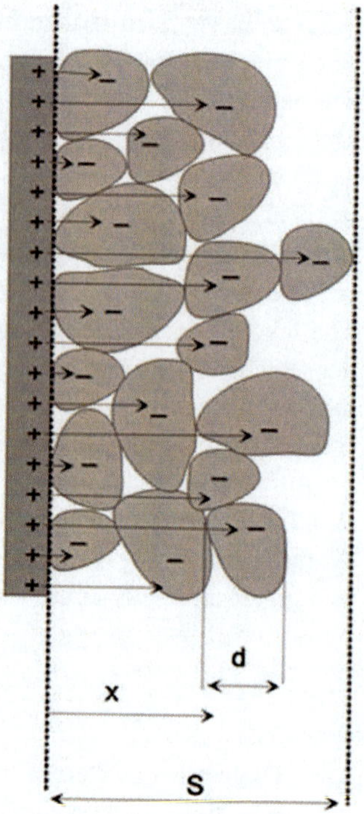

Due to the reionization from the already applied powder layer, only a coating of limited thickness can be applied during electrostatic powder coating. On the one hand, the powder must have Q^*_{min} a certain minimum charge in order to adhere to the non-preheated workpiece, but on the other hand, this charge already causes such a high field strength in the air pockets at relatively low coating thicknesses s that ionisation points are formed which prevent the application of further powder (Fig. 2.8). The collision of the positively charged ions with the negatively charged powder particles leads to a loss of charge in the powder particles. They can no longer be deposited on the workpiece [29]. On the other hand, the supposed disadvantage of the layer limitation causes the advantage of a largely constant layer thickness on the entire workpiece surface.

The maximum coating thickness to be achieved depends on the powder properties. As a rule, it is 100–200 μm for a single application, in exceptional cases (PVC, polyamide) up to 500 μm. In the case of thin-film powder coatings, the coating thickness is approximately 50 μm for a single application.

Re-ionization is associated with the formation of back-spray craters. However, since the ionization channels in the powder layer are not stable, but are constantly formed and extinguished again at different points, the craters are often covered again by subsequent particles. However, larger air pockets remain, which have a negative effect on the coating

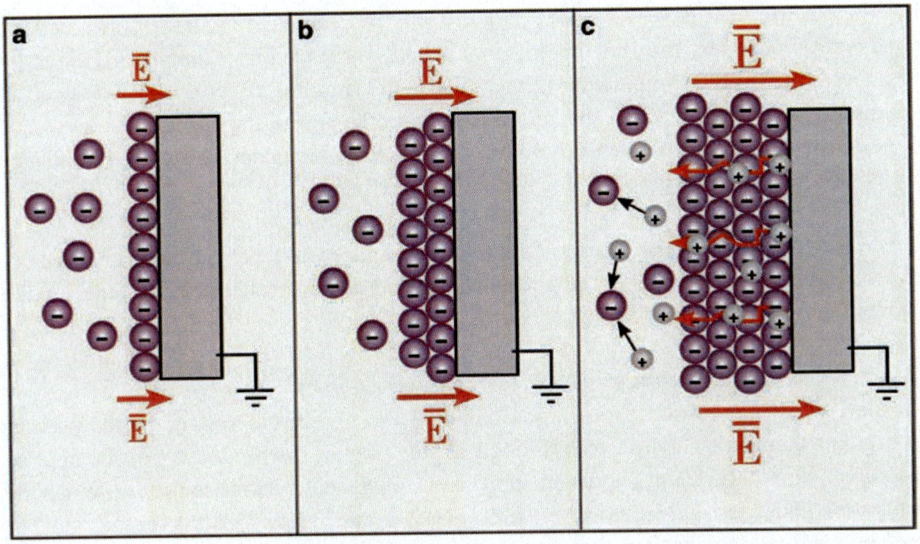

Fig. 2.8 Deposition of the powder particles on the metal surface (**a**); growth of the powder layer (**b**); maximum achievable layer thickness is reached and the back-spray effect starts (**c**)

result, especially in the case of powder coatings with poor flow properties, as they lead to voids and pores in the melted film. Therefore, it is better to set the system parameters so that the coating thickness is not limited by the saturation effect, but by the coating duration.

The backspray is particularly strong when the powder feed fails with high voltage applied.

This is because the field-weakening influence of the space charge of the sprayed powder is missing. In this case, one can speak directly of a destruction of the powder layer.

The backspray effects occur more strongly with exposed corona tips than with retracted ones. In the case of positive voltage polarity, they already start at lower layer thicknesses and are stronger in terms of current (workpiece acts as a negative electrode with lower corona insertion voltage). With tribo devices, on the other hand, backspraying occurs later and in a weakened form [30].

2.3.8 Technological Comparison of Sprayers

On the one hand, the use of electrical forces in the process technology results in the advantages of electrostatic painting processes, such as high automation, handling and good coating uniformity, but inevitably the laws of the electrical field also cause some difficulties or the limits of the application:

- The electric field is very weak in partially shielded cavities (e.g. back-drawn edges), therefore only very little coating material reaches these cavities. Compressed air assistance brings only a limited improvement due to the jamming effect as well as the partial discharge of the powder by the reionization from the protruding edges.
- At sharp edges and corners the field lines bundle, here it can come to an overcoating or to the appearance of the so-called picture frame effect.

Since the formation of the electric field in front of the workpiece surface is essentially determined by the selected charging mechanism, this also has a decisive influence on the technological properties of the spray device.

2.3.8.1 Coating Efficiency

In the case of electrostatic spraying equipment, the higher the voltage or the powder charge, or the smaller the powder throughput, the greater the proportion of the applied powder in the total amount of powder sprayed during the layer formation phase. When the saturation phase is reached, however, it decreases very sharply. For this reason, too, the system parameters voltage, powder throughput and conveyor speed in technical coating systems must be matched to each other in such a way that the thickness of the applied layer is limited by the coating duration and not by the saturation effect. In the case of electrokinetic (tribo) charging equipment, the result is very much influenced by the powder charge achieved. If this is at or above the level of the charge in the devices operating with high voltage, the coating efficiencies are also equivalent. Under constant kinetic conditions in the charging zone (air volume and velocity), the specific powder charge decreases sharply with increasing throughput [24]. In the same way, the coating efficiency also decreases.

2.3.8.2 Coverage (Electrostatic Wrap Effect)

Due to the supporting effect of the encompassing “far field” of the corona electrodes connected to the high voltage, the coverage effect of corona guns is significantly better than that of tribo devices, where mainly the “near field“of the space charge is effective. A retracted electrode arrangement reduces the influence of the high voltage. Their polarity, on the other hand, does not cause any noticeable differences.

2.3.8.3 Penetration Capacity

Even under the same application conditions (distance, spray head shape) a greater penetration depth can be achieved with electrokinetic coating equipment.

The particular advantage of these devices, however, is that the spray head can even be inserted into small openings, making it much easier to coat electrically shielded areas (cavities, backed edges, so-called “shadow areas”, etc.).

If, on the other hand, the workpiece surface is approached too closely with corona spraying equipment, back-ionization sets in there, especially at sharp edges or in the already applied powder layer, which leads to a discharge or reloading of the sprayed powder particles.

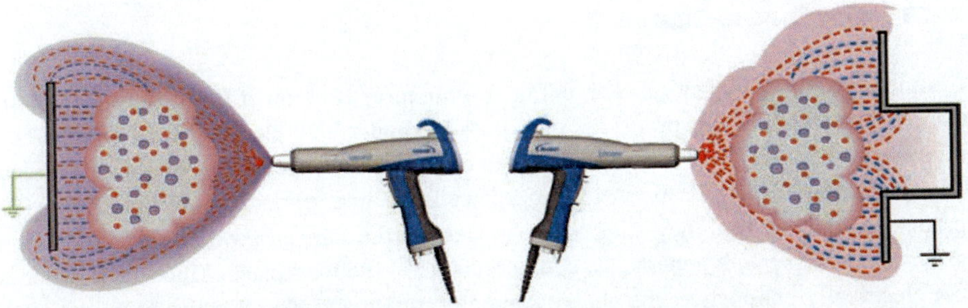

Fig. 2.9 Influence of the Faraday cage on the field lines and coating result. Left: flat workpiece the field lines reach the workpiece and a certain "wrap" of the coating is given. Right: the field lines cannot enter the depression (Faraday cage), build-up of the powder layer is restricted [29]

The high current in the bipolar ionized field also causes a sharp drop in the voltage at the corona electrodes and thus a weakening of the electric field strength. The powder particles mechanically blown into the cavities are practically uncharged and do not adhere. An improvement in penetration is achieved by flat spray nozzles. On the one hand, the retracted electrode arrangement minimizes the flow of free ions; on the other hand, the higher mechanical flow component also leads to a deviation of the powder paths from the field lines leading to the edges.

Modern corona spray devices allow the spray stome to be limited to <0.5 μA or the characteristic curve of the high-voltage cascade to be changed. This can significantly reduce the disadvantages of coating cavities, recesses and edges (Fig. 2.9). Today, corona charging dominates the market.

2.4 The Charging Systems in Practice

The following Sects. 2.4, 2.5, 2.6, 2.7, 2.8, 2.9 and 2.10 were compiled primarily with the help of the extensive writings and information in [30–32].

Various systems are used to charge the powder. The use of the different processes depends on the application and the wishes of the user. Basically, a distinction is made between three different charging methods: electrostatic charging, low-ion charging and tribocharging. Air ion reduction is achieved by practically all suppliers by means of a special attachment to the gun. Other systems, which have not yet found widespread use, are the so-called hybrid guns, in which electrostatic and tribo are combined, and the internal charging systems. They complement the main processes mentioned.

2.4.1 The Corona Charge

Corona charging or ionization charging is the unhindered escape of free electrons from an electrical conductor. This process is only possible under high electric field strength. The ionized powder particles, like the free ionized air particles, are attracted to all grounded objects. In practical applications, the workpiece is the first grounded part that the powder particles encounter and they stick to it. As soon as the ionized powder particle hits the grounded workpiece, it generates a counter charge in the workpiece. Since the powder is not electrically conductive, the charge cannot dissipate and the attractive force remains. The current flowing away from the grounded workpiece consists of the impacting free electrons, of air ions and of the influence current of the countercharge.

A very good grounding of the workpieces is an essential factor for a good coating. Active measures such as paint stripping of the hangers or special hanger constructions with shielded support points are necessary. The grounding of all system components itself is again an essential safety aspect, see Sects. 2.9 and section 3.1.

Corona guns consist of a voltage transformer, a high voltage cascade and a thin wire as electrode. In order to ensure an optimal and constant charging of the powder, the electrode is rinsed with a small stream of air. Without this rinsing, the powder would sinter on the electrode in a very short time and its electron flow would be significantly reduced. The air flow also increases the corona discharge at the electrode, as the electrons are washed away. Of the free electrons, 1–3% strike powder particles, the rest ionize in the ambient air or remain free. These free electrons and the ionized air, together referred to as space charge, have an influence on the coating quality (back spray effects).

The air speed used also has a great influence on the coating quality. The air is mainly used to convey the powder from the powder container to the gun, through the nozzle to the object. High air velocities support penetration into recesses; however, they also increase the tendency to blow off. Low air velocities increase the degree of deposition, but can also have a negative effect on penetration. Only the combination of the right charge with the appropriate air velocity will give optimum coating results.

The charging principle with constant supply of electrostatic charge allows the use of this type of gun, Fig. 2.10, for all applications and for practically all types of powder available on the market.

2.4.1.1 Low Ionic Charging

Low-ion charging is used wherever thicker powder layers and a high optical coating quality are required at the same time. With conventional electrostatic charging, it is not always possible to achieve a high surface quality when depressions have to be covered and the front edges have to be smooth at the same time. Either the recesses cannot be coated sufficiently or the layer thicknesses at the edges are too large, whereby orange peel can occur from layer thicknesses of 60–70 µm. This results from an excess of free ions that can no longer discharge over the object due to the excessively thick powder layer, leading to an

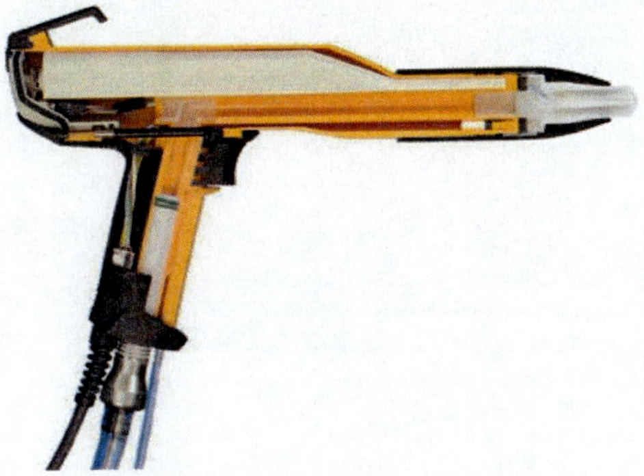

Fig. 2.10 Structure of a corona gun

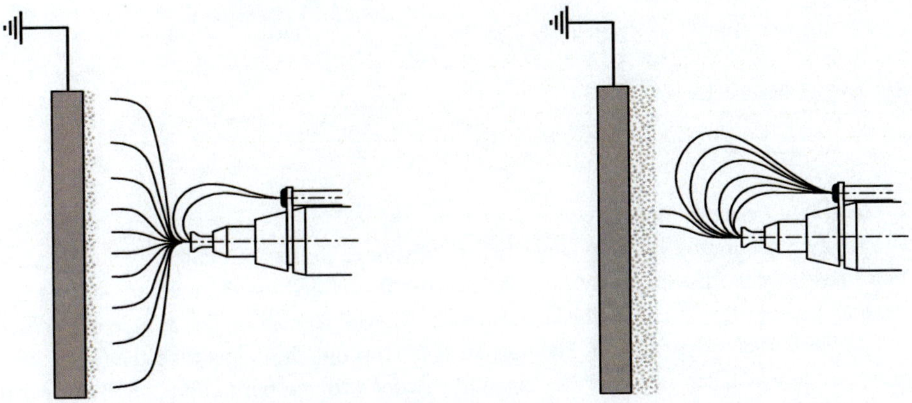

Fig. 2.11 Functional principle of the low-ion charging. The powder layer is not yet homogeneously closed at layer thicknesses of less than 60–90 μm (powder dependent), so that the air ions can flow off over the ground, see figure on the left. In the figure on the right, the powder coating layer is homogeneous, i.e. air ions can no longer flow off via the grounded workpiece. With the additional grounding, the main part of the ions moves to this grounding and not to the workpiece. This means that the workpiece remains largely free of free air ions [33]

irregular coating. An attachment of a grounded metal ring or tip prevents the majority of the free ions (space charge) from reaching the workpiece by returning them to the grounding of the gun (Figs. 2.11 and 2.12). Typical applications for this type of gun are the coating of drawers, rims and profiles. Another area of application is coating with textured powder, where a much more uniform pattern or surface texture is achieved. In addition, these guns reduce the powder recoil when coating by hand. A disadvantage is that the electrode(s) of the attachment are wearing parts and make handling more difficult with manual guns.

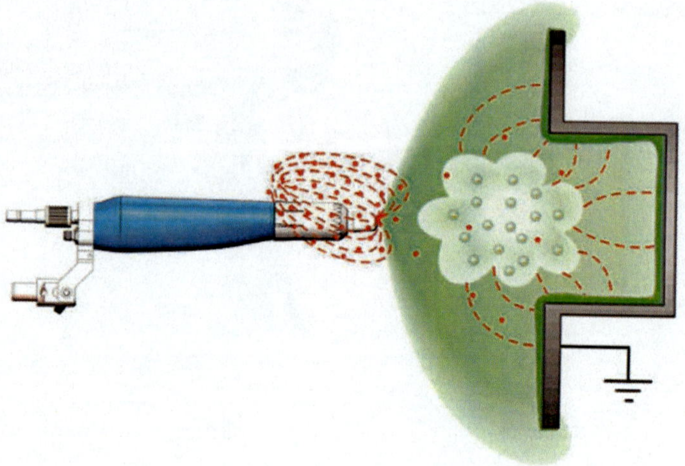

Fig. 2.12 Low-ion charging [29]

2.4.1.2 The Powder Bell

The mode of operation of the powder bell corresponds to that of the bell, which is very well known from the wet paint area. The powder is charged via the loading edge of the rotating impact plate. The rotation speed can be changed depending on the type of powder and the desired cloud formation. The powder is fed through the center tube. The high voltage cascade was taken from the electrostatic powder gun and built into the bell body. An auxiliary air allows the jet to be adjusted and the forward speed to be increased in the direction of the object (Fig. 2.13).

With the use of the powder bell, an extremely uniform coating application with a high application efficiency is achieved. Through its use, the number of atomizers can be reduced depending on the application and the average layer thickness can be reduced due to the high uniformity of the powder discharge. This offers a high powder saving potential. Powder bells only achieve a very limited penetration depth and are therefore primarily suitable for coating large, flat parts. When selecting a booth, it is important to note that powder bells build up large, soft powder clouds and therefore react very sensitively to air currents. The maximum powder output for powder bells is 600–700 g/min, whereby this amount of powder is generally not well chargeable and therefore the application efficiency is low and more overspray occurs.

2.4.1.3 Internal Charging

In the early days of powder coating, most corona guns had an external charge. The electrode is located at the far end of the gun. The closest grounded part is the workpiece. The electrostatic field builds up between the workpiece and the electrode. The charge on the

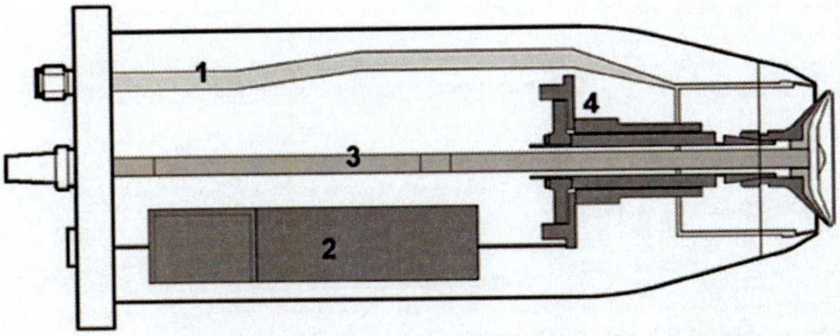

Fig. 2.13 Schematic diagram of a powder bell, 1 = forming air, 2 = cascade, 3 = powder tube, 4 = high voltage

powder builds up near the electrode. In the case of internal charging, the ground is in the gun itself. The electric field builds up inside the gun. The charging of the powder occurs inside the gun and near the electrode. Most of the free ions are conducted directly to earth. The result is that the space charge is smaller than with external charging and that the penetration capacity is increased. The problems lie in the sintering of powder and fogging of the grounded electrode. Due to the very short charging distance, the powder throughput is limited.

2.4.2 Tribo Charging

The Tribo process makes use of friction charging. The process requires a comparatively high air velocity, as turbulence is then created in the tube, which increases the number of contacts. The plastic material used is mainly Teflon. Tribo charging can only be regulated to a limited extent. Since the conveying speed of the powder alone is not sufficient to generate the required charge on the powder in the gun, the speed at which the powder passes through the gun is increased with an additional air supply. In this way, regulation of the desired charge on the powder can be controlled to a certain extent (Fig. 2.14).

The advantage of tribo charging is the good penetration depth. This charge does not produce free air ions. Therefore, the penetration of powder into recesses, such as letter boxes and rims, is better than with the corona gun. The coated surfaces are smoother and orange peel is less pronounced. The construction of the gun is simple. The nozzles for atomizing the powder can be formed as desired. Disadvantages of tribo charging are its low flexibility, higher wear and recovery of the powder. Not every powder can be processed with Tribo, it needs special additives or a special formulation. Fresh powder can be charged best, powder recovered several times cannot be used again. The higher the fines content of the powder, the less tribo-capable it is.

The main advantages of the Tribo coating can be summarized as follows:

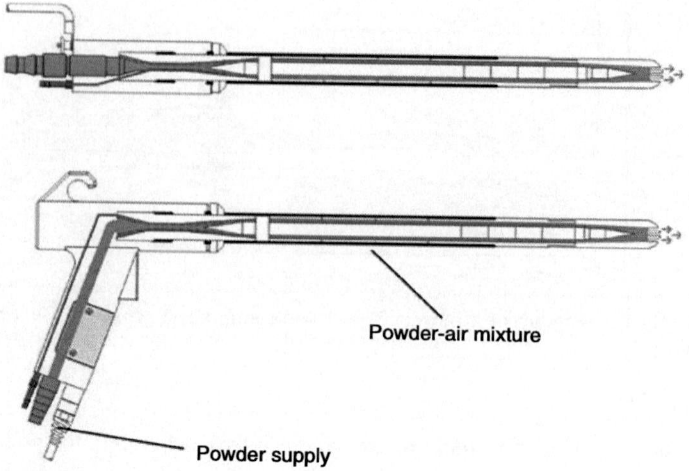

Fig. 2.14 Schematic diagram of a tribo gun assembly

- Overcoats and second coats are possible without any problems.
- More uniform layer thickness distribution for complicated part geometries. Especially for tube constructions with small construction depth, practically no Faraday cages
- The surface finish can be improved with tribo-coating as there are no free field lines from high voltage generators.
- Higher degree of automation for filigree workpieces. Manual coating may not be necessary at all.
- Tribo coating systems are cheaper to purchase than corona systems with the same number of guns.
- Often fully automatic coating with lower control effort

2.4.3 Comparison of Charging Systems

2.4.3.1 Tribo or Corona?

The question in practice is usually under which conditions or workpiece geometries a tribo coating is more advantageous than a corona coating.

This applies in the case of:

- rough welded constructions (e.g. formwork elements)
- internal coating (e.g. for pipes)
- deep parts with Faraday cages
- very complicated part geometries

- rough surface (e.g. cast iron), here processes such as sanding and filling can be partially dispensed with by applying a relatively thick layer.
- extreme levelling requirements (of the paint film)
- the requirement for very uniform layer thicknesses
- multiple coating – in addition to tribo charging, a combination of tribo and corona or low-ion charging with corona can also offer advantages here

The disadvantages of Tribo over corona are:

- the greater wear in the gun and therefore higher costs
- a partly higher air consumption
- not all powders are to be used for tribo application
- a stronger layer structure on hangers
- a lower powder throughput, thus often more guns are necessary, especially for flat parts

2.4.3.2 Processing Guidelines for Tribo Coating

For an optimum Tribo coating, some requirements must be fulfilled with regard to the powder and the grain spectrum. The processor should also observe some rules in order to achieve an optimum result with the Tribo coating.

Requirements for the Tribo-Powder The promise that the powder is tribo-capable is not sufficient. The powder must also be circulation stable. Effect powders in particular can lead to problems here. The powder grain content of less than 8 µm should be as low as possible.

Reasons for a Low Fine Grain Content A grain content <8 µm of max. 6% can only be used with cyclone separation. With filter separators, there is a risk that the recycled powder will be too fine. From an economic point of view, the grain content <8 µm should also be as low as possible with cyclone separation, as the cyclone does not separate these particles for the most part. Applying this fine grain to the part to be coated is very difficult. Especially with drawers/cabinets, the so-called blow-out effect occurs i.e. the fine fraction is blown out by the air vortex in the housing and sucked out by the recovery. This fine fraction is for the most parts lost and should be avoided.

2.4.4 The Nozzle Systems

Guns can be equipped with different nozzle geometries, depending on the coating requirements (Fig. 2.15). In order to obtain optimum coating results, the choice of nozzle must be adapted to the effective coating requirements. The nozzles perform two tasks: they atomize the powder and produce a homogeneous powder cloud (Fig. 2.16). Depending on the charging process, they are also carriers of the electrode.

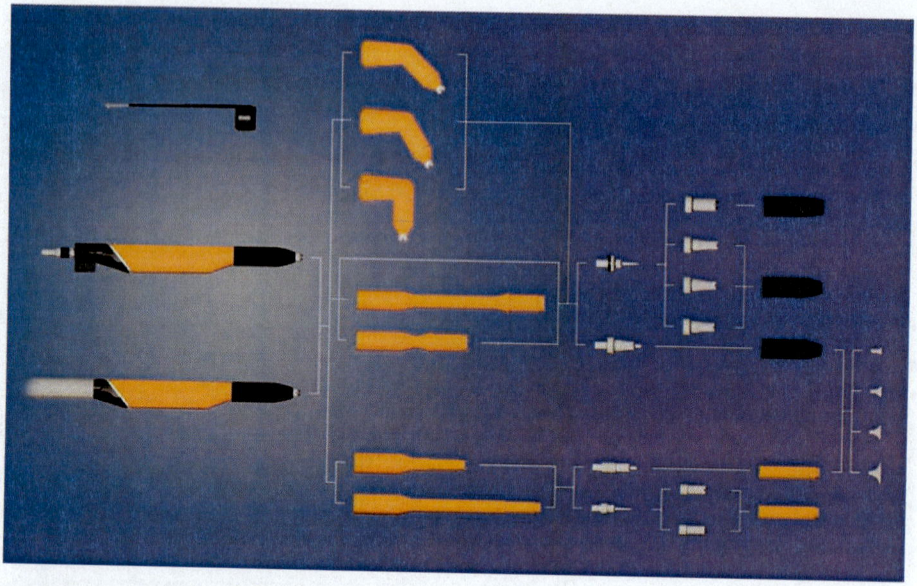

Fig. 2.15 Nozzles and extensions for electrostatic charging

2.4.4.1 The Baffle Plate

The impact plate has been known since the beginning of powder coating. The strongly bundled powder jet hits an impact cone and is thereby torn apart. A homogeneous powder cloud is created, which slowly approaches the grounded workpiece. The cloud can only be controlled to a limited extent and the penetration capacity is small. It is used for flat parts and high surface powers (Fig. 2.17).

2.4.4.2 The Flat Spray Nozzle

The flat jet nozzle has a nozzle geometry with a slot. The powder in the tube is accelerated before the slot and decelerated after the slot and thus put into high turbulence. The powder cloud has an elliptical cross-section and can be easily aligned with the object. It is used for more complex parts with recesses (Fig. 2.18).

2.4.4.3 The Swirl Nozzle

It attempts to tear apart the powder jet with a tangentially entering air jet. However, the additional air introduced into the coating chamber disturbs the coating quality. Therefore, this nozzle is not widely used.

2.4.4.4 The Metallic Nozzle

Today, special nozzles are used for metallic powders to avoid deposit effects. These deposits lead to an increased risk of short circuits. With the help of an improved geometry of the nozzles and an optimal air flow in the nozzle, deposits are avoided as far as possible.

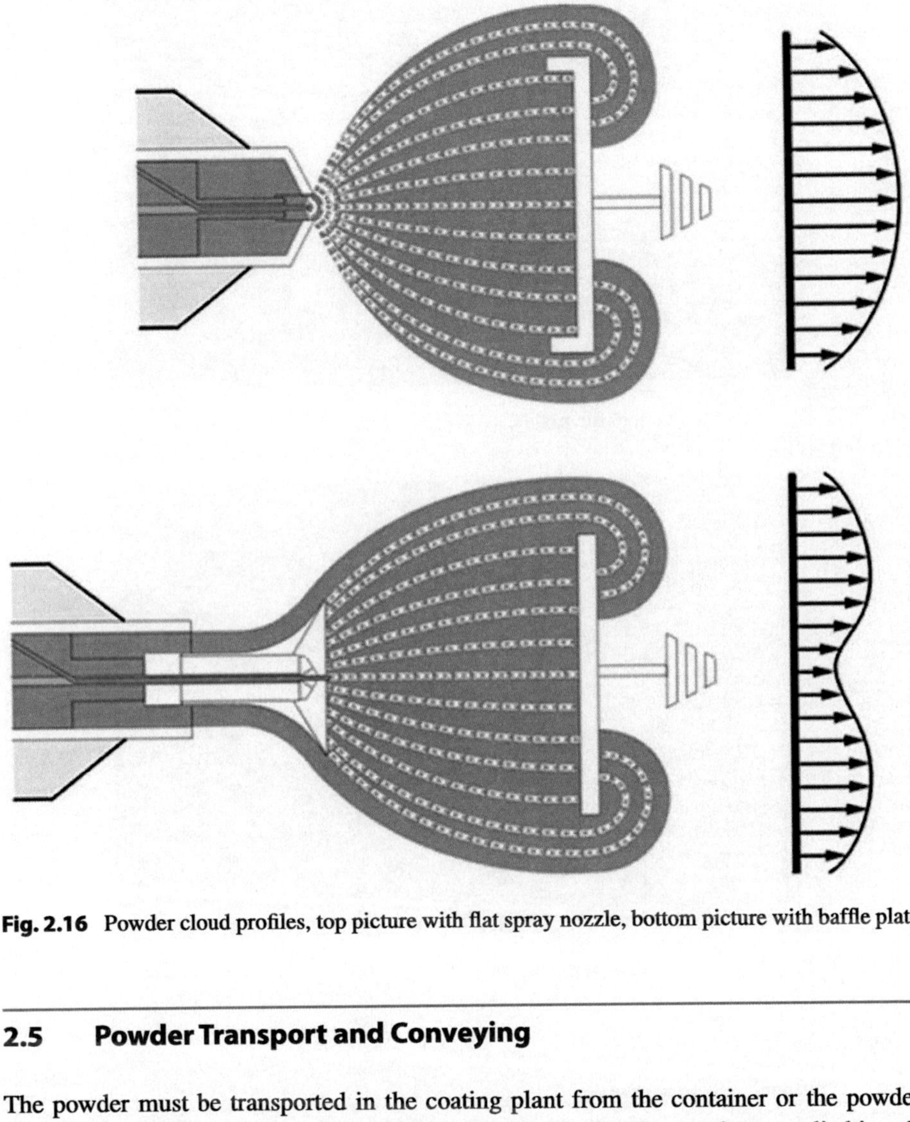

Fig. 2.16 Powder cloud profiles, top picture with flat spray nozzle, bottom picture with baffle plate

2.5 Powder Transport and Conveying

The powder must be transported in the coating plant from the container or the powder center to the spraying elements. In addition, the powder that has not been applied is collected in the booth, recovered and returned to the intermediate container. The powder is therefore always in motion. During transport both from container to container and from the container to the spraying guns, it is important that the structure, e.g. the particle size, is maintained. Basically, a distinction is made between pneumatic flow and dense flow conveying (Fig. 2.19).

In flight conveying, a powder/air mixture is conveyed to the powder guns at a speed of 10–15 m/s. The powder content is very low, corresponding to the set powder quantity, e.g.

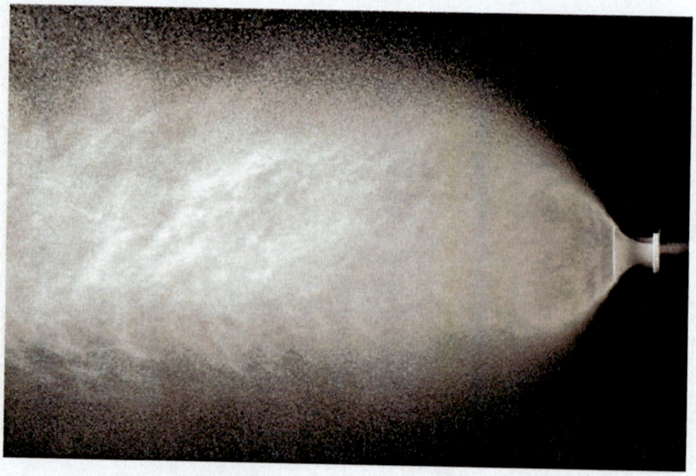

Fig. 2.17 Powder cloud with baffle plate

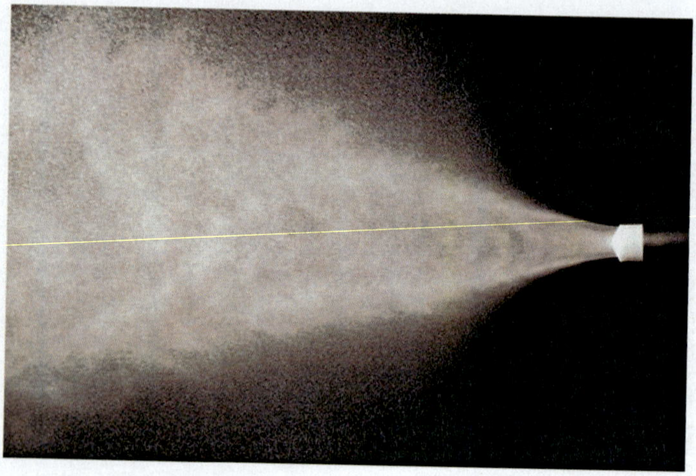

Fig. 2.18 Powder cloud with flat spray nozzle

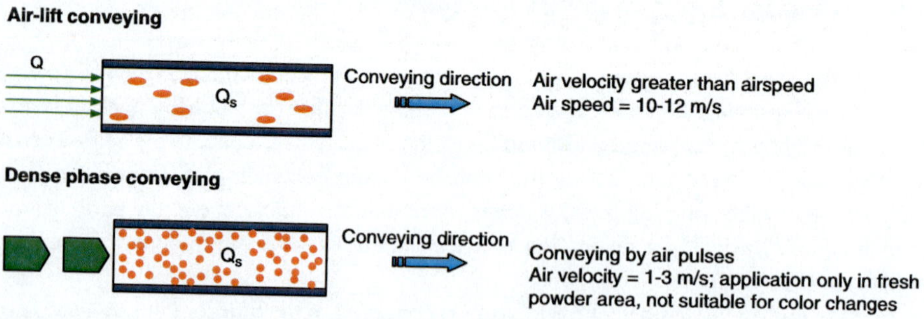

Fig. 2.19 Principle of the various powder conveying options [34]. Pneumatic conveying is mainly used in powder extraction; dense-phase conveying is suitable as injector conveying or in the powder hose

180 g/min. The powder content is very low, according to the set powder quantity, e.g. 180 g/min. The dense phase conveying works with a considerably higher proportion of powder and a slower speed in the hose between 2 and 4 m/s. Depending on the application, this powder/air mixture must be accelerated again at the outlet (e.g. gun) in order to achieve atomization.

Characteristics of flight feeding are:

- Stress on the powder due to kinetic energy – resulting from the conveying speed
- Increase in the proportion of fines
- Relatively complex output regulation
- Limited conveying length.

Characteristics of dense phase feeding are:

- Easier and more accurate dosing
- No mechanical stress on the powder due to low conveying speed
- Low conveying air
- If necessary, acceleration of the air/powder mixture at the guns for sufficient atomization.

Properties of dense phase feeding:

+ Less mechanical stress on the powder, uniform and gentle transport
+ Greater flexibility in adjusting the intensity of the atomization (for guns)
+ Low conveying air
+ Low abrasion (hose – slow transport of the powder, therefore less abrasion)
  – Tendency to pulsation with shorter hoses
  – Long hoses from 20 m necessary
  – Slow response time to desired change in flow rate
  – Higher risk of hose clogging (given by hose length)

Characteristics of flight feeding:

+ Uniform powder flow, no pulsation
+ Shorter hoses possible from 5 m
+ Fast response time to desired change in flow rate
+ Minimal risk of hose clogging
  – Greater mechanical stress on the powder due to the higher speed
  – Basic intensity of atomization (for guns) defined by air volume for delivery

2.5.1 Mechanical Properties of the Powder

The mechanical and partly also the chemical properties influence the conveyance of the powder. Powder coatings have substantially different properties than liquid coatings. They are dry and often poorly pourable. Depending on its additives, the powder has poor sliding properties and thus a high tendency to sintering. Its essential properties are:

- Dry, grainy
- Particle sizes from 1 to 150 µm
- Wide repose angle
- Tendency to sinter at too high speeds
- Different chemical compositions
- Temperature-sensitive from approx. 50 °C
- Bulk density approx. 1.4 g/cm3

2.5.2 Requirements for the Support System

The requirements on the conveying system are diverse. Depending on the application, a distinction is made between precision conveying and mass conveying. The precision conveying is used in the powder feed to the gun, in the remaining transports no high consistency of the mass flow is required.

Precision conveying requires high consistency of powder flow and repeatability of the flow rate setting on the equipment. Uniformity must be maintained at all times to avoid pulsing or "spitting" of the powder at the guns. The required powder flow rate is 50–500 g/min. Possible conveying systems for this are injectors or powder dosing systems. In order for the powder to be drawn in continuously and homogeneously, it must be fluidized and/or vibrated or stirred.

In mass conveying, the powder must be transported from one container to another. The conveying should be economical and gentle here. The required powder throughput is 2–5 kg/min. Suitable conveying systems are injectors, powder pumps, plug conveyors, vacuum conveyors, screw conveyors and vibrating chutes.

2.5.3 Powder Feeding Systems

2.5.3.1 Powder Feeding to the Gun: Precision Feeding

Today, three systems are in use for powder feeding, injector feeding, pump feeding and hybrid feeding, where elements from different systems are combined.

Typical conveying systems used in flight conveying are venturi systems, often referred to as "injectors". Here, powder is sucked from a container by the negative pressure in the venturi suction area and accelerated at the outlet and thus transported to the guns. New on the market is the Smart Feeding technology, which works with a combination of pressure tank and external powder dosing. The provision of the powder quantity is separated from the dosing. This gives flexibility in the following type of powder conveying, whereby the flight conveying is favored here.

Dense phase feeding based on the plug conveying principle has been in use for many years. Here, the powder is conveyed in so-called plugs at a speed of 2–4 m/s by means of two alternately operating pinch valves and an intermediate chamber. Larger quantities, up to 4 kg/min, can be conveyed here. Due to the functional principle, however, a discontinuous powder flow is produced, which is not suitable for conveying to the guns, but is good for fresh powder metering or intermediate conveying from container A to B.

Rarely used, because of the missing possibility of a color change, are double piston diaphragm pumps, which suck in and eject powder via two alternately working diaphragms. Dense phase conveying systems, which are also used for powder supply, are relatively new.

The DP conveying system (digital dense phase conveying) was introduced to the market in 2003. It works like a motor via a piston, which creates a vacuum during start-up (suction) and transports the sucked powder quantity further during shut-down. The piston movement is controlled by compressed air. The powder quantity is adjusted by the stroke frequency, since the piston space is constant. There are stroke limiters with which one mechanically sets a working range as a basic setting. In addition to the drive air, the conveying process works with very little air and leads to a discontinuous conveying process (small powder plugs). In order to achieve a homogeneous powder flow at the end of the conveying distance as a prerequisite for a good coating result, the pump is designed as a double piston pump. Nevertheless, a minimum conveying distance is required until the small powder plugs are evenly distributed in the hose. However, due to the low air volume (approx. 1–2 m^3/h), the flow rate is very low and does not yet allow atomization at the gun outlet. Therefore, at the gun inlet, an additional air is used to bring the powder back to a higher velocity in order to obtain a powder cloud. In addition to the low-air conveying, the lower wear compared to the injector technology was mentioned as an advantage, whereby the advantages mainly lie in longer maintenance intervals and not in the overall expenditure, as the maintenance is quite complex.

DP conveying has raised awareness in the market for the influence of air during powder application. The air in the powder cloud has an enormous influence on the coating behaviour, as the air must escape when the powder is deposited on the workpiece. Thus, there are greater advantages when coating workpieces with depressions, grooves and undercuts, since here the air interferes more strongly with the deposition behavior.

DP was the starting point for further developments of dense phase conveying systems for gun conveying, such as the HDLV, AP01 pumps and others. Here, a combination of vacuum and plug feeding is used, whereby powder is sucked in by means of vacuum in a first step, and is conveyed further and blown out in the next step. To reduce a pulsating powder cloud with negative influence on the layer thickness distribution and application

Fig. 2.20 IPS coating center for powder preparation

efficiency, these systems work with two chambers, which work alternately or with only one channel, but higher frequency of the cycle between suction and conveying.

New on the market is the so-called smart feeding system (SFT). This differs from the previous injector or pump solutions. It is a system consisting of a pressure hopper, powder preparation, powder dosing and powder transport, which is used as a component in the coating center (Fig. 2.20). The coating center designed in this way is a complete solution for powder preparation, powder feeding and color change and thus differs from previous powder centers. All process steps can be fully automated and centrally controlled. Powder preparation with fluidisation, ultrasonic sieving and vibration, conveying and dosing are all housed in a closed system, thus preventing contamination of the environment or, conversely, the ingress of dirt from outside. In order to prevent the fluctuations of manual activity and operating errors, the colour change is carried out fully automatically at the push of a button and autonomously. Table 2.5 shows a comparison of the different conveyor systems.

The Injector Feeding The powder is sucked in here according to the Venturi principle via the suction pipe. The powder particles are transported at a speed of approx. 15–20 m/s. The transport is carried out with air. The necessary air volume is composed of the conveying air and dosing air.

Generally speaking:

- Conveying air high/dosing air low → large powder quantity
- Conveying air low/dosing air high → small powder quantity

Table 2.5 Conveying system for gun feeding in comparison

Criterion	Injector	Injector with control circuit	Rod injector	DDF	HDLV	Powder pump AP01	Powder pump SFT
Size	Small	Small	Small	Large	Large	Large	Large System concept
Air consumption in the delivery duct (m^3/h)	4–5	2,5–4,5	3–5	1–2	3–4	3–4	3–4
Air consumption operating air (m^3/h)	–	–	–	2–3	2–3	2–3	–
Setting range (g/min)	50–300	20–300	20–300	50–300	50–350	50–350	50–400
Delivery accuracy (%)	±10	±3	±3	±3	±3	±3	±2
Volume fluctuation due to powder level	Yes	Yes	Yes	Yes	Yes	Yes	No
Typical service life	2–12 weeks	2–12 weeks	2–12 weeks	3–6 months	6–12 months	6–12 months	8–15 months

[a]Depending on powder type and flow rate, 1-shift operation

The requirements for injector feeding are:

- High consistency of the powder flow
- Repeatability, reproducibility
- no pulsation
- Throughput 50–500 g/min.

The conveying is influenced by the powder hose length, powder hose routing, powder hose diameter, filling level and fluidizing air in the powder hopper, suction height in the hopper and the total air volume in the powder hose (Figs. 2.21 and 2.22).

The Injector The function of the precision injector is to suck out and convey a predetermined quantity of powder from the container to the gun and to add metering air to keep the total air quantity constant. By changing the nozzle pressure, the vacuum in the injector changes. Depending on the vacuum, the powder quantity is varied. High pressure means high vacuum at the nozzle and more powder is sucked in. Due to the flow changes, the air volume also increases/decreases. With smaller powder quantities (smaller air quantities), the air quantity must be increased via the dosing air, so that the powder can be continuously conveyed in the powder hose without pulsation.

The advantages of this system are its mechanically simple design and easy maintenance. The disadvantages are the low control accuracy (±10%), a regularly necessary maintenance in relatively short intervals, due to the wear of the nozzles. *The rod injector:*

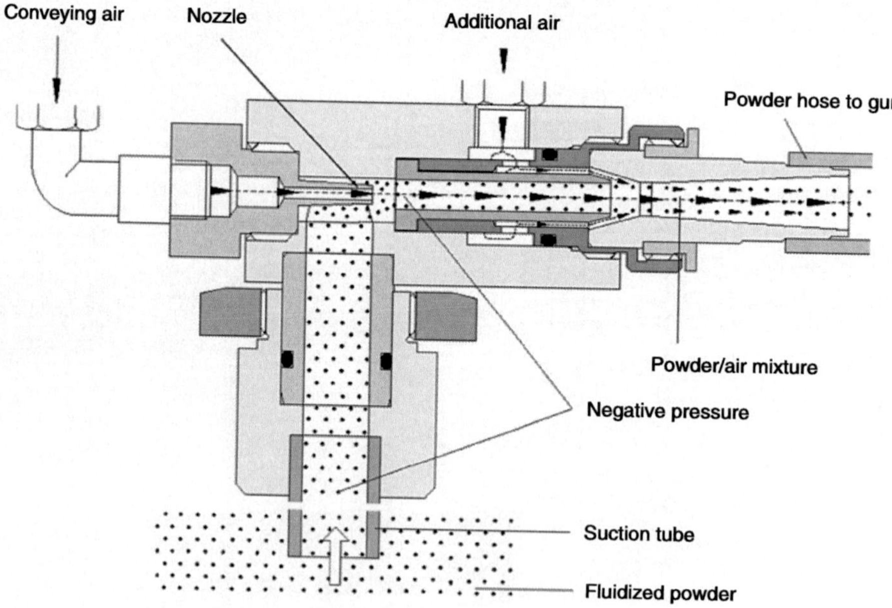

Fig. 2.21 Cross-section through an injector

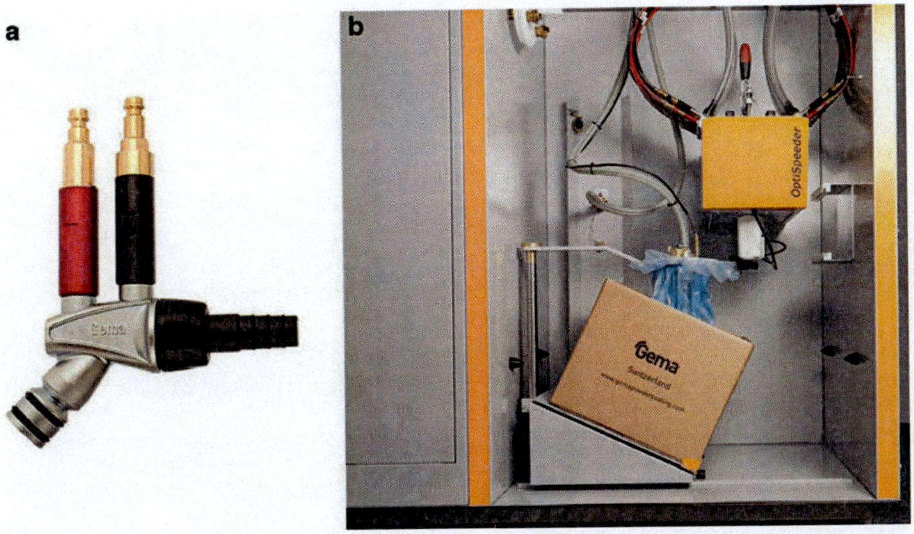

Fig. 2.22 Injector (**a**); injector delivery from the container (**b**)

Here the powder is sucked in at the bottom of the container. The characteristics of the rod injector are:

- small dimension, diameter 20 mm
- no fluid air necessary, therefore no dust development to the outside
- streamlined transport of the powder, almost straight, radii only occur in the connection from the hose to the gun
- uniform powder feeding
- versatile use (manual systems, powder center, powder container)
- automatic cleaning
- an inexpensive system when used in the original container
- somewhat more expensive compared to simple injectors.

The powder pump (Fig. 2.23): It is suitable for both precision and mass conveying. The great advantage here is that the air velocity is many times lower than with the previously known systems. With this conveying technique, long transport distances (>10 m) are not a disadvantage, but rather an advantage, so that a quiet powder cloud can be achieved.

Features:

- low wear
- constant, reproducible conveying
- constant output
- clean environment due to closed system

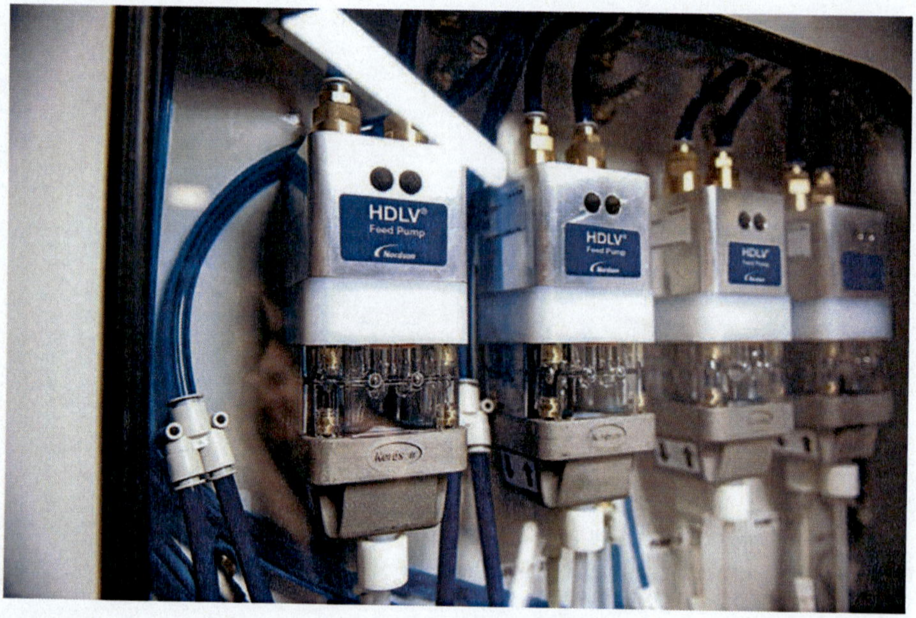

Fig. 2.23 Powder pumps for conveying the powder coating to the gun

- long conveying distances possible, up to 40 m
- complex technology
- higher costs

2.5.3.2 Conveying the Powder to the Booth/Powder Centre: Bulk Conveying

Injectors or dense phase pumps can be used to convey powder from the delivery container to the powder hopper and return it from the cyclone or filter separator to the hopper.

The advantage of injector conveying is the low purchase price. However, the systems are subject to high wear, require a lot of air and are not suitable for conveying over long distances.

Nowadays, only low-air systems, feed pumps, are generally installed in mass conveyance.

The plug conveying is used for the return conveying from the cyclone/filter separator and the conveying from the large container with suction at the bottom. Diaphragm pumps are rather seldom in use, as no colour change is possible. Dense phase pumps are commonly used, which have advantages in the further processing of the powder due to their low air content in the delivery flow and are designed in such a way that a colour change is possible. The available pumps differ from each other in detail, but they are all active pumps with one suction and one delivery cycle (DDF, HDLV, PP06, IP5000).

The powder quantity is 3–5 kg/min. For powders with poor flowability, depending on the type of suction, vibration alone is not sufficient and the powder must also be fluidized.

2.5.3.3 The Powder Pump

In mass conveying, the powder must be transported from one container to another. An economical and gentle conveying of the powder is decisive.

The task of the pump is to suck in and convey as large a quantity of powder as possible with the smallest possible expenditure of energy, e.g. from the fresh powder container into the central powder container, from the cyclone collecting container to the powder container.

2.5.3.4 Dense Phase Conveying Systems for Bulk Conveying (Intermediate Conveying)

Plug conveying (thrust conveying): In plug conveying, the available pressure is used for conveying. Plug conveying in the pressure chamber is terminated by two valves. The conveying process takes place in different steps. In the first step, the lower valve is closed and the upper valve is open. The powder falls into the pressure chamber. Now, in the second step, the upper valve is closed and the lower valve is opened. In the third step, a pressure pulse is fed into the pressure chamber. In the fourth step, the lower valve is closed again to prevent the pressure in the line from decreasing. In the fifth step, the upper valve is opened and the process starts again. If the conveying system is well adjusted, only a small amount of air is added to the powder, so that no separating elements are necessary at the end of the line. The powder is pushed and not entrained by an air stream. The push conveying is used when powder quantities of 4–6 kg/min are to be conveyed with little air. The conveying distance can be up to 30 m. Separation of the powder-air mixture at the point of consumption is no longer necessary. Figures 2.24, 2.25, 2.26 and 2.27 show how plug conveying works.

Fig. 2.24 Initial situation

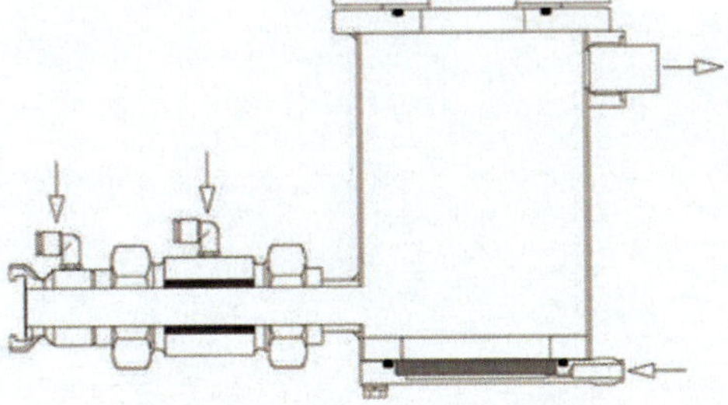

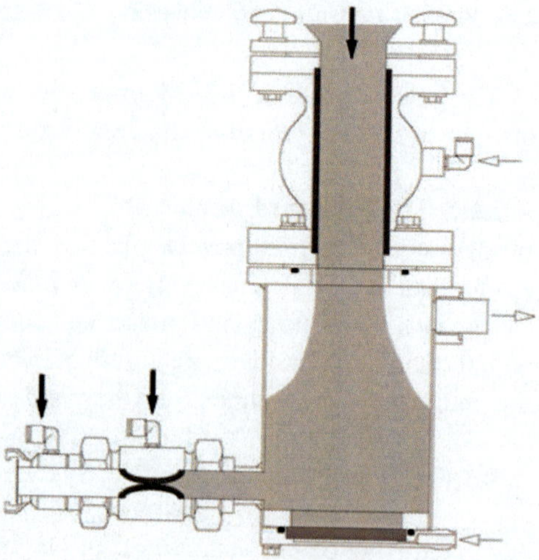

Fig. 2.25 Filling

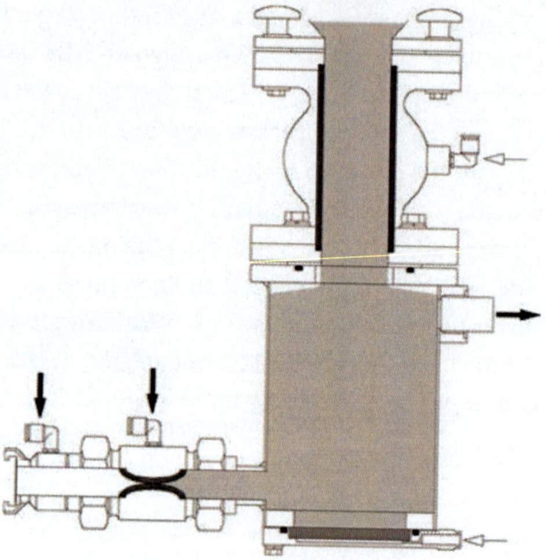

Fig. 2.26 Intermediate tank is filled

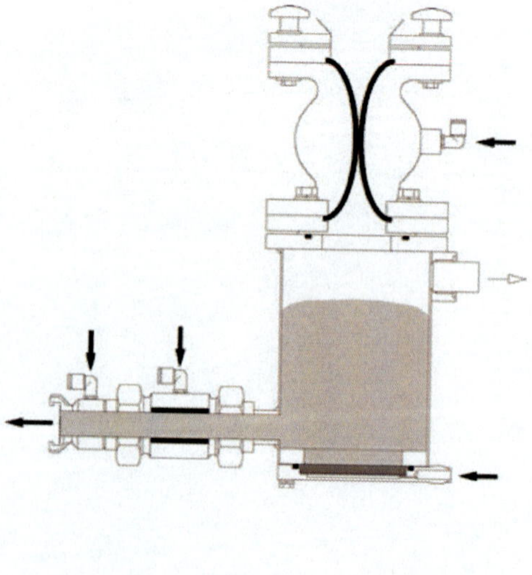

Fig. 2.27 Conveying the powder

Dense phase conveying: Dense phase conveying is a system with low-air conveying of the powder with low flow velocities in the transport line. This has decisive advantages. The principle of the pump is based on counter-pressure conveying, which allows considerably lower transport speeds of the powder. The pump consists of two cylinder units which alternately generate a vacuum when the pistons are raised. The powder is thus sucked in a defined manner into the dosing chamber from the powder container. Fluidization of the container is not necessary. The two cylinders are precisely counter-controlled. Once the dosing chamber is filled with powder, the conveying air is blown in. Then the piston moves back to the lowest position in order to generate a vacuum or powder suction again. The alternating suction and blow-out is controlled by pinch valves, which are opened or closed accordingly. By changing the frequency, a powder quantity adjustment is possible.

The DP pump for mass transfer works on the same principle as described in Sect. 2.5.3.2. However, the cylinder space is considerably larger to allow larger powder quantities to be processed.

The majority of manufacturers use a porous fluid tube as a powder chamber, to which vacuum (negative pressure) and positive pressure are alternately applied. Before and after this, pinch valves are used, which alternately close off the powder chamber on both sides. Two chambers work alternately to increase the powder quantity. One manufacturer uses a closed plastic hose instead of a porous tube as the powder chamber. This reduces the risk of contamination and clogging (Figs. 2.28 and 2.29) (Table 2.6).

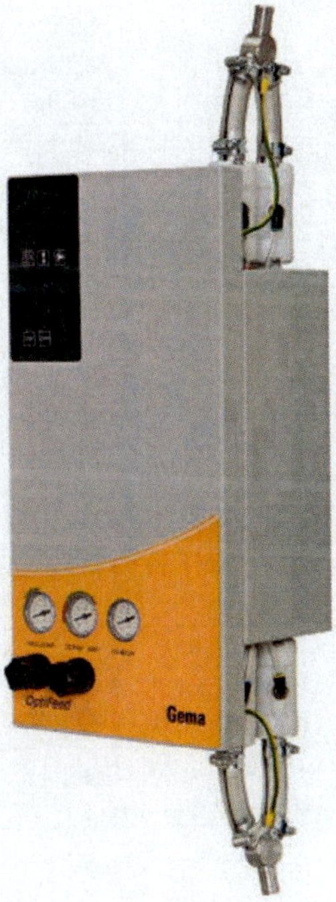

Fig. 2.28 Example of a powder pump for mass transfer

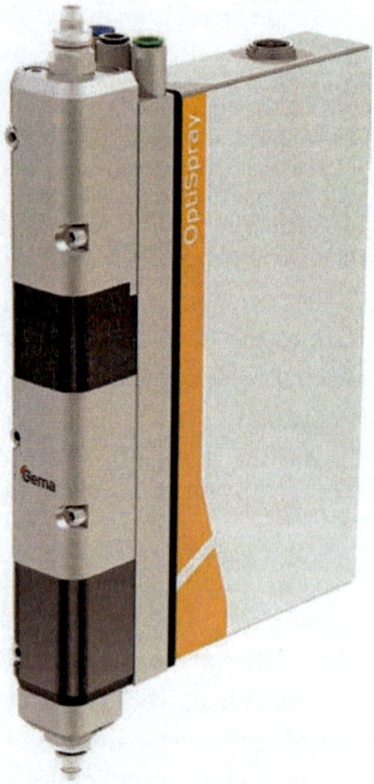

Fig. 2.29 Example of a powder pump for delivery to the gun

Table 2.6 Comparison of mass conveyors

Criterion	High-performance injector	Plug Conveyor	DP	HDLV	Pump PP06	Pump IP5000
Size	Small	Small	Large	Very large	Large	Large
Type of installation	Directly at the container	Below tank, cyclone hopper	Near powder container <1 m	Near powder container <1 m	Near powder container <1 m	Near powder container <1 m
Suction principle	Active	Passive, by gravity	Active	Active	Active	Active
Separation powder-air-mixture	By mini cyclone	Not required	Not required	Not required	Not required	Not required
Air consumption (m^3/h)	4–15	4–7	10–25	13–16	Approx. 12	10–25
Color change	Yes	Yes	Yes	Yes	Yes	Yes
Delivery rate* (kg/min)	1–1,5	2–3	4–6	3–5	3–5	3–5
Loading of the powder	High	Low	Low	Low	Low	Low

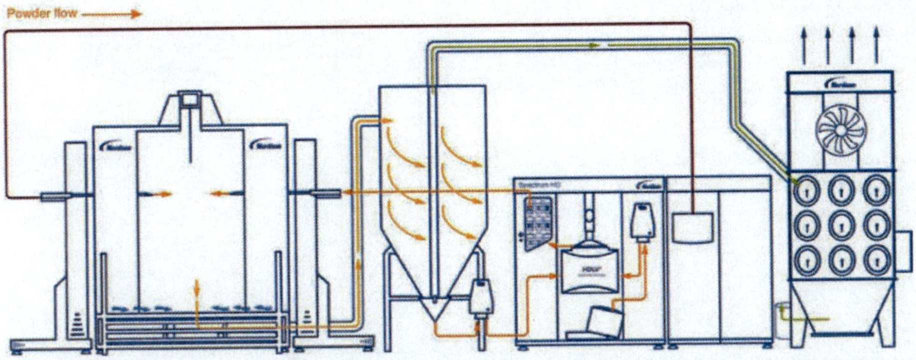

Fig. 2.30 The powder flow diagram: 1 – coating booth, 2 – gun lifting device, 3 – bottom suction of overspray powder, 4 – powder centre with powder sieve (5); gun pumps (6); transfer pumps, transport of powder from the container to the powder container, as well as recycled powder back to the powder container (7), control (8), 9 – cyclone, 10 – final filter, 11 – filtered out powder residue (waste)

Figure 2.30 shows the powder flow diagram in principle. The powder is transported with precision conveying from the powder center to the guns. The overspray from the booth is transported to the cyclone. In the cyclone the powder-air mixture is separated. Fine particles are extracted via the filter system, the prepared, sieved powder is transported back to the working hopper.

2.5.4 Separation of the Powder-Air Mixture

The extraction of the powder from the booth is normally carried out in accordance with the safety regulations, i.e. with large quantities of air. Before the powder can be returned to the powder hopper, it must be separated from the air. Separation systems such as monocyclone, multicyclone, cartridge, plate or bag filters are used for this process.

The monocyclone (Fig. 2.31): The cyclone makes use of the physical law of centrifugal force. The powder particles entering the cyclone tangentially are thrown against the outer wall by the rotation. With the help of the spirally moving air, the powder particles are conveyed towards the cyclone outlet and collected. The air is extracted via the central pipe. Basically, the higher the inlet velocity, the higher the loss of fine powder particles and the higher the probability of sintering. The smaller the cyclone diameter, the higher the resistance, which requires a larger motor and more energy. Too low an air velocity will cause more powder buildup. And with powder, the larger the powder particles the higher the collection efficiency. False air attracted via leaks also leads to poorer separation efficiencies (Fig. 2.32).

Fig. 2.31 Construction of a cyclone

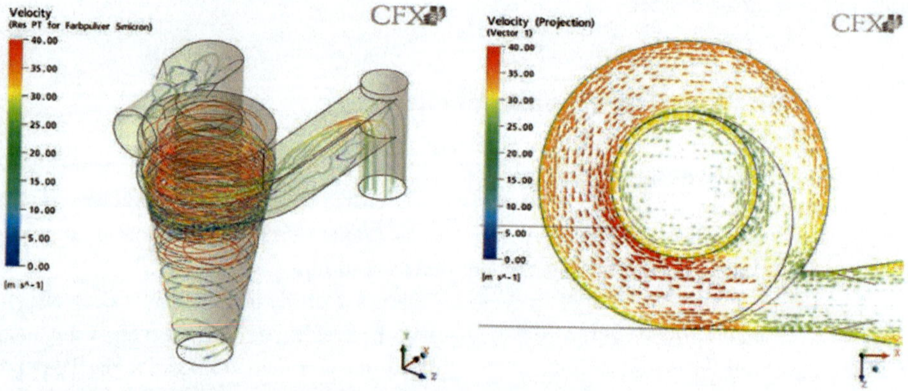

Fig. 2.32 Velocity of 5 μm-sized powder particles in the cyclone

Features:

- low-cost separation system
- suitable for colour change
- large particles are separated better than smaller ones
- change of the grain spectrum
- separation efficiency 90–97

- possibility of powder sintering (max. Air speed 20–22 m/s)
- voluminous, large overall height
- piping required between booth and cyclone inlet.

The Plate Filter Sintered porous plastic materials are used for powder separation in the plate filter (Fig. 2.33). To reduce the porosity, the surface is coated with a Teflon coating. The powder is deposited on the filter and the clean air is extracted through the fan via a final filter. During operation, the filter cake on the plate filter increases continuously, thus increasing the pressure resistance and reducing the air volume. The plate filter is cleaned from the inside by cyclic pressure surges and the filter cake is thrown off.

Features:

- robust, durable filter material → low maintenance requirements
- stable suction volume
- constant differential pressure
- reduced compressed gas consumption due to differential pressure-dependent control
- very high separation efficiency, almost 100
- compact design
- suitable for a wide range of materials, including abrasive materials
- not suitable for colour change, generally no final filter is suitable for colour change → colour change only possible with cyclone
- energy consumption depending on installed filter area
- somewhat higher investment costs

Fig. 2.33 Structure of a sinter plate filter (rigid body filter)

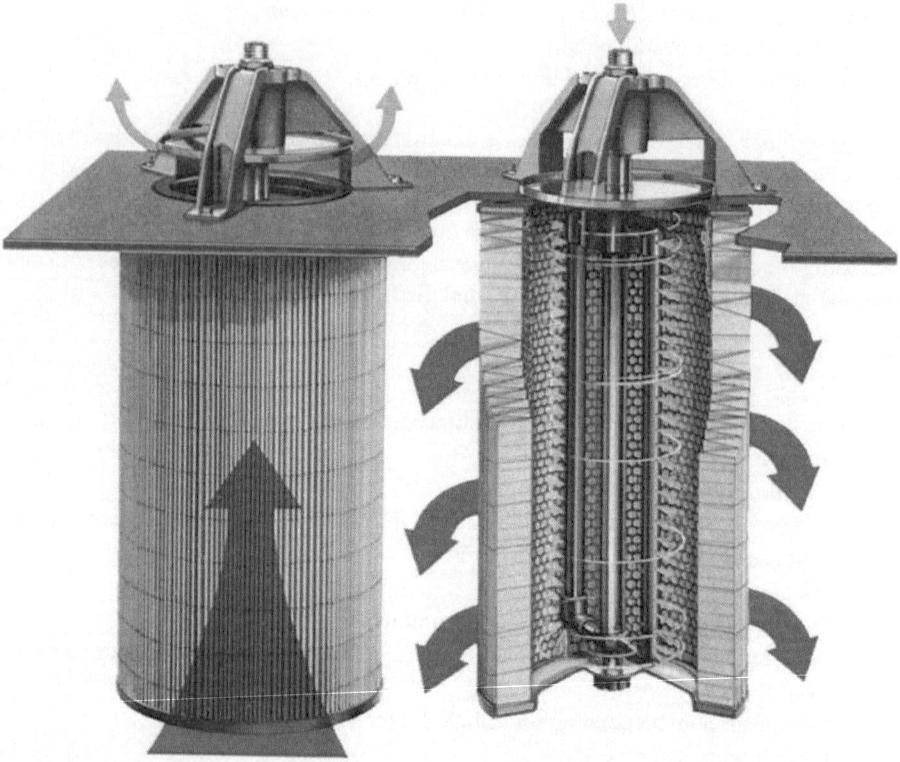

Fig. 2.34 Design of a cartridge filter

The Cartridge Filter Cartridges are cylindrical filters with endlessly folded filter materials, Fig. 2.34. For filtration, a filter cake is built up which is blown off cyclically. Compressed air jet systems or rotary vane cleaning systems are used for filter cleaning. The cleaning effect in the depth of the pleat is problematic (Table 2.7).

Features:

- high efficiency
- low pressure resistance
- high air performance
- cartridges are susceptible to wear (abrasion → "fluff")
- not suitable for colour change
- sensitive to moisture
- danger of clogging with fine powders

Table 2.7 Comparison of cyclone and filter separation

Criterion	Monocyclone with final filter	Plate filter	Cartridge filter
Separation efficiency (%)	90–97	>99	>99
Color change	Yes	No	No
Air flow rate	Large	Large	Large
Air resistance	Medium – high	High	Medium
Absolute filter	Necessary	Recommended	Recommended
Lifetime	Years	>5000 h	2000 h
Susceptibility to interference	Low	Small	Medium
Power – volume ratio	Large	Small	Medium
Acquisition price	High	High	Medium

2.5.5 Powder Preparation

The recovered powder must be cleaned of fibres, large particles and dirt before use. Mechanical sieves and/or ultrasonic sieves with different mesh sizes are used for this purpose. The following requirements should be met by screening machines:

- Separation of powder and contaminants
- Dissolving or sieving out powder lumps
- Discharging the contamination
- Loosening of fresh powder

Depending on the quality requirements, different types of screening machines are used.

The Vibrating Sieve The powder is guided onto a vibrating sieve mesh. The combination of horizontal and vertical movements causes the powder to fall through the sieve. The dirt particles are retained (Fig. 2.35). Most vibrating screens do not have a device for automatic dirt removal. The dirt must be extracted manually.

Features:

- inexpensive, simple screening device
- easy to clean
- low-wear
- no automatic dirt discharge
- mesh size 200–750 µm; standard 300 µm
- not suitable for high quality demands

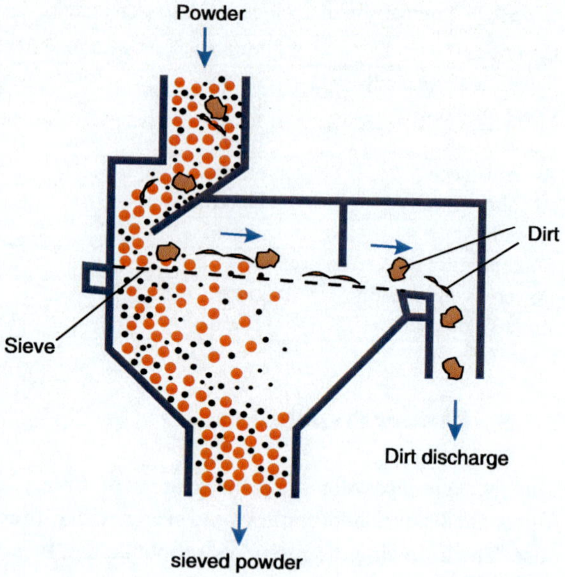

Fig. 2.35 Schematic diagram of screening with dirt discharge [35]

Rotary Sieve (Centrifugal Device) In the rotary sieve, Fig. 2.36, the powder particles are centrifuged through the sieve basket by means of centrifugal force. The powder is fed continuously via a screw to the centrifugal blades. The dirt particles are automatically conveyed to the dirt discharge opening.

Features:

- automatic dirt discharge
- mesh sizes from 200 μm possible
- color change complex
- many moving parts
- high-maintenance
- continuous supply required

Vibratory and Tumbler Screening Machines The tumbler screen, Fig. 2.37, is used when high quality is required. Due to the clearly defined adjustable movements of the sieve machine, the powder collects in the centre of the sieve and the dirt particles are conveyed to the sieve edge where they are automatically separated out.

Vibratory tumbler sieves are screens with an unbalance motor drive. The material is fed to the centre of the sieve. The circular tumbler vibrations cause the powder to move outwards in a spiral and the fines are separated downwards through the sieve.

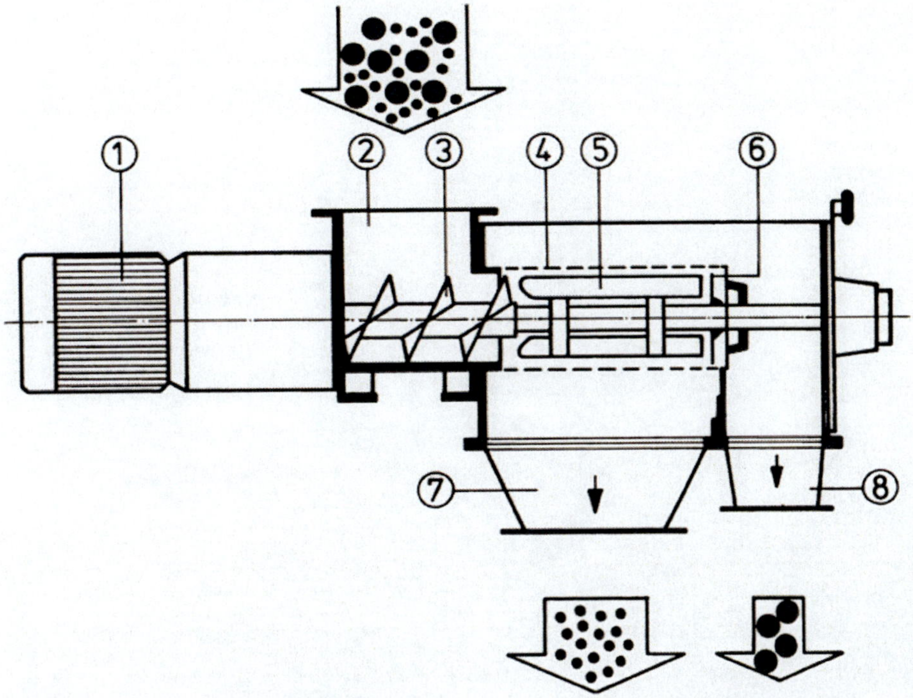

Fig. 2.36 Structure of a rotary sieve. 1 = motor, 2 = incoming powder + dirt, 3 = screw conveyor, 4 = screen basket, 5 = centrifugal vane, 6 = dirt discharge opening, 7 = screened powder, 8 = dirt

Features:

- automatic dirt discharge
- mesh sizes of 100 μm possible, gentle powder conveying
- color change very complex
- vast engine
- high investment costs

Vibratory and tumbler sieves can be additionally equipped or retrofitted with an ultrasonic cleaning device. In this way, the mesh size can be reduced and the throughput increased.

2.5.5.1 The Ultrasonic Sieve

The ultrasonic sieve [36], Fig. 2.38, is used when high quality is required. The ultrasonic oscillating head keeps the sieve or the powder in motion, thus enabling significantly faster sieving with smaller mesh sizes. Ultrasonic sieves can be used practically everywhere and usually lead to a higher throughput. In addition to the sieve parameters, the sieving

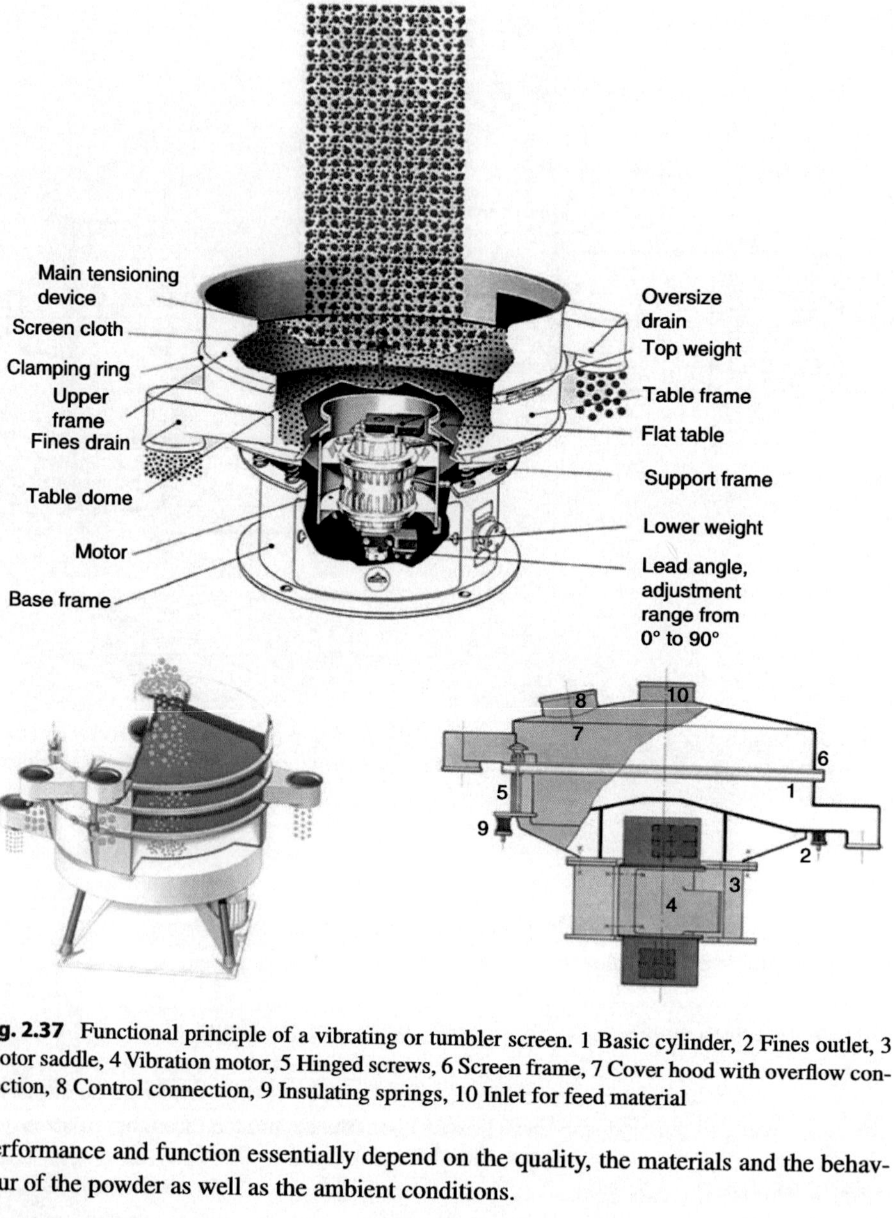

Fig. 2.37 Functional principle of a vibrating or tumbler screen. 1 Basic cylinder, 2 Fines outlet, 3 Motor saddle, 4 Vibration motor, 5 Hinged screws, 6 Screen frame, 7 Cover hood with overflow connection, 8 Control connection, 9 Insulating springs, 10 Inlet for feed material

performance and function essentially depend on the quality, the materials and the behaviour of the powder as well as the ambient conditions.

Features:

- high screening capacity
- short cleaning times, automatic dirt discharge possible
- mesh sizes of 80 μm and smaller theoretically possible, but throughput low

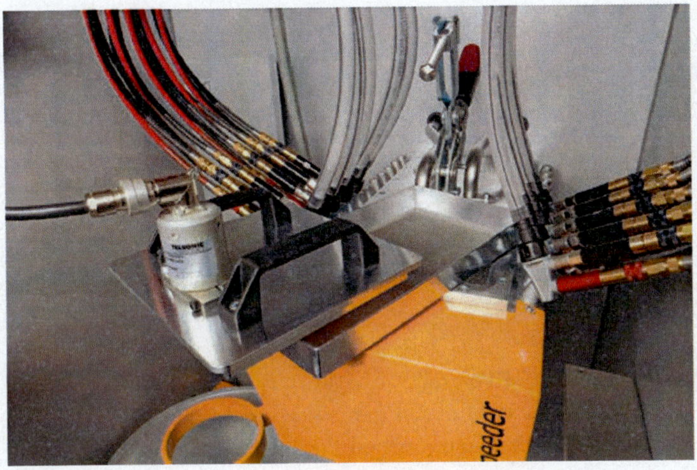

Fig. 2.38 Ultrasonic sieve

- standard mesh size 250 µm
- homogeneous mixing of overspray and fresh powder possible
- low heat generation
- the finer the sieve, the greater the risk of fabric tears
- higher costs in acquisition and maintenance

The ultrasonic sieve has the task of loosening the powder, crushing loose powder agglomerates and sieving out dirt particles such as fibres, fluff and smaller dirt particles. It is not suitable for screening out large particles in the sense of changing the particle size distribution. Larger or heavy particles, such as hooks, which mistakenly get into the sieve, can damage the sieve cloth.

The powder feed to the sieve should be as uniform and low in air as possible. Suitable conveying systems are, for example, plug feeding with pumps. The feed rate into the sieve must be adjusted so that the maximum sieve capacity cannot be exceeded, otherwise the sieve will become clogged. The smaller the mesh size, the lower the sieving capacity. Manufacturers recommend a mesh size of 200 µm as standard, with which most powders can be sieved well and in sufficient quantity. When using acrylic, low temperature and effect powders, smaller mesh sizes are not recommended. As a general rule, light powders with specific weights <1.4 g/cm^3 are more difficult to screen, as gravity has a supporting effect on screening performance.

The fluidizability and flowability of the powder is measured with a fluidized measuring vessel in standardized tests according to DIN ISO 8130-5 or AFNOR T30A. The fluidizing capacity is calculated from the difference between the fluidized and non-fluidized height of the powder column. The flowability results from the quotient of the two heights multiplied by the discharge quantity in a defined time.

The tricle factor R is often specified by the powder supplier in the powder data sheet. Here there are the following quality classifications:

- $R > 140$ = very good
- $R = 120–140$ = good
- $R < 120$ = moderate to poor

When using an ultrasonic sieve, it is recommended to use only powders with a factor $R > 120$. Powders with poorer flowability values can also cause problems during conveying and fluidization.

Particle Size Distribution of Powder

The mesh size of the sieve must match the grain spectrum of the powder. In general, it should be noted that grain sizes in the range of the mesh size can clog the sieve, Figs. 2.39 and 2.40. Sieving in powder production must be carried out with at least ¼ smaller mesh size. A high proportion of fine grain powder (Fig. 2.41), e.g. particles smaller than 10 μm with a proportion greater than 8%, can also lead to problems, as agglomerates form and clog the sieve.

Wax Content in the Powder

Powders with high wax contents are often difficult to sieve and lead to clogging of the sieve. Since wax contents are not specified, a simple manual test provides indications. Fresh powder is pressed together by hand. A solid powder lump indicates high wax contents and probably difficult conditions for powder conveying and sieving (Fig. 2.42).

Sieve Maintenance

The screen mesh is subject to wear and must be replaced regularly. Preventive maintenance can significantly extend the service life. For this purpose, the mesh must be inspected daily for adhesions and sintering. These must be removed in order to maintain the

Fig. 2.39 Sieve mesh clogged by fine powder grain

Fig. 2.40 Fine grain deposits on the screen mesh

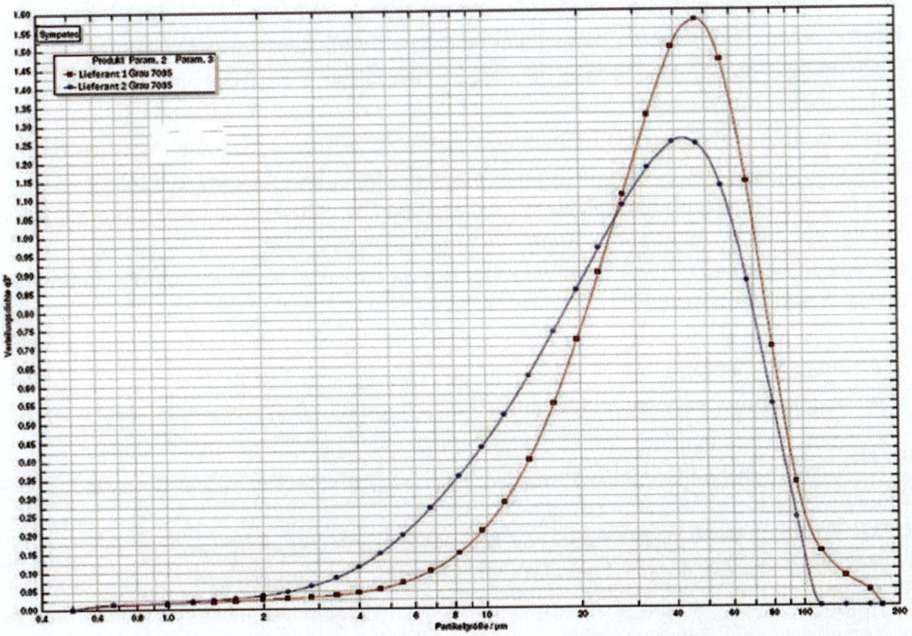

Fig. 2.41 Diagram of the particle size distribution of two powder coatings of normal (red) and high (blue) fines content

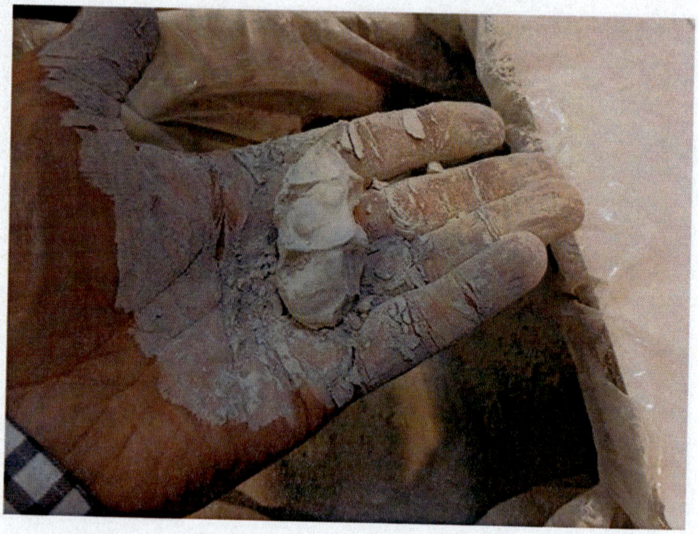

Fig. 2.42 Simple test of the wax content in the powder

screening performance. This is best done with solvent and a brush or by cleaning in an ultrasonic bath. The sieve mesh must not be cleaned with steel brushes or similar.

Cleaning is recommended after one week at the latest, more frequently if necessary. A replaceable sievemakes the maintenance timing flexible and in case of damage to the screen mesh, a replacement is immediately available. Screen tensioning may only be carried out by specialised service providers. The sieve must be subjected to a quality control with measurement of the temperature increase and the contact resistance from the sieve fabric to the frame, in order to comply with the ATEX suitability.

Humidity and heat have a negative influence on the sieve performance. The following values should be observed:

- Ambient temperature in powder storage and coating area T < 30 °C
- rel. Air humidity in powder storage and coating area < 50%.

General recommendation and advice:

- An ultrasonic sieve cannot improve a bad powder into a good powder.
- In general, the use of tested powder with LEL >40 g/m3 is assumed.
- A replaceable sieve enables the quick replacement of contaminated or defective sieves.
- The pre-screening at the powder supplier should be at least 1/4 smaller than the selected mesh size for coating. Example: Sieving for coating with 200 μm → Pre-sieving for powder production <150 μm.
- Only use powders with good flowability and a flow factor R > 120.

- Fine grain content <10 μm and less than 8%.
- Regular maintenance and cleaning of the screen mesh and the screening machine.
- The sieving capacity of powders can only be reliably determined in production use at the customer. For this purpose, the customer must observe the sieving effect when using a new powder for the first time.
- Use a suction gun or vacuum cleaner for cleaning when changing colors.

2.5.5.2 The Choice of Mesh Sizes

The quality of the screening depends on the choice of mesh size and the screen bit. Vibration sieves require larger screening areas than rotary sieves for the same capacity. Table 2.8 shows a summary of the most important characteristics of the different screening machines.

2.5.6 Powder Preparation in the Hopper

Powder is difficult to trickle and cannot be easily transported out of the container. The powder tends to agglomerate, which must be prevented by suitable measures. The preparation of the powder applies to both manual and automatic systems. In automatic systems, fluidization and vibration are a must.

The following techniques are used today:

- Preparation by agitators
- Preparation by fluidisation
- Preparation by a combination of fluidization with vibration.

The powder container has the task of providing the paint powder for the spraying elements. In the process the powder should be handled as gently as possible. In addition, the

Table 2.8 Comparison of the screening machines

Criterion	Vibration sieve	Rotary sieve	Vibrating tumbler sieve	Ultrasonic sieve
Mesh size (μm)	300–500	200–300	120–250	80–260
Sieve quality	Sufficient	Good	Very good	Very good
Relative screening capacity[a]	Low	Good	High	High
Dirt discharge	No	Yes	Yes	Possible[b]
Size	Small	Large	Large	Small
Color change	Yes	No	No	Jab
Maintenance costs	Low	Medium	Medium	Medium
Integration in plant engineering	Simply	Elaborate	Elaborate	Simply

[a]Sieve power in relation to size
[b]Design conditional and depending on system integration

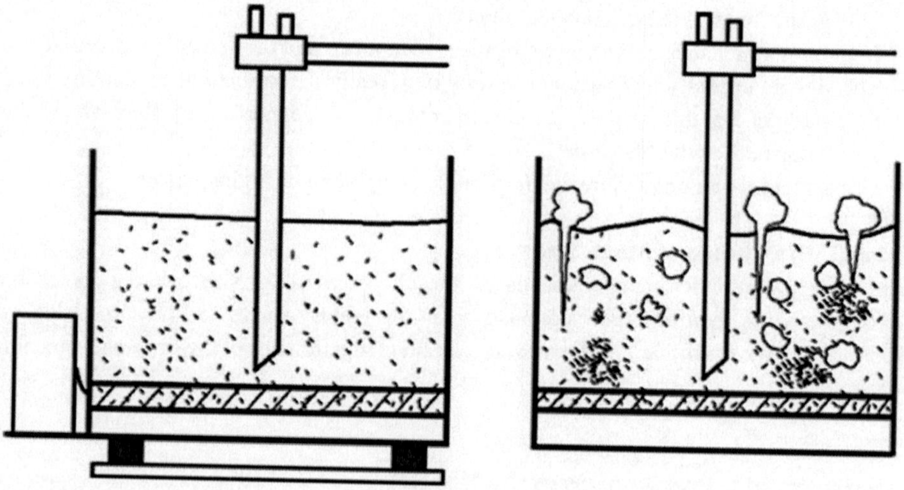

Fig. 2.43 Right: formation of air channels and craters in the fluidized powder container, left: uniform homogeneous powder/air mixture [33]

fresh powder and the recovered recirculating powder must be mixed as uniformly as possible. A uniform powder cloud depends largely on the powder feed.

Figure 2.43 shows the problem of the formation of craters and air channels in the powder. Suitable measures (fluidization/vibration) must be taken to keep the powder moving uniformly.

The Stirring Hopper The agitator prevents the powder from agglomerating. The stirring hopper can be used for all powder types. It is particularly suitable for powders that are difficult to fluidize. However, these hoppers are rarely used in Europe today. Figure 2.44 shows the structure of an stirring hopper. The injectors for the spraying elements are attached to the sides at the bottom, and the bottom has a discharge flap.

Features

- suitable for all types of powder, small and large quantities
- regular powder feed (suction directly under the container)
- no change of the grain spectrum
- container can be easily emptied and cleaned
- additional agitator mechanism
- powder is loaded by agitator, not suitable for fine and metallic powder
- expensive to buy
- limited number of pistols (12 pieces)
- injectors difficult to access, difficult to dismantle when tank is full

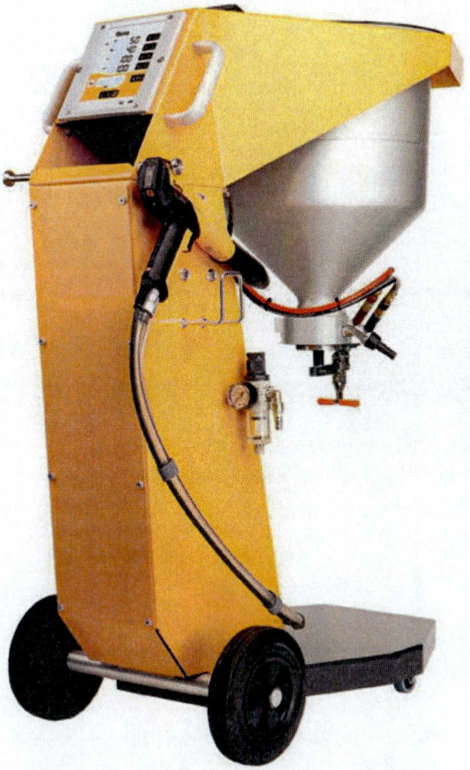

Fig. 2.44 Manual coating system with stirring tank

Fluidized Feed Hopper Fluidization is the best known method of powder preparation. Compressed air flows through a uniformly porous bottom and makes the powder flowable. Depending on the powder quality, additives and particle size distribution, the powder tends to form lumps to a greater or lesser extent. As a result, channels and craters form, which affect regular powder discharge. The tendency to irregular fluidization is increased when the powder sinks during a plant stop and the plant is restarted. These effects can be prevented by suitable prefluidization. The volume of the fluidized powder increases by a factor of 2, this means that the containers must only be half filled when at rest, in order to avoid overflowing of the powder during working operation.

Features:

- simple low-cost design
- can be cleaned for colour change
- injector connections easily accessible on the cover
- powder demix
- powder output is level dependent

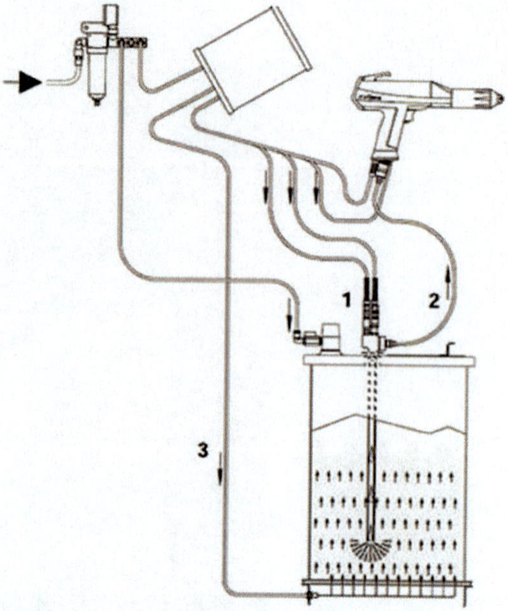

Fig. 2.45 Functioning of a fluid container, 1 = transport air, 2 = powder hose, 3 = fluidizing air – drum container

- regular fluidisation difficult for large containers
- powder build-up at the edge
- fluid air consumption 20–40 m3/h per m2 fluid floor
- more difficult cleaning of the powder container

Combination Container Fluidization/Vibration To eliminate the disadvantages of the fluidized container, the powder preparation is supported with additional vibration. The vibration prevents the formation of lumps and craters in the container, air channels (see Fig. 2.45) are broken immediately, the powder can flow evenly in the container.

Features:

- constant regular fluidization
- also suitable for powders with fine grain spectrum
- approximately 30% lower air consumption compared to fluidisation without vibration
- no margin
- optimum powder, air mixture
- increased investments, as additional equipment is necessary

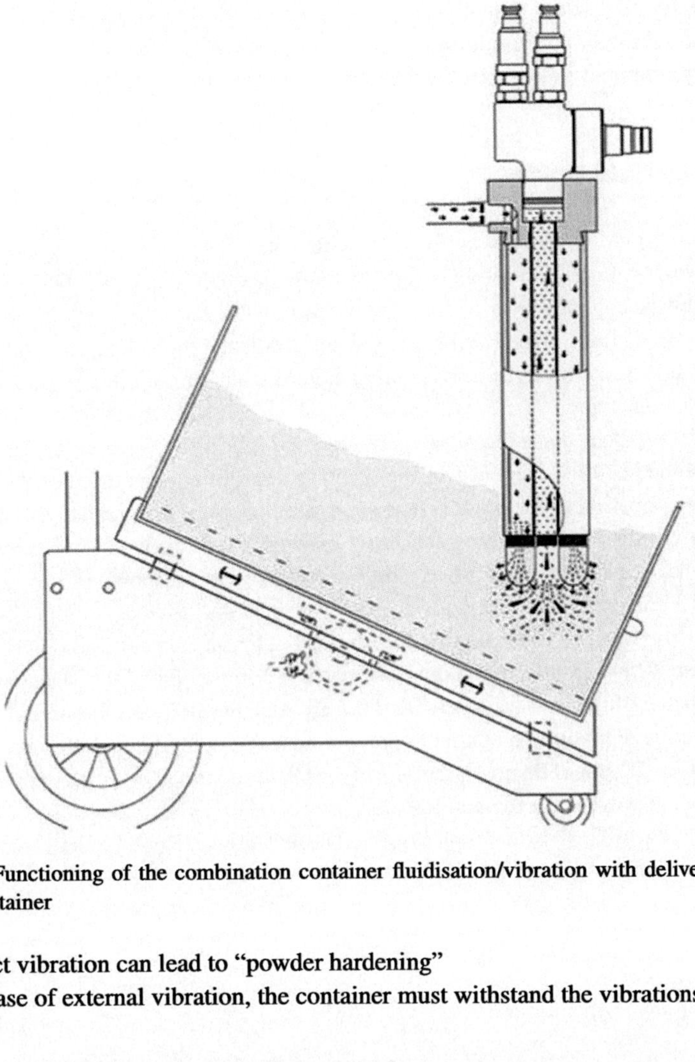

Fig. 2.46 Functioning of the combination container fluidisation/vibration with delivery from the original container

- incorrect vibration can lead to "powder hardening"
- in the case of external vibration, the container must withstand the vibrations

Powder Delivery Directly from the Original Container- Vibratory Box Feeder The delivery container is used as powder container. Fluidization takes place locally at the suction pipe. The flowability of the powder is supported by the vibrating floor on which the container stands, Fig. 2.46.

Features

- fast colour changes, ideal for frequent colour changes
- no powder transfer and cleaning of the powder container

- suitable for all powder systems
- minimum fluid air consumption
- delivery container must resist the vibrations

2.5.7 The Hose Guide

The following guidelines only apply to systems with injector feeding. When using dense phase conveying, the hose routing – apart from the fact that kinks must be avoided – is not critical. However, in order to obtain a uniform powder output, a certain hose length is necessary. Here, the longer the better. For an efficient color change, bends should be avoided or at least the hose should be routed in 90° bends. This applies to both injector and dense phase conveying.

Immediately after the injector, a "calming section" in the form of a straight powder hose should be installed. Already during the planning of the plant, attention must be paid to the correct location of the powder hopper, in order to be able to realize the shortest possible hose lengths. When routing the hose, care must be taken to ensure that it is laid without kinks and that the prescribed radius are adhered to, as shown in Fig. 2.47. Only in this way is a uniform and continuous powder transport guaranteed.

An optimum powder hose length is less than 10 m for an internal diameter of 10–11 mm and less than 12 m for an internal diameter of 12–13 mm. Table 2.9 gives an overview of the dependence of the powder output and the air consumption on the internal hose diameter and recommendations on suitable powder hose diameters, depending on the desired powder output. It should be noted that the higher the powder quantity, the more difficult it is to charge the powder for the application.

Figure 2.48a and b show examples of good hose routing, Fig. 2.49a and b are examples of how powder hoses should not be routed.

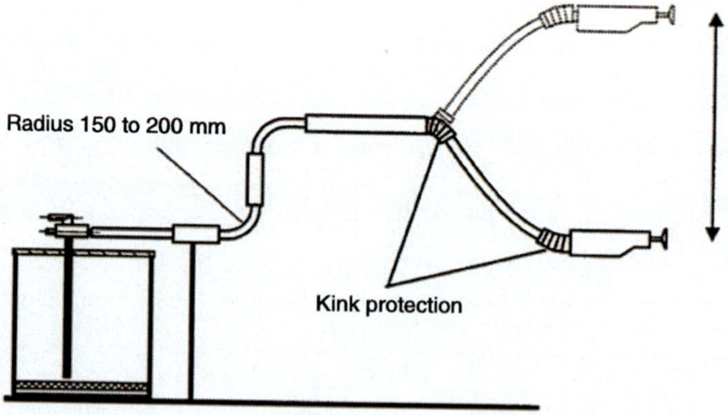

Fig. 2.47 Example of an optimal hose routing [33]

Table 2.9 Information on hose dimensioning with recommended powder output quantity by injector delivery at the gun. The values are valid for a max. Hose length of 12 m [35]

Inner diameter of powder hose (mm)	Cross-sectional area (mm^2)	Compressed air consumption in (m^3/h)	Powder output (g/min)
8	50	Approx. 2.0–3.0	<50
10	78	Approx. 2.8–4.0	30–200
11	95	Approx. 3.3–5.5	50–300
12	113	Approx. 3.7–6.8	100–400
14	154	Approx. 5.7–7.8	400–700[a]

[a]The problem of large powder quantities is not necessarily the transport itself. The limitation here is the efficient loading of the powder quantity

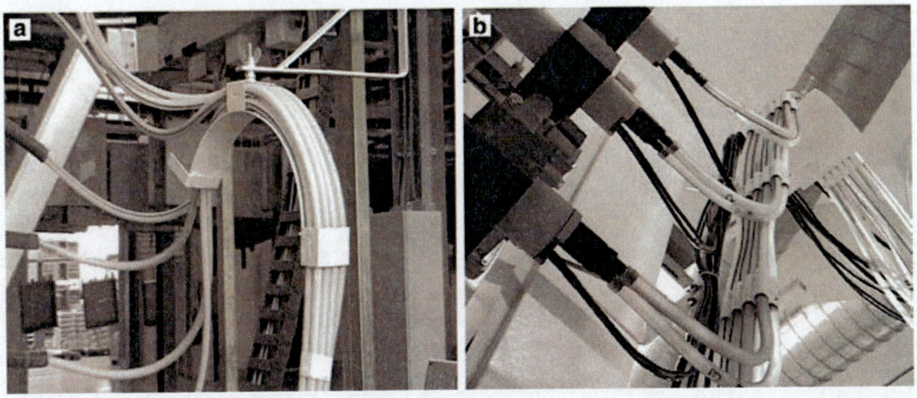

Fig. 2.48 (a,b) Examples of optimum hose routing. Hoses are free of kinks and clearly arranged. Changing the hose is relatively simple and can be carried out quickly

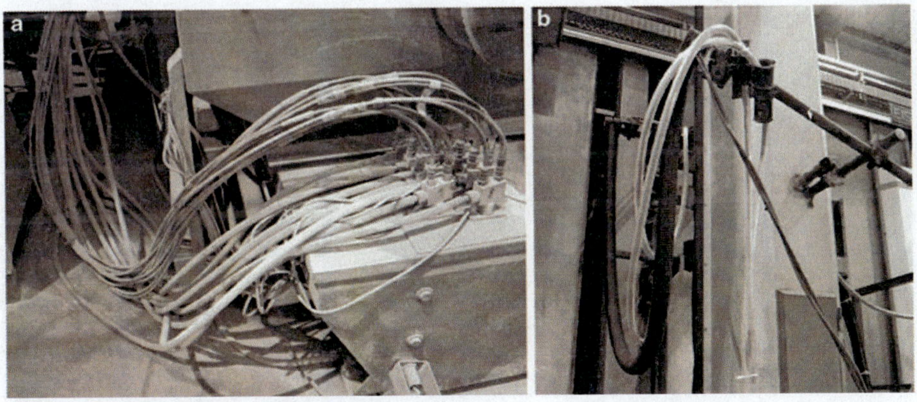

Fig. 2.49 (a,b) Example of how powder hoses should not be installed hoses laid too long and unclear

The hoses recommended by the equipment manufacturer should always be used as powder hoses. PVC hoses or material containing plasticizers are not suitable.

2.5.7.1 Conductive Hose

Conductive powder hoses are increasingly being used. The advantage over conventional hoses is, a lower tendency to powder deposits in the hose. This means that cleaning can be achieved quickly and easily by blowing out the hose when changing colours. Color carry-over can also be reduced in this way. With the antistatic property of the hose, electric shocks caused by static charge during powder transport (friction) are avoided.

Features:

- antistatic
- prevent powder deposits
- reduce color carryover
- wear resistant – longer service life is achieved
- stable and kink resistant

In order to comply with the ATEX directive, only conductive powder hoses have been available for some years now. On the one hand, these are safe with regard to electrostatic charging and the associated risk of high-energy discharges. In addition, their flake-like surface offers advantages during color changes due to the low adhesion tendency of the powder in the hose. In the past, foam plugs were shot through the hose for better cleaning during color changes. This leads to a high effort during the color change. With the conductive hoses, blowing through with compressed air pulses is usually sufficient. (Fig. 2.50).

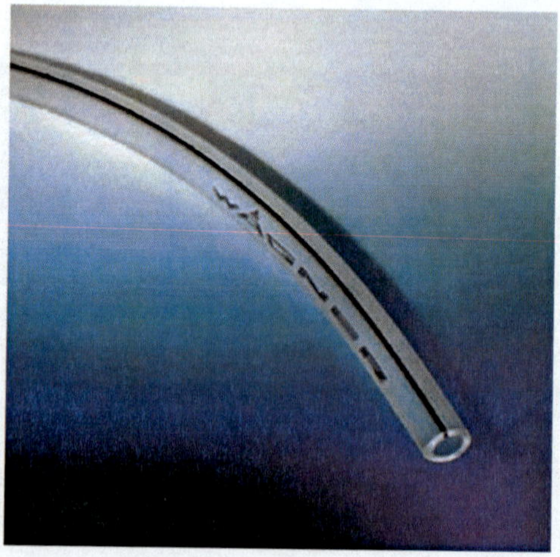

Fig. 2.50 Powder hose with conductive carbon fibre

2.6 System Planning and Design

If a new coating plant is projected at the beginning often stands the wish or even the obligation to use an environmentally friendly coating process and/or to automate the process. The range of parts to be coated and the desired plant technology already result in many relevant parameters that dictate a choice between different plant concepts, system, booth and recovery types. The degree of automation to be selected is also determined here.

Specifications from the Range of Parts to Be Coated

The first step is to determine the main parameters that have an impact on the design of the system. These are:

- logistic conditions
- budget guidelines
- the quantity of parts to be coated
- the workpiece geometry
- the required layer thicknesses
- the required coating quality
- number of colors, frequency of color changes with recovery
- the percentage distribution between main and spot colours
- use of the workpieces (inside, outside, aggressive environment such as marine climate or chemical laboratories)

Further parameters that have an influence concern the specifications from the plant technology

- Feeding system
- Suspension gear and hanger design
- Pre-treatment

and the structural conditions, such as

- space conditions
- floors
- other disturbing influences such as pillars, room height or ceiling structures

The floors must be electrostatically conductive and there must be no influences such as air turbulence or solvent vapors in the vicinity of the planned booth location that could interfere with the coating quality. Excessively high room temperatures must also be avoided.

2.6.1 System Concepts

The most common system concepts are briefly presented below. They can all be realized with normal booths as well as with the so-called quick color change systems.

2.6.1.1 Two Recovery Units

This concept is suitable for conventional color change with additional recovery. After booth and gun cleaning, the recovery systems are changed. This can be done by repositioning the powder hoses or by an automated changeover (Fig. 2.51).

2.6.1.2 Mobile Cabins

By means of movable booths, which are operated alternately, booth cleaning is possible outside the chain conveyor (Fig. 2.52). Today, this system technology is being replaced by alternating booths with a powder centre.

2.6.1.3 Booths with Separate Conveyor Lines

Two completely separate conveyor lines are fed through two coating booths. While one booth is coating, the other booth can be cleaned (Fig. 2.53).

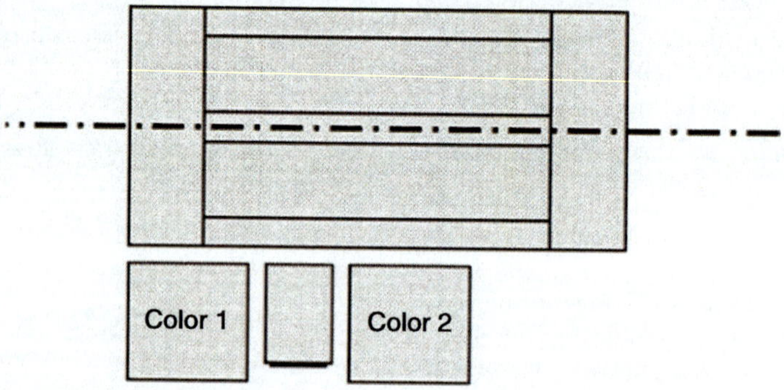

Fig. 2.51 Cabin with two recovery units

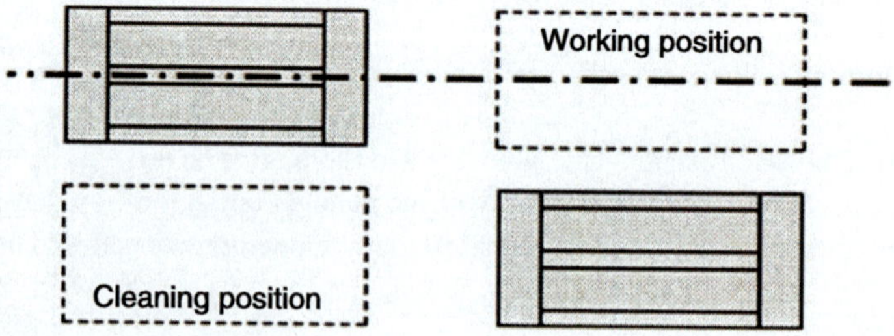

Fig. 2.52 Example of movable cabins

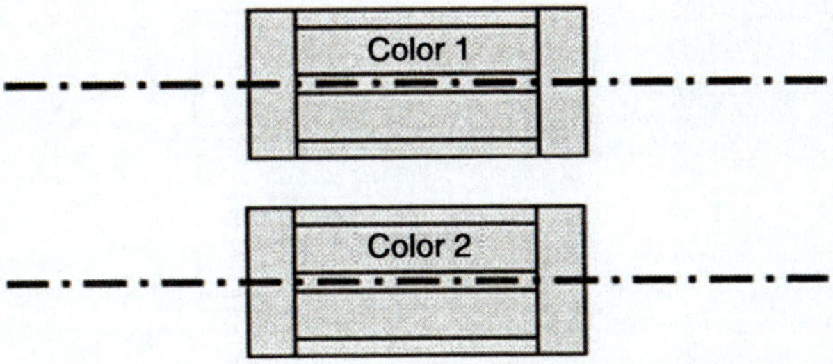

Fig. 2.53 Booths with separate lines

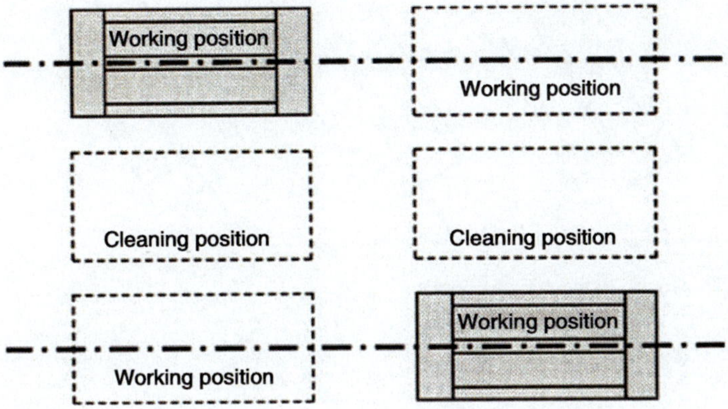

Fig. 2.54 Mobile booths with two conveyor lines – no longer built today

2.6.1.4 Mobile Booths with Two Conveyor Lines

Very high flexibility is achieved by using two conveyors in combination with movable cabins. The disadvantage of this concept is the large space requirement (Fig. 2.54).

Booths with Power&Free Conveyor Here, the parts pass through the pre-treatment and the oven together, but pass separately the specific booth. In this way, one booth can be cleaned while the other is in operation. In most cases, Power&Free conveying systems can only be used economically if other conveying tasks can be solved by this conveying system too. (Fig. 2.55).

The *multi-colour system* uses a booth with a quick colour change system. The quick change is achieved by a compact design of the booth and a minimum amount of powder in the circuit. The booth construction is designed for short cleaning times due to a flat booth floor, flow-optimized suction and a non-conductive booth material with low tendency to powder adhesion.

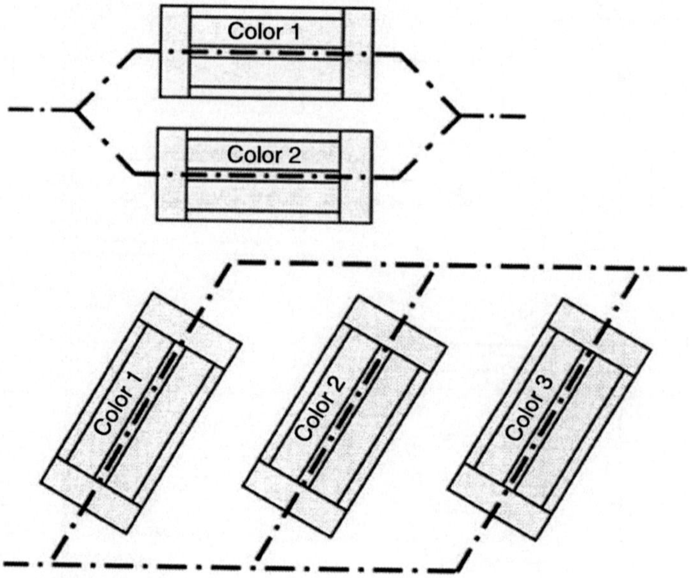

Fig. 2.55 Cabins with Power&Free conveyor

2.6.2 Coating Booth

A coating booth is a closed space on all sides except for the openings for the part entrance/exit openings, automatic electrostatic spraying devices and/or the working openings for manual coating.. All coating booths must be technically ventilated during the spraying of the coating material in order to largely exclude explosive mixtures, to avoid hazards from inhalation of the particle-air mixtures and to minimize contamination of the booth environment with powder. Metal, glass or plastic are used as materials for booth walls. The distances to the actual coating area must be large so that the walls themselves do not become a coating part, because all conductive parts must be grounded like the workpiece. Today, plastic booths are often used. As an electrical non-conductor, plastic has the advantage of charging with the first powder approach, repelling the subsequent powder particles and thus itself acting as a trigger electrode on the spray cloud. The material used must be flame-retardant, and the outside of the plastic wall or parts of it may only be backfilled with a conductive material at a distance of >9 mm in order to exclude sliding brush discharges.

An important point is the air flow in the booth. The task of the booth exhaust, the exhaust air is to maintain a uniform air sinking velocity at the booth openings, not to allow powder to escape, and no turbulence may occur.

The air volume flow is calculated from the amount of coating material sprayed per time unit. Air velocities of 0.4 to 0.7 m/s must be met at the booth openings [37]. As a rule, an inflow velocity of 0.5 m/s is aimed for.

DIN EN 16985:2019–04 describes the safety requirements for spray booths for powder coating [38].

2.6.3 The Choice of Cabin Type

Basically, a distinction is made between booths for single-color, multi-color and quick color change applications. Today, the system elements are usually modular in design. This means that the individual system elements are interchangeable, depending on requirements and application. Depending on the manufacturer, powder booths can be made of coated or stainless metal, single- or double-walled plastic or glass.

For the booth construction materials steel and plastic, Fig. 2.56 shows the cleaning times and powder accumulation on the booth walls in comparison with uncoated sheet steel.

In the case of plastic booths, the booth wall is also negatively or positively charged by the negatively or positively charged powder particles and these are thus repelled. The difference in the course of the electrostatic field lines in a steel and plastic booth is shown in Fig. 2.57. The electric field lines always start at the electrode and end at a grounded part.

In the following, the advantages and disadvantages of the different cabin materials and different cabin designs will be shown.

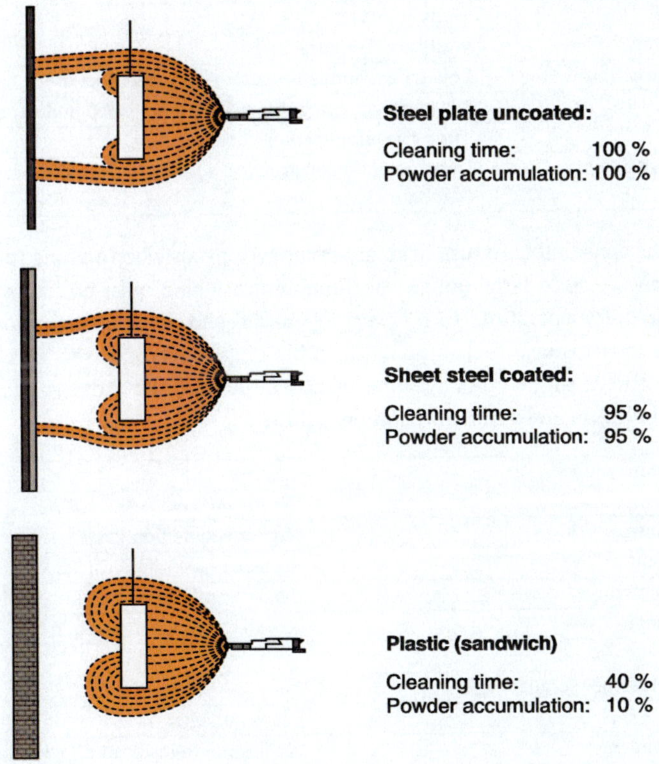

Fig. 2.56 Cleaning times and powder accumulation on the different booth materials. State of the art today is the use of plastic sandwich elements

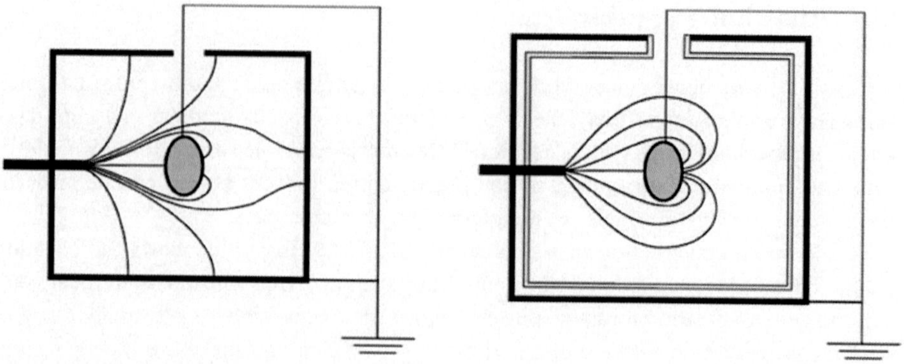

Fig. 2.57 Left: Course of the field lines in a steel cabin, right: Course of the field lines in a plastic cabin

The steel cabin	
Advantages	Disadvantages
Cost-effective	Many joints
Simple assembly	Walls powder attracting
Element construction	Powder accumulation after prolonged operation
Any shape	Cleaning (scraping, blowing) complicated and expensive
Hard surface	Time consuming color change
Compliance with safety requirements	Depending on construction additional floor cleaning necessary (wiper, scraper)

Steel booths are suitable for use in single-color systems or when coating to loss. Longer production times lead to large powder accumulations, which must be removed regularly even in single-color operation. In recovery operation and color change, booth cleaning must be done in two steps: first scraping, then blowing. This increases the cleaning time considerably. Today, single-color systems are also built in plastic. Advantages are the minimal powder build-up and a high circulation stability.

The plastic cabin	
Advantages	Disadvantages
Practically joint-free	Higher acquisition costs
Powder repellent	Surface not scratch resistant
Cleaning by blowing	Safety regulations require complex design
Fast color change	Regular maintenance required

Glass cabin with floor scraper	
Advantages	Disadvantages
Scratch resistant	Additional floor cleaning necessary
Cabins bright, better visibility when cleaned	Powder residues in corners

Glass cabin with floor scraper	
Advantages	Disadvantages
Work friendly, especially with pale colours	Longer color change times
	Regular maintenance required
	Risk of glass breakage during installation

2.6.3.1 Comparison of Plastic Cabins

With plastic booths, the basic advantages compared to a sheet metal booth are always the same. However, the design of the booth leads to differences, e.g. in the cleaning time. In general, a distinction is made between the classic rectangular booths, round booths and drop-shaped booths [39].

Cabin with cleaning machine	
Advantages	Disadvantages
Automatic interior cleaning	High investment costs
Shortest cleaning time independent of cabin height	large construction heights require a pit
Extraction along the entire length of the cabin	High compressed air and energy consumption
Simple pipe connection	
No scratching of the walls	
Low powder adhesion	
Cabins bright, better visibility when cleaned	
Work friendly, especially with pale colours	

Plastic cabin with fixed floor and inserted exhaust air duct as well as front-side suction unit	
Advantages	Disadvantages
Simple construction	Scratching walls and floor
Low construction height	Powder deposits in the floor covering
Fast color change	Cabin length and height limited
	Only limited accessible

Round cabin	
Advantages	Disadvantages
Simple construction	Scratching walls and floor
No moving parts	Powder deposit on the insertion wedge
No cavities that cannot be controlled	High energy consumption
Acceptable airflow	Cleaning of large surfaces
Simple pipe connection	

Round cabin with hand openings	
Advantages	Disadvantages
Small cabin surface, quick cleaning	Size, height and width

Round cabin with hand openings	
Advantages	Disadvantages
Cleaning possible without inspection	Two pre- or post-coating stations each not possible
Walls far away, therefore little powder deposit	
Simple pipe connection	
Hand coating without additional openings	
Low energy consumption	
Additional, lossless main color recovery	

If the highest coating quality is required, the following conditions should be met:

- The booth must be equipped with a separate coating room. This must be operated in positive pressure with purified air.
- The screening machines must be equipped with a dirt discharge system.
- The cabin walls must be powder-repellent.
- For multi-colour operation, an additional microfilter must be installed after the filter unit.
- During cabin cleaning, personnel must wear special fibre-free clothing.
- Filter materials must not cause additional contamination (degradation of fibres).

Contamination smaller than 70/80 μm is difficult or impossible to remove even with very good screening machines. This can lead to disturbance effects on the surface.

Fresh and recirculated powder must be sieved. Mesh sizes of more than 120 μm are not sufficient. Depending on the powder and quality requirements, screening should be between 90 and 120 μm. The trend is clearly towards ultrasonic sieving with automatic dirt removal (Figs. 2.58, 2.59 and 2.60).

2.6.3.2 The Importance of Powder-Repellent Walls

Steel, glass and plastic walls are covered with the overspray powder. So-called "powder cakes"form on the walls and ceilings of the booths.

From time to time, powder deposits may detach from the booth ceiling, causing powder spills on the coating objects.

Powder accumulation on the booth walls must be wiped off the walls after a few hours, but at least once a day, and returned to the running powder circuit. This not insignificant amount of powder, which is conveyed back into the powder hopper at one time, can negatively influence the powder quality and thus the coating. Here is an example:

- 3 mm dry powder corresponds to approx. 1 mm powder coating film
- 100 μm powder coating film = 120–160 g powder/m2
- 1000 μm (1 mm) powder coating film = 1.2–1.6 kg powder/m2

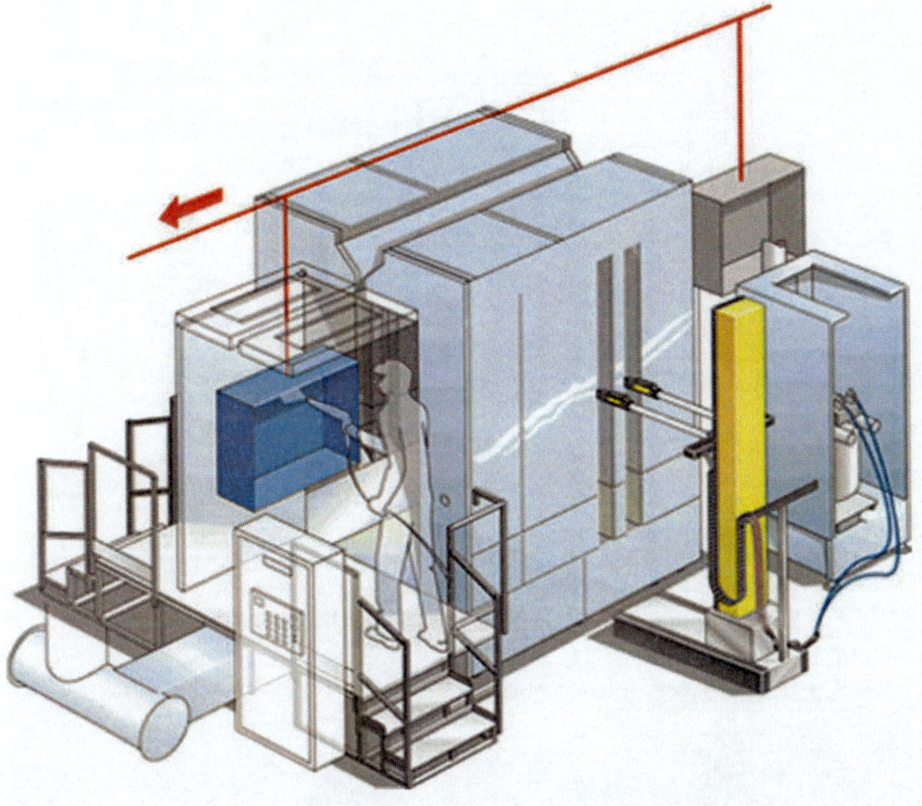

Fig. 2.58 Example of a high-speed cleaning booth with external manual coater station

Disadvantages may include:

- Change in grain size distribution
- Reduction of the charging behaviour
- higher load on the screening machine

The two diagrams Figs. 2.61 and 2.62 show the influence of the seasons winter and summer on the amount of adhering powder on the booth walls as a function of the coating time. The Y-axis shows the amount of adhering powder and on the x-axis the application time is plotted. Green = PVC cabin walls, yellow = glass and red = steel.

PCV sandwich shows the lowest adhesion tendency over the entire season. Stainless steel shows very strong adhesion throughout and glass shows slightly higher adhesion tendency in winter than PVC, which increases significantly in summer due to higher humidity.

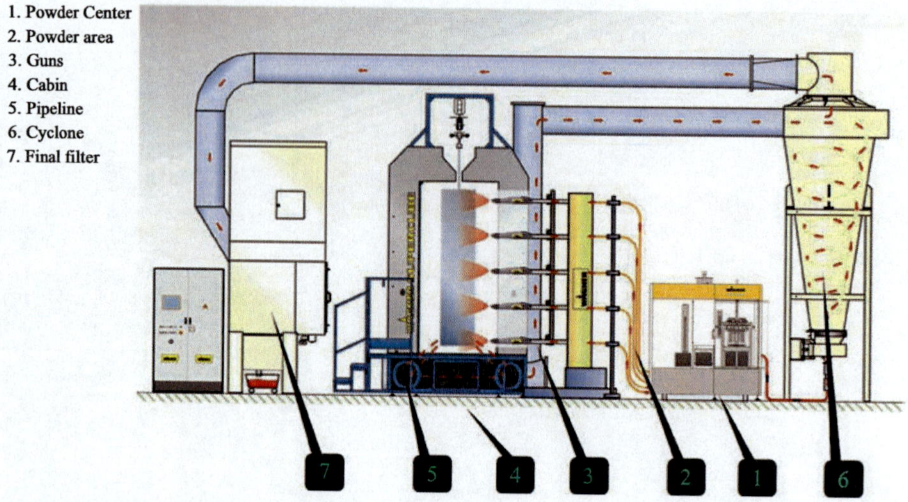

Fig. 2.59 Schematic representation of a booth concept with powder supply and recovery for a quick-change booth. 1 Powder center, 2 Powder transport/hoses, 3 Guns, 4 Booth; 5 Piping; 6 Cyclone: Overspray recirculation/powder-air mixture separation, powder sieving and return transport to the powder center, 7 Final-Filter: Filtration of the air in the absolute filter

Fig. 2.60 Example of a coating booth made of glass. The advantage of this booth material is the good visibility into the coating chamber [40]

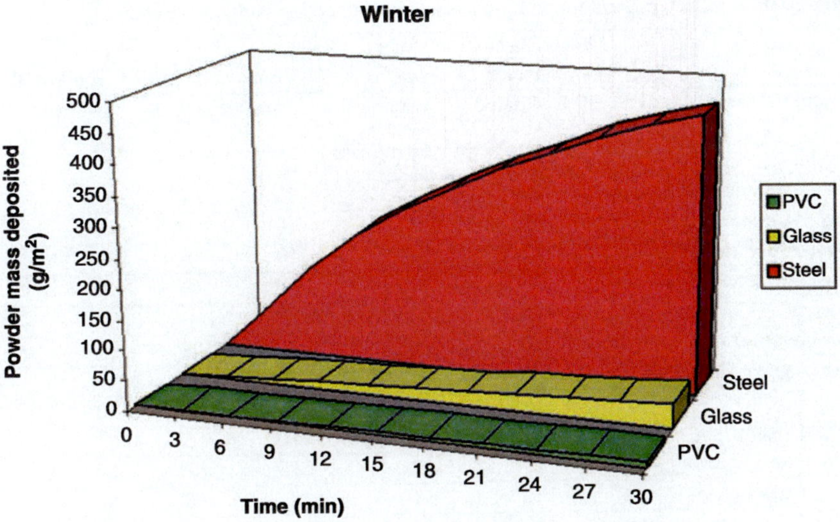

Fig. 2.61 Diagram of powder build-up on cabin walls (PVC, glass and steel) in winter

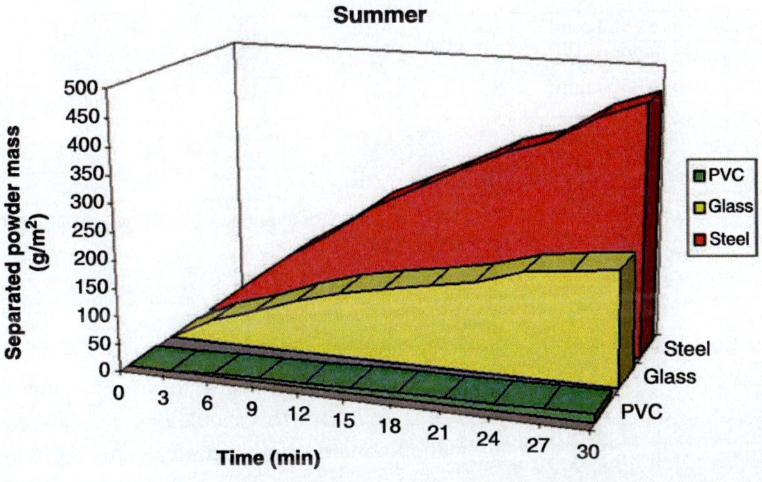

Fig. 2.62 Diagram of powder build-up on cabin walls (PVC, glass and steel) in summer

2.6.3.3 Quick Colour Change Booths

The question today is color change yes/no. If the decision is yes, a quick color change system with a booth specially designed for this application is usually used. These are optionally equipped without or with up to two manual coating openings, which are designed as pre- or post-coating stations. A vertical gun arrangement simplifies object-related adjustment and saves space. The plastic walls of the booths prevent powder

Table 2.10 Selection criteria for booth materials in connection with colour change

Selection criteria:	Sheet metal Coated	Sheet metal Stainless	Plastic Sandwich	Plastic Slide	Wood/MDF Sandwich panels
Durability	8	10	8	–*	7
Scratch resistance	8	10	7	–*	6
Easy to clean	5	5	10	–*	10
Low powder accumulation	4	4	10	8	10
Low powder adhesion	5	4	10	8	10
Security	10	10	8	8	7
Low investment	10	8	6	8	9
Up to 6 guns	10	10	10	10	10
Up to 10 guns	8	8	10	10	10
Up to 15 guns	5	5	10	5+	6
More than 15 guns	2	2	10	2	6
Up to 1 color change/day	10	10	8	8	8
Up to 2 color changes/day	8	8	8	8	8
3–4 colour changes/day	5	5	10	10	10
5 and more color changes per day	2	2	8	6	8
Object height up to 1500 mm	10	10	10	10	10
Object height up to 2000 mm	8	8	10	8	8
Object height up to 2500 mm	5	5	10	5	5
Object height up to 3000 mm	5	5	8	5	4
Object height < 3000 mm	5	5	6	5	4
Expandability	10	10	5	2	5

10 = very suitable, 1 = not suitable, * foil is renewed with every colour change, + foil width limited to 2.5 m

accumulation, which is supported by a central exhaust system. The booth is usually cleaned manually with an air jet lance; systems with automatic cleaning machines are also possible (Table 2.10). If a colour change is necessary, this should be carried out as quickly as possible. Today, booths with automatic floor cleaning are used for this purpose.

2.6.3.4 Cabin with Automatic Floor Cleaning System

These booths are suitable for single colour operation. Due to their simple construction, they are extremely reliable and easy to maintain, and do not place great demands on the operator. The powder paths are short and therefore also very gentle on the powder. The floor is kept free by automatic blowing off (Fig. 2.63).

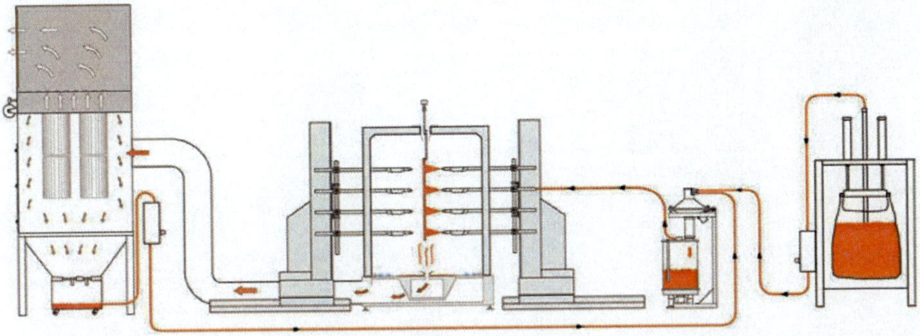

Fig. 2.63 Schematic diagram of a booth with floor cleaning by blow-off for single-color application

2.6.4 The Recovery Systems

Cyclones have the task of separating the powder recovered from the booth from the air, which makes the powder usable again. The recovered powder is returned from the cyclone directly to the powder center or to the powder hoppers. The air leaving the cyclones still contains fine powder particles and must therefore be cleaned again in a post-filter.

The monocyclone uses the physical law of centrifugal force. The powder-air mixture extracted in the booth is introduced tangentially at the top of the cyclone. As the air spirals downward, the powder particles are deposited on the wall. From there they follow the boundary flow to the tapered part of the cyclone and are collected in a bin. The air flow cannot be completely separated from the powder, as the air carries along fine particles. These are separated from the air by the final-filter.

The multicyclone has the same task as the monocyclone and works in the same way. The only difference is that it uses several small cyclones instead of one large one. Accordingly, it requires a smaller overall height than the monocyclone, but is more complex to clean. The multicyclone works in suction mode, its performance is therefore strongly dependent on absolute tightness. Poor separation rates are usually caused by leaks. Today, the multicyclone is only rarely used.

2.6.4.1 Monocyclone

All colour change booths, both sheet metal and plastic booths, are equipped with monocyclones. The separation efficiency of a monocyclone is approx. 95% (Fig. 2.64).

2.6.4.2 Recovery for Quick-Change Cabins

For quick-change booths, monocyclones with vibrating screen, discharge cone and push feeding are usually used. They are practically "self-cleaning". Cleaning is limited to the discharge section and the conical discharge section. The degree of separation varies between 95% and 97% depending on the powder loading. The differences between the individual suppliers are to be found in the details. The cyclones differ in height and

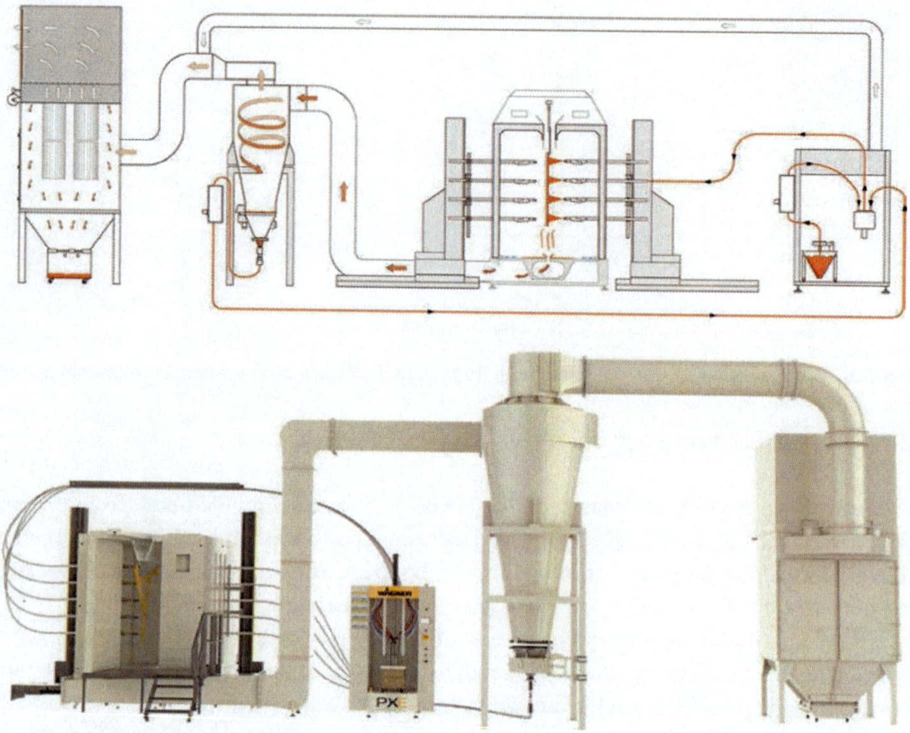

Fig. 2.64 Illustration of a plant with monocyclone

diameter. For reasons of space, compromises have to be made in some cases. With larger diameters, the peripheral speed decreases and the separation efficiency deteriorates. Lower speeds prevent the sintering of powder particles. Smaller diameters increase the separation efficiency. The higher peripheral speed can lead to sintering. The air movement close to the screening machine should be kept as low as possible to prevent sintering on the screen fabric. If the overall height is short, the air is left with too short a settling distance, which can lead to powder losses as well as powder sintering on the sieve and in the conical outlet section (Table 2.11).

2.6.4.3 The Efficiency of Powder Recovery Systems

Depending on the recovery system, the powder separation efficiency is approx. 90–98%. Of the sprayed powder, approximately 15–80% remains on the coating object, depending on the type of part. The following worksheet provides guidance on the criteria to be considered when selecting a recovery system (Fig. 2.65).

However, more important than the degree of recovery is the coating efficiency of the application/plant.

Table 2.11 Comparison of the advantages and disadvantages of the recovery components

	Advantages	Disadvantages
Monocyclone with final filter	Wide range of applications multicolor operation Simple construction Colour changeable	large construction height Efficiency losses Powder loss due to efficiency <100
Filter system	Simple system with high separation efficiency compact design Space-saving No pipe connections	Single color operation only Long service life of the cartridges or plates One filter unit required for each color

The choice of the recovery system

The following points should be clarified when choosing a recovery system:

1. what is specifically emphasized?

a) High degree of separation: _______ c) Fast color change: _______

b) Minimum space requirement: _______

2. system cost

max. cost: __________

3. number of colors

3.1. __ 1 colors __ 2-5 colors __ 5-50 colors __ over 50 colors __ over 200 colors

3.2 Are there main colors? If yes, indicate main colors in %:

1 color 2 colors 3 colors 4-10 colors

_______% _______% _______% _________%

3.3 Is it possible that two large orders have to be coated at the same time?

Yes ___/ No ___

4. number of color changes

How many color changes must be performed?

On average: _____ x per day _____ x per week
_____ x per month

In extreme cases: _____ x per day _____ x per week
_____ x per month

5. location of the powder booth

Hall height: ________ m

Max. Floor space: ________ m^2 (if no planning is available yet)

6. Powder consumption per year: _____t

7. Utilization of the plant (hours per day/week)

_______________ / _______________%

Fig. 2.65 The worksheet is intended as a basis for the plant manufacturer/planner to select the correct system

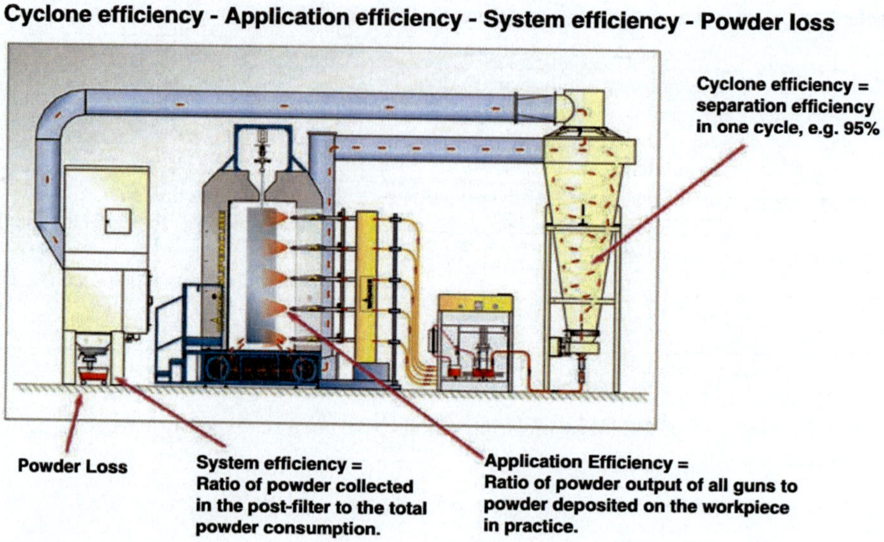

Fig. 2.66 Explanation of the individual process steps that have an influence on the overall efficiency

The different efficiencies of a plant are shown in Fig. 2.66. In practice, the overall efficiency and thus the powder loss can be significantly influenced by the application efficiency. A high technical standard with efficiencies of 95–97% is achieved with the cyclone technology of the well-known manufacturers.

The application efficiency is significantly influenced by the type of workpieces. With flat parts, efficiencies of 70% can be achieved in practice. However, when coating small parts on a collecting hanger or frame parts, the area to be coated is very small in comparison with a lot of space in between, so that a high overspray proportion is generated per se. The correlation between application efficiency, recovery efficiency of the extraction system and the resulting powder overspray is shown in Table 2.12. This can be minimized by control functions such as gap and height control. The workpiece is measured at the booth inlet by an optical system and then the guns are only switched on if a surface passes in front of the guns. The functions are explained in Figs. 2.66 and 2.67a to d.

The separation efficiency of a cyclone depends on the size of the powder particles. Particles smaller than 10 µm are more difficult to separate.

2.6.5 Dimensioning of the System

The different types of booths and recovery units are all available in different sizes. Their correct sizing is important, as a booth chosen too large will only add unnecessary cost, while one chosen too small will quickly be at the expense of coating quality and safety. The essential parameters for sizing the booth and its recovery system are:

Table 2.12 Powder loss table as a function of cyclone separation efficiency and application efficiency [32]

Table of powder loss	Cyclon separation efficiency (%)							
Application efficiency (%)	99,0 (%)	98,0 (%)	97,0 (%)	96,0 (%)	95,0 (%)	94,0 (%)	93,0 (%)	92,0 (%)
90,0	0,11	0,22	0,33	0,44	0,55	0,66	0,77	0,88
80,0	0,25	0,50	0,74	0,99	1,23	1,48	1,72	1,96
70,0	0,43	0,85	1,27	1,69	2,10	2,51	2,91	3,31
60,0	0,66	1,32	1,96	2,60	3,23	3,85	4,46	5,06
50,0	0,99	1,96	2,91	3,85	4,76	5,66	6,54	7,41
45,0	1,21	2,39	3,54	4,66	5,76	6,83	7,88	8,91
40,0	1,48	2,91	4,31	5,66	6,98	8,26	9,50	10,71
35,0	1,82	3,58	5,28	6,91	8,50	10,03	11,50	12,94
30,0	2,28	4,46	6,54	8,54	10,45	12,28	14,04	15,73
25,0	2,91	5,66	8,26	10,71	13,04	15,25	17,36	19,35
20,0	3,85	7,41	10,71	13,79	16,67	19,35	21,88	24,24
15,0	5,36	10,18	14,53	18,48	22,08	25,37	28,40	31,19
10,0	8,26	15,25	21,26	26,47	31,03	35,06	38,65	41,86
5,0	15,97	27,54	36,31	43,18	48,72	53,27	57,08	60,32

The separation efficiency of a cyclone depends on the size of the powder particles. Particles smaller than 10 µm are more difficult to separate

- maximum part weight
- the calculated exhaust air volume
- the quantity of powder used
- the number of guns
- the quality requirements
- the space conditions (Fig. 2.68).

Figure 2.68 shows an example of a quick colour change booth and the arrangement of the individual system components.

2.6.6 The Coating Equipment: Guns

Tests in the technical center of the plant manufacturer or powder manufacturer, the gun types to be used and their number must be determined. The corona gun is a high-voltage gun which can be used for all powder types. With the low-ion or ion-reduced charging form, guns are available which allow thicker powder coatings to be applied without the

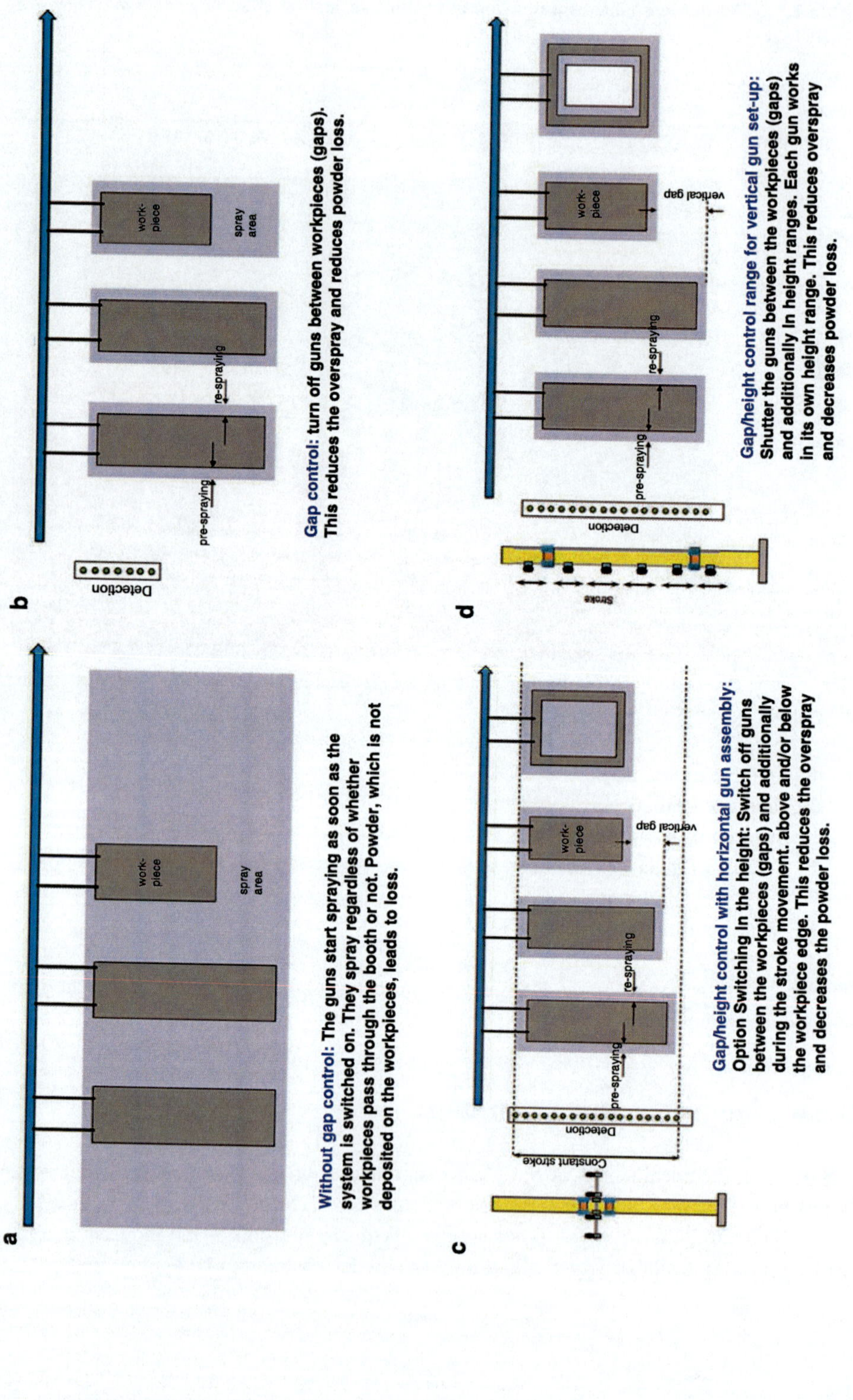

Fig. 2.67 (**a** to **d**) Examples of gap and height control to reduce overspray

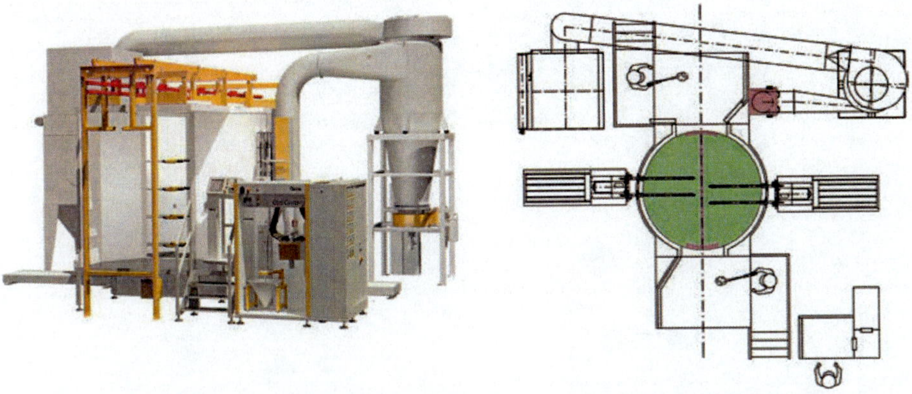

Fig. 2.68 Quick colour change booth with round booth and powder recovery, control cabinet, powder centre and two gun racks on the opposite sides

formation of the orange peel. These types of guns are also used for coating enamel powders, but because of the abrasive properties of the enamel powder, they are made of different materials. The tribo gun needs a specially treated powder. The powder bell is sensibly only used if only flat parts are coated in a booth. Different nozzles are also available for the guns for the various applications, depending on the requirements that the object to be coated places on the coating.

2.6.6.1 Arrangement of Application Equipment and Determination of the Number of Guns

For automatic painting, the guns are arranged fix or on reciprocators. Since the powder output of a spray gun is only variable within certain limits, several guns must be used simultaneously. The variable arrangement of the units relative to each other is advantageous, whereby care must be taken that the spray clouds of the same pole repel each other. A staggered arrangement and/or oscillation transverse to the direction of movement of the workpieces to be coated prevents the formation of color streaks. Since the size and shape of the spray clouds, their space charge, the distance between the spray heads, the distance to the workpiece and the surface profile facing the guns all have an influence on a multiple arrangement, as a rule only the experimental test provides sufficient information about the result to be expected. The prerequisite for a consistent painting result is a constant dosing of the powder to the guns and a sufficient charging. A measurement or control of the powder feed is so far costly [41, 42]. An exception is the measurement of the charging current for guns with tribo charging, since here, with a constant air volume flow, the leakage current of the tribo gun is a powder-specific measure of the powder output.

For processes with recovery, the change from fresh to recycled powder must also be taken into account in the coating tests. Figure 2.69 shows possible arrangements of sprayers [39].

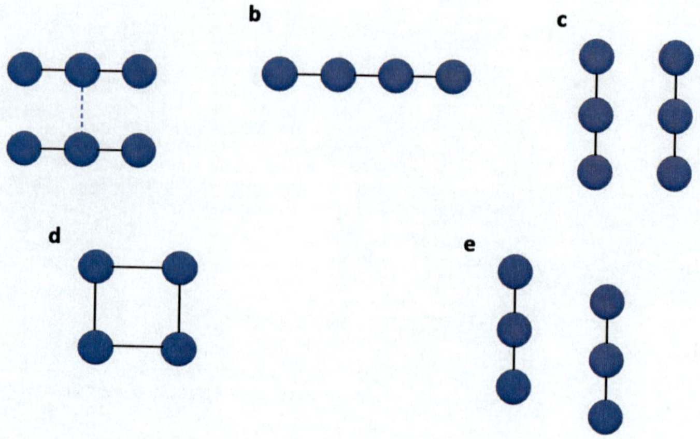

Fig. 2.69 Possibilities for the arrangement of guns for powder application

a) Optimization of the distance between the guns easily possible, low lifting height (approx. 950 mm), painting height approximately 850 mm,
b) Largest stroke required, therefore insensitive to output fluctuations of individual guns, powder oversupply at the reversal points, stroke height approximately 1500 mm, painting height approximately 300 mm,
c) Lowest stroke, danger of streaking with uneven dosing, stroke height approximately 600 mm, painting height approximately 1200 mm
d) Frequent middle way for a reduced lifting height and simple spray pattern overlapping in horizontal and vertical direction, lifting height approximately 1200 mm, painting height approximately 600 mm
e) Lowest stroke, avoidance of stripe formation with uneven dosing, stroke height approximately 600 mm, painting height approximately 1200 mm

2.6.6.2 Characteristics for Coating with Several Guns Mounted Rigidly or on Automatic Moving Equipment

The coating window A designates the area which can be coated from one side with approximately the same coating thickness in the specified time unit. The coating height h is the height that can be powdered with approximately the same coating thickness with fix arranged guns or guns moved with reciprocators. The transport speed v is the speed at which the parts to be painted are conveyed through the booth.

$$A = h \cdot v$$

In order to achieve a desired coating thickness s, the quantity $_{mm}$ of powder must be sprayed per time t. This results in the theoretically required spraying performance, i.e. output of powder per time unit. This must be increased taking into account the application efficiency η for a surface coating and corresponding to the workpiece format n. To calculate the necessary volume output quantity (powder output), the density δ of the coating material is necessary. n is the correction factor of the part type: surface n = 1; profile n = 2.

$$m_m = \frac{s \cdot A \cdot \delta}{\eta \cdot n}$$

m_m Mass output
s Layer thickness
A Painting window
δ Density of the powder coating
n Correction factor Part type
η Order efficiency

$$\eta = \frac{m_{FL}}{m_F} \times 100\%$$

η Order efficiency [%]
m_F Applied powder coating quantity [g]
m_{FL} Amount of powder deposited on the workpiece/surface [g].

The necessary minimum number of guns P_{min} results from the necessary mass output m_m divided by the output quantity m_P of the gun. The usual output quantities of the guns are m_P = 100 to 250 g.

$$P_{min} = \frac{m_m}{m_p}$$

P_{min} minimum number of pistols
m_m Mass output
m_P Output quantity at the gun

In the experiment, the arrangement must now be designed in such a way that a high layer thickness uniformity is achieved. Experience shows that this can be solved more easily with a moving arrangement than with a fixed arrangement, unless it is already clear in advance that the required height can only be achieved with a moving arrangement of the guns. Suggestions for the arrangement of the equipment are shown in Fig. 2.69 [37].

Table 2.13 and Fig. 2.70 give guide values for determining the number of guns. In general, it should be noted that with a powder output of more than 200 g/min, the deposition efficiency decreases. The chargeability of the powder particles decreases and the proportion of overspray increases.

Table 2.13 Suitability of automatic guns for different part profiles depending on gun attachment, powder output and visible surface performance

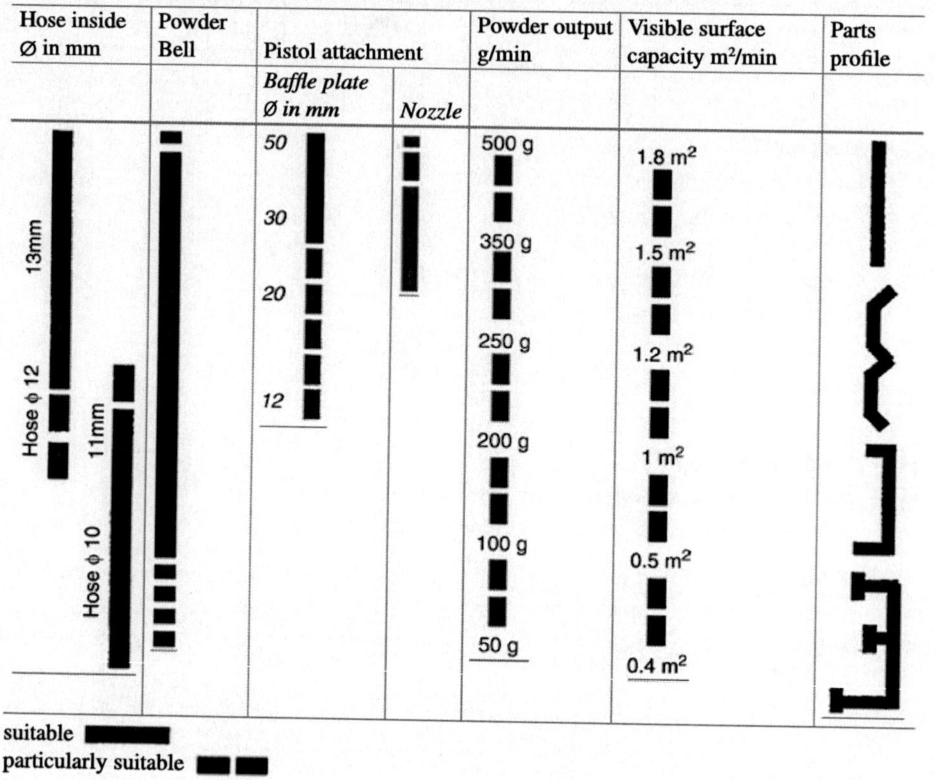

2.7 The Powder Centre

The task of the powder center is to feed the powder recovered via the cyclone, including the fresh powder supply, to the coating guns via powder lines. The color change here should be carried out as quickly, cleanly and fully automatically as possible. In order to be able to carry out a clean colour change, nozzle blocks must ensure automatic internal cleaning of suction pipes, injectors, powder hoses and guns. A return hose directs the recovered powder directly back into the container. Dust formation or contamination is prevented by an exhaust air unit that is either directly installed or routed through the cyclone's final-filter. Ideally, it should be possible to control all functions from the powder center. If the powder center is designed without an independent exhaust system, the correct dimensioning of the cyclone final-filter must be ensured.

The powder center is an essential part of the color change system and contributes greatly to the quality of the end products. Optimum vibration and local fluidization in combination with a level-controlled immersion tube immersion depth are prerequisites for

Korona and Tribo baffle plate nozzle

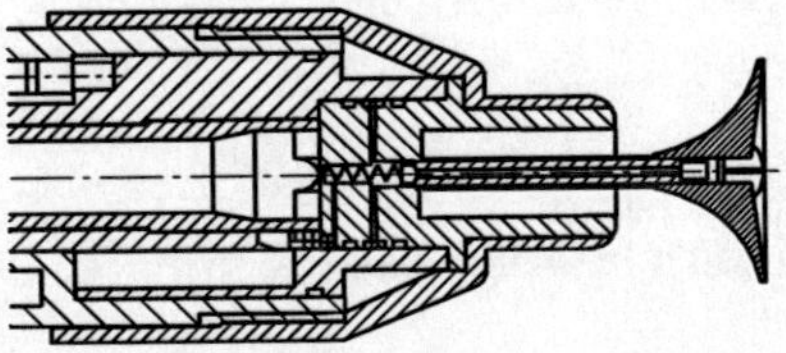

For surface coating
Powder output max 200-300 g/min (Corona)
Powder output max. 150-200 g/min (Tribo)

Gap nozzle Korona and Tribo

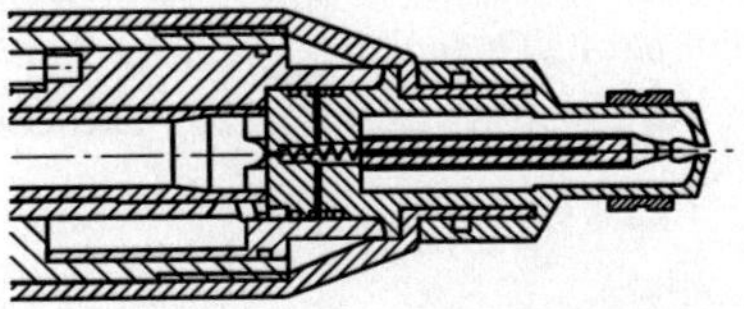

For complicated and deep objects
powder output max. 150–200 g/min (Corona)
powder output max. 100–150 g/min (Tribo)

Powder bell Corona

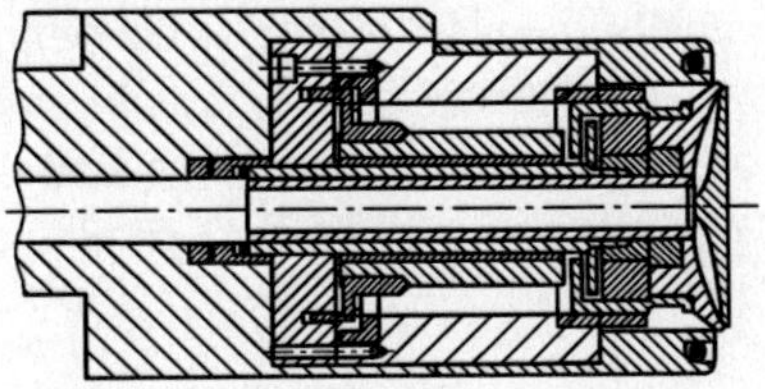

For large area parts
Excellent area performance and high application efficiency
Powder output max. 500-600 g/min

Fig. 2.70 Powder output for corona and tribo guns with different nozzles

good coating results. A self-contained, generously designed exhaust air prevents the escape of powder particles during the working process and especially during cleaning during color change.

Function of the powder center:

- Transport of the recovered powder via the cyclone into the original container,
- Fresh powder dosage into the container, as continuous as possible and appropriate to the consumption,
- to convey the powder to the guns via hoses or pipes using appropriate injectors
- When changing powder or color, the following assemblies must be cleaned:
  - the conveying paths of the powder from the cyclone to the powder centre
  - the powder dosage
  - the injectors, dip tube, guns, hoses
  - the screening machine
  - the powder centre itself

As a rule, only the injectors with suction pipe, hoses and guns can be cleaned automatically on the inside of the powder centers. All other areas must be cleaned manually or with manual assistance.

2.7.1 Powder Centre with Injector Technology

Concept Powder Center with Rod Injector

With this concept, the bar injector tubes and the level probe on the outside, as well as the bar inectors, hoses and guns on the inside are automatically extended at the push of a button and cleaned in pulses. Depending on the version, e.g. with monocyclone, the powder roller is also cleaned automatically via the final-filter by separation to the cyclone. Manually the plastic plate at the bottom with injector tips, as well as the intermediate space, if necessary with sieve and frame must be blown out. The suction is carried out via the cyclone to the final-filter. Finally, the powder carton must be replaced. The color change takes approximately 3–5 min, see Fig. 2.71.

Fig. 2.71 Powder center with rod injectors, container in automatic application with fluidization and vibration

2.7.2 Powder Centre with Pump Technology

Here the powder is transported to the guns or into the powder hopper with dense phase pumps. The cleaning takes place automatically. At the end of the production with one color, the powder hoses and the guns are cleaned with powerful air blasts. An automatic switching device disconnects powder hose "A" and connects hose "B" to the gun. The cleaning of area "A" takes place in offline mode and is prepared for the next colour, while the system continues coating with colour "B".

2.7.3 Integrated Powder Centre

The increasing automation also has an effect on the design of the powder centers. The new systems have a high level of integration of the powder supply, powder preparation, powder conveying and application functions. Only minimal manual activities need to be carried out, which increases the availability and quality of the coating. An example is shown in Fig. 2.72, a coating center with high integration of various functions and a high degree of automation.

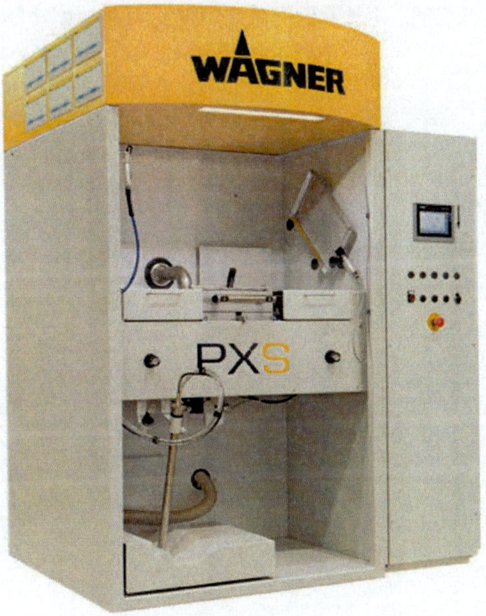

Fig. 2.72 Example of a powder center with a high degree of automation

2.8 Plant Control and Automation

The simplest, but also efficient system control is the manual coater. He recognizes the workpiece at a glance, knows what is to be coated and how to coat, and he sees the powder application immediately and can take corrective action. Increasing demands on process reliability, output, automated documentation, increasing shortage of skilled workers and cost pressure lead to increasing automation in the area of coating systems.

While a simple hand-held unit gets by with a single gun control, an automatic system has higher requirements, up to a total system control. The following automation levels are possible in the coating sector today.

2.8.1 Automation Level 1: Gap Detection

Automation level 1 is the gap control for gun groups. With this function, the object to be coated is detected at the booth entrance by a light barrier and all guns are switched on with a slight delay. In case of a gap, the guns are again switched off with a delay (Figs. 2.73 and 2.74).

2.8.2 Automation Level 2: Height Detection

The gap control for single guns. With this function, the object height is also detected and the guns are switched on individually. In case of a gap, they are switched off again with a delay. From this automation level onwards, the control system can be integrated into a higher-level booth control system (PLC) (Fig. 2.75).

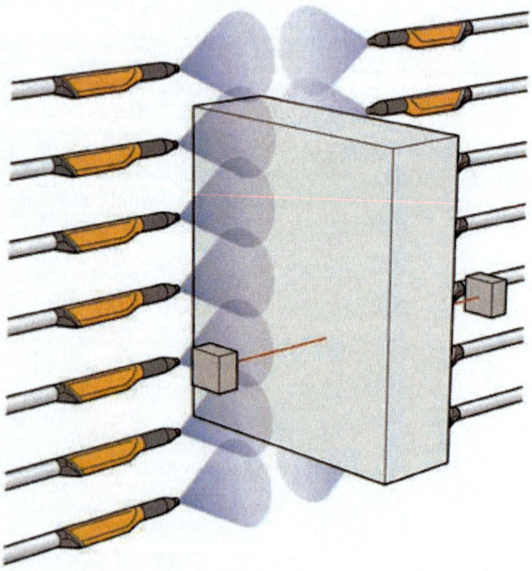

Fig. 2.73 Example of automation level 1

Fig. 2.74 Automated workpiece recognition with light barrier

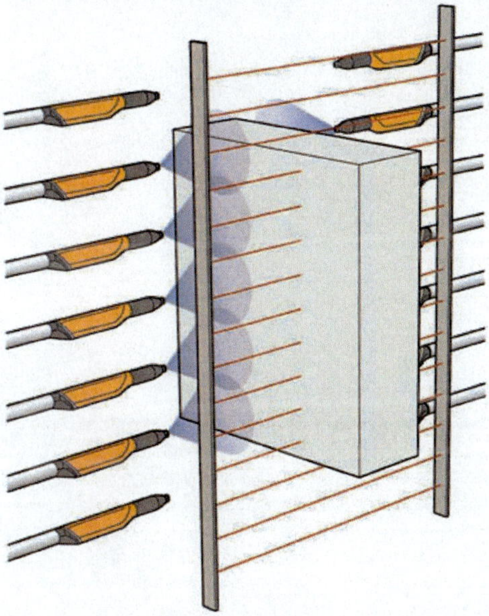

Fig. 2.75 Example of automation level 2

2.8.3 Automation Level 3: Width Detection

The gap control and depth feed. The gap control functions as already described. With depth infeed, the reciprocators are retracted to the width of the widest workpiece at the start of coating and extended again from the booth after the coating process is complete (Fig. 2.76).

2.8.4 Automation Level 4: Vertical Gun Arrangement

The height detection and depth infeed with vertical gun arrangement. The width infeed is again performed on the widest object, whereas here the coating heights are detected and only as many guns are switched on as are necessary for coating the respective workpiece (Fig. 2.77).

Fig. 2.76 Example of automation level 3

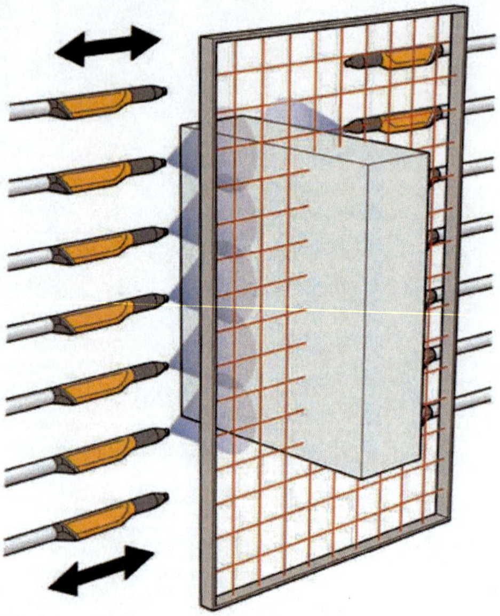

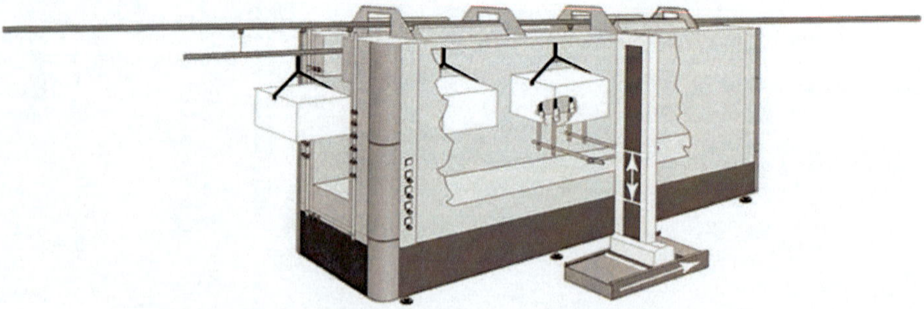

Fig. 2.77 Example of automation level 4

2.8.5 Automation Level 5: Dynamic Contour Detection

Fully automatic operation. The objects are no longer coated on the basis of the detected geometry, but the object outer contour is detected via laser scanning and the coating program belonging to the respective object is set in motion. Here, the width adjustment is also carried out to the corresponding workpiece width and no longer only to the widest object (Fig. 2.78). This stage also includes synchronized coating, in which the reciprocators travel with the object and several axes must be moved together (Fig. 2.79).

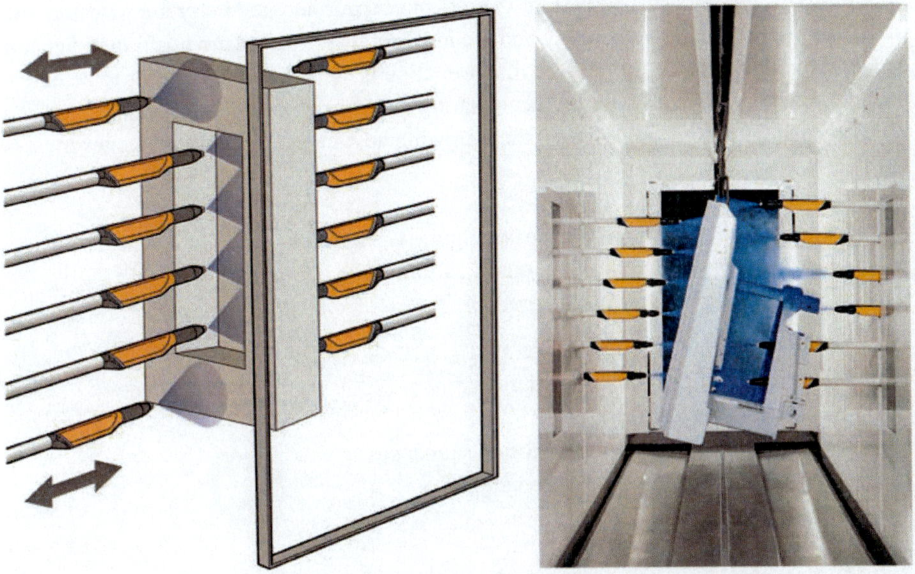

Fig. 2.78 Example of dynamic contour detection and individual gun positioning

Fig. 2.79 Example of automation level 5

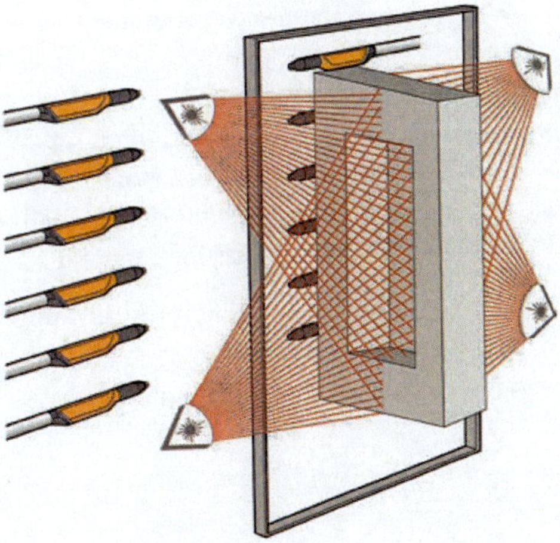

Gap, Height and Width Detection The **gap control** switches the guns depending on the presence of objects. The workpiece is detected at the booth entrance by a light barrier and the guns are switched on shortly before the object. If there is a gap, the guns are switched off.

A vertically arranged light barrier is used for **height detection.** Based on the detected object height, only the required guns are switched on. With a horizontal gun arrangement, the lifting movement is automatically adapted to the object height.

The width detection is an additional automatic adjustment of the horizontal gun positions to the object width.

With **dynamic contour detection**, the object contour and geometry are detected with the aid of laser scanning technology and the individual spray guns are positioned dynamically and individually. A high degree of automation is achieved here in order to be able to paint complex workpieces with uniform coating quality.

When using guns on single axles, the following advantages and disadvantages must be considered:

- Sensor technology required for measuring the workpieces
- Increased to high demands on the qualification of the operating personnel
- Stable workpiece suspension required

  - + High area output
  - + Reduction of overspray/powder loss
  - + 3D coating of the workpieces
  - + High reproducibility of the motion sequences
  - + Increase in the degree of automation

2.8.6 Robotics

Robots are increasingly used in the area of pre- and post-coating of complex workpieces..

When using a multi-axis robot, the following points must be observed:

- Sensor technology required
- Increased space requirement, often separate cabin
- Low area output as only 1–2 guns
- - Increased to high demands on the qualification of the operating personnel
- Stable workpiece suspension required

  - + High application efficiency, low powder loss
  - + High reproducibility
  - + Highest degree of automation

2.8.7 The Lifting Equipment

Lifting devices have the task of moving the guns. Depending on the application, different versions can be selected. While a simple short-stroke device may be sufficient for the simplest case, a reciprocator with travel capability must be used for a more complex coating (Fig. 2.80). Today, manufacturers offer a wide range of possibilities, from simple vertical strokes to complex multidimensional motion sequences. Important for good coating results is, the distance between the spray gun and the workpiece a precisely running motion unit.

The reciprocator and its optimum settings must be determined and documented during the coating test. Only through the customer-specific design of the reciprocators is it possible to achieve an exactly reproducible coating performance of uniform uniformity and layer thickness. The reciprocators can work in up to three dimensions, depending on the effort required. The reciprocator control is located in the control cabinet of the powder coating unit. A microprocessor control regulates the constant lifting speed as well as the exact adherence to the reversal points. The setting adjusted to the conveying speed and lifting height can be stored and recalled at any time.

Fig. 2.80 Reciprocator y, r-u axis with one gun

Axes of Motionand their Task

Z-axis: the vertical axis movement is the basis for automation. With the vertical pendulum stroke, uniform coating results are achieved.

X-axis: it moves the reciprocator horizontally to the desired position for width adjustment.

Y-axis:. the reciprocator moves along the booth synchronously with the conveyor speed. The movement of the Y-axis is synchronized with the speed of the conveyor chain, which enables precise coating of edges and recessed surfaces

U-axis: this feed axis positions individual guns in object recesses

R-axis: the rotary axis swivels the spray guns around their own axis. Rotary axes are suitable for tracking inside edges, recessed geometries or rotating parts with angled gun nozzles (Fig. 2.81).

2.8.8 Reciprocators or Coating Robots

The easy-to-operate and quiet cost-effective 1–5 axis automatic reciprocators can still be found in most powder coating plants today. With one device, several powder spray guns arranged horizontally next to each other or vertically above each other can be moved simultaneously. With the help of a workpiece recognition via light barrier or RFID, stored, workpiece-specific coating programs can be called up, with settings for: powder output, spraying distance, spraying current, gap control, etc. The exact adaptation of the lifting movement to the workpiece height and the conveying speed is important in order to obtain an optimum overlapping of the coating passes and thus an even coating thickness distribution over the entire workpiece.

Increasingly, coating robots are being used for the pre- and/or post-coating of complex components. The precisely reproducible movement of the guns, optimally adapted to the workpiece, reduces powder consumption and coating thickness fluctuations. The prerequisite for this is a precisely defined part position on the product carrier. However, it must be taken into account that a new coating program must be defined for each new workpiece. This requires appropriately qualified personnel. Figure 2.82 shows the influence of the important components in the choice of the degree of automation of the coating plant.

2.8.9 Cabin Systems for Automatic Coating

The topic of Industry 4.0 develops solutions to be able to manufacture customized products with proven methods of mass production. If necessary, down to batch size 1. In the coating industry, individual colouring is of great importance in many markets and poses great challenges for coaters. For the coating plant this means:

Fig. 2.81 Reciprocator z and x-axis with four guns

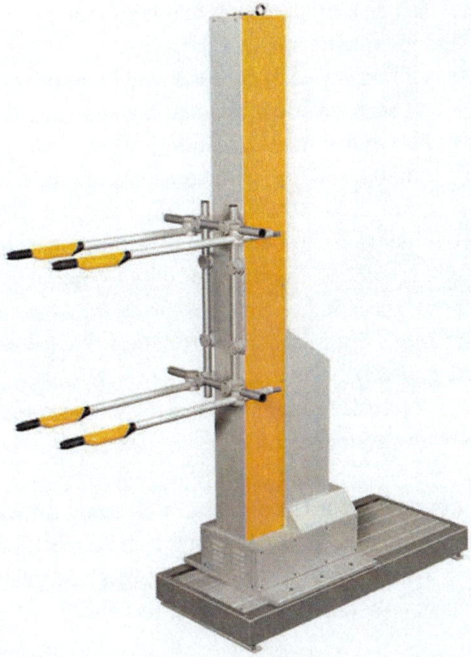

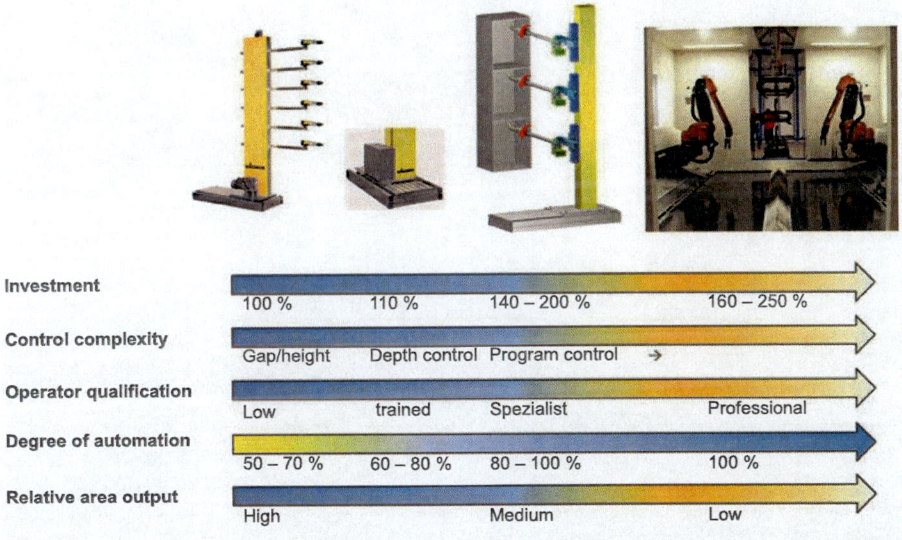

Fig. 2.82 Important criteria and their influence on the choice of the degree of automation of the application

- fast coating processes and high output
- high quality, no rework
- uniform layer thicknesses and layer thickness distribution
- resource-efficient application
- Fast and effective cleaning
- durable and low-maintenance components

In order to be able to realize these points and the expectations of the coaters and customers, it will be important not only to optimize individual components, but rather the entire powder coating system must be coordinated and networked. This means: the powder material, the lines, conveyor systems (powder and workpiece), guns, extraction and booth (Fig. 2.83).

A powder coating system for semi-automatic or automatic coating consists of the following system elements:

Final-filter: It has the task of cleaning the exhaust air of the coating booth from powder particles. The maximum values specified by the legislator must not be exceeded. Various systems with cartridges and plate filters are available. Since the powder is mainly fine dust particles, the filter must be generously dimensioned. The exhaust air

Fig. 2.83 Powder coating plant with modern control system and integrated network interface for production data acquisition, system status monitoring and remote diagnosis

volume is designed on the basis of the booth openings in such a way that an average air intake velocity of 0.5 m/s for sheet metal booths and 0.5–0.75 m/s for plastic booths is guaranteed at all openings. If the exhaust air from the powder center is also processed in the final-filter, the exhaust air volume must be increased by approx. 3000 m^3/h· Lower values for powder center exhaust lead to contamination of the workplace.

Powder center: In a color change system, it facilitates the processing of a wide variety of colors without risk of contamination. The powder is processed either from a powder box or directly from the container. As a rule, the powder center has a compartment. In case of very high quality requirements or extreme color changes, two powder centers are used, each of which has its own exhaust air system and separate powder hose routing.

Application equipment for automatic coating: While a simple hand-held unit gets by with a single gun control unit, an automatic system has greater needs. At its simplest, each gun and reciprocator needs its own control unit (slide-in). With each further level of automation, additional control plug-in units are needed to control the individual steps and to coordinate them with each other.

Controls: Basically, a system control consists of a control cabinet with various slide-in modules that control the individual functions of the guns, reciprocators, etc.

Gun plug-in units: They are used for the electrostatic and pneumatic control of the guns and injectors. The high voltage, current as well as conveying, supplementary and rinsing air can be set on the control plug-in unit. The flow measurement allows an exact adjustment of the optimal powder output. Different degrees of automation are possible. The pneumatic/fluidizing units distribute the compressed air to the gun control inserts, control the fluid air and air mover pressure as well as the prefluidization. A powder hose flushing unit can be used to clean the powder hoses. Depending on the application, the guns can be flushed individually or in groups.

Monitoring controllers: Interlock control modules are used to control and enable interlock functions in systems with fluidized powder hoppers and to display powder hopper status. Diagnostic systems are used to monitor the function of guns and reciprocators.

Reciprocator controls: These controls are used for free programming of reciprocator and axis movements in different travel positions.

Fresh powder supply controls: The fresh powder control inserts monitor and control the powder transport from the container into the powder hopper.

Fresh powder supply: Fresh powder systems are used when powder consumption is high and quality requirements demand a constant admixture of fresh powder to the recovered powder. Depending on the size of the plant (number of guns) and the delivery container of the powder suppliers, different systems are used. Fresh powder supply from the original small container (cardboard box) or the original big bag of the powder manufacturer.

Automatic guns: The prerequisite for automatic external cleaning is a smooth surface. Hose and cable connections must be located outside the booths.

Object height detection and object depth detection: Due to the use of monocyclones for powder recovery, powder losses of 3–5% are to be expected. Therefore, it makes sense to operate the guns only when objects pass the booth. With the object height detection this requirement can be fulfilled. The object width detection is used where clearly defined shapes are to be coated and the automatically preselected gun distance also produces the best coating results. For variable object shapes, an individual feed of the individual axes by the system operator is preferable.

Automatic cleaning of the cabin interior: Automatic cleaning of the cabin interior is used where the cabin dimensions from approximately 3 m object height no longer permit manual cleaning with air lances or where entering the cabin is not desired for safety or quality reasons.

2.9 Piping from the Cabin to the Cyclone

The professional execution of the piping is of importance and influences to a great extent the quality of the colour change. Each cabin manufacturer has its specific piping, and there are differences in the materials:

Galvanized ventilation pipes	
Advantages	Disadvantages
Cost-effective, simple assembly	Many connection points
All shapes possible	Part segments in sheets (bends)
Sheet available	Danger of powder accumulation in seams/bends
Safety requirements can be easily met	

Stainless seamless welded tubes	
Advantages	Disadvantages
No powder deposits	High costs
No seams	
Single piece sheet	
Safety requirements are easily met	

Plastic pipes	
Advantages	Disadvantages
No seams	Compliance with safety regulations more difficult
No powder deposits	Additional earthing necessary to prevent sliding brush discharges
Single piece sheet	

Square tubes should not be used for fluidic reasons. The air velocity profile in rectangular pipes ensures turbulence in all cases and thus uncontrollable powder deposits. In general, it can be said that stainless tubes meet all the requirements of an optimum pipeline. Of course, the dimensioning of diameters and the design of transition pieces round or rectangular must be carried out according to the rules of fluid mechanics.

2.10 Plant Technology for the Processing of Effect Powder Coatings

Most commercially available application systems take the special properties of metallic powder into account quite well in the electrostatic and mechanical design of their guns and gun controls. Thus, most of the metallic powders in use today can be processed without problems and without special precautions. Nevertheless, the following problems can occur during processing:

- Recovery problems
- Charging problems (short circuits)
- Spray pattern changes.

2.10.1 Recovery Problems

Here, problems arise primarily in the processing of the non-bonded powders, which tend to segregate. From a plant engineering point of view, bonded powders should be used if possible, these should be processed in systems with little powder in the circuit (approx. 1–2.5 kg powder in the fluid pot) and a continuous supply of fresh powder should be ensured. Causes for the recovery problems when processing non-bonded powders can be:

- frequent cleaning
- too much powder in the circuit
- no continuous fresh powder supply
- powder preparation in agitator vessel
- fluidization and vibration lead to segregation.

2.10.2 Charging Problems

Charge problems are most often caused by short circuits, especially inside and outside guns and nozzles. Due to the formation of a short circuit, the entire charge current flows off in the direction of the earth. As a result, the powder particles are insufficiently charged or not charged at all. Coating is impossible. The following causes can be the source of short circuits:

- Deposits on the gun body
- Short circuit formation in the area between gun and nozzle
- Layer formation in the powder tube or powder hose
- Short circuit formation due to powder deposits on the outside of the gun body.

The insulation distance between the nozzle and the conductive powder layer is too small, electrical flashover occurs. The accumulation of powder on the guns represents a considerable potential danger and can lead to uncontrollable discharges and sparking. Only regular blowing off of the gun body can help here. The type of nozzle used has a major influence on the surface effects. Deposits in the hose can be avoided with conductive powder hoses.

2.10.3 Spray Pattern Changes

A well-known phenomenon of metallic and effect coating is that these powders, processed with different guns or controls, show uneven surface effects. This is due to the different charging and the resulting layer structure on the object.

However, spray pattern changes can also occur during coating. This is then due to the change in the powder-pigment mixture that occurs during the preparation and conveying of the powder. Only a dosed and continuous supply of fresh powder can help here. In general, it can be said that the higher the pigment content of a powder, the more critical its preparation and atomization.

Errors can also be avoided when preparing the powder for transport. Gentle transport with powder pumps is advantageous. When using powder containers with fluid bottom, the simultaneous use of vibration and fluidization must be avoided, as the vibrations increase the separation of the powder particles. Dense phase pumps can be used to transport the powder evenly and gently. There should only be approx. 5–6 kg of powder in the circuit. It should be noted that with long series on recovery, the mixing ratio between recovered and fresh powder must be closely observed.

2.10.4 Influence of the Nozzle Systems

2.10.4.1 The Flat Spray Nozzle

The advantage of the flat spray nozzle lies mainly in the high charging of the powder and the associated high application efficiency. For flat parts, an application efficiency of 60–70% is achieved. The flat jet nozzle is mainly used in automatic systems with recovery. Here, a continuous supply of fresh powder is an absolute necessity in order to be able to achieve uniform coating results without effect changes.

If higher pigmented powders are processed, a separation of the powder mixture can occur due to the electrode holder integrated in the flat jet nozzle. The metal pigments then deposit inside the nozzle. The agglomerations break up into clumps and cause spitting. Since these accumulations of metal particles are conductive, they cause short circuits after only a very short time of operation by striking back at the powder deposits on the nozzle and on the gun.

The use of the flat spray nozzle with hidden electrode can provide a remedy in many cases. Here, the rinsed electrode is guided inside the body up to the powder outlet opening. In this case, the application efficiency is 10–15% lower than with the normal flat spray nozzle, but coating can be carried out without spitting. If care is also taken that the rinsing air of the electrode is adjusted so that it does not split the powder jet, very good, stable results can be achieved with these nozzles.

2.10.4.2 The Round Spray Nozzle

Due to the larger insulation distance between the electrode and the powder deposits on the gun or in the powder pipe, the danger of short circuits is considerably smaller than with the flat spray nozzle. Nevertheless, only flushed nozzles should be used to avoid deposits on the baffle plate. These immediately lead to a short circuit again. The round jet nozzle gives good to very good results in most applications. Especially on flat parts the surface quality is very good. Due to the slow travel of the cloud, the penetration is not very pronounced. In difficult cases a small baffle plate of 15–20 mm can improve the penetration. The charging efficiency of the gun is much lower compared to the use of a flat fan nozzle, it drops to less than 50%. Accordingly, less powder adheres to the object and correspondingly more powder enters the recovery circuit. There is a tendency for the round spray nozzle to separate the metal pigments from the powder, which can lead to irregular metallic effects (cloud formation). However, this disadvantage can be corrected in most cases by a continuous supply of fresh powder (Fig. 2.84).

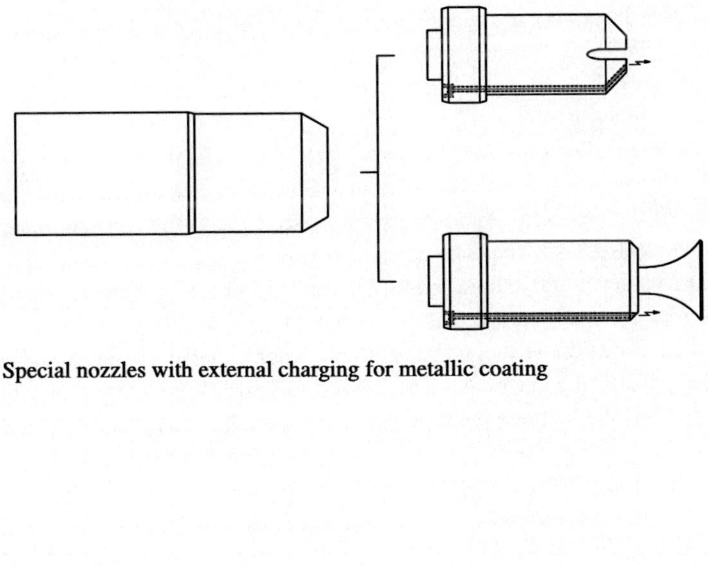

Fig. 2.84 Special nozzles with external charging for metallic coating

2.10.5 Short-circuiting Between Gun and Nozzle

When using guns, which are not specially designed for processing metallic powders, short circuits can occur in the area of the high voltage transfer to the nozzle.

The use of a gun with an extended powder tube is ideal. If a regulargun is used, make sure that the contact surface between the gun body and the nozzle is carefully cleaned when changing the nozzle. The union nut must also fit well and be tightened firmly, so that the nozzle sits snugly on the gun.

2.10.6 Short-circuit Formation Due to Layer Formation in the Powder Tube or Powder Hose

Some effect powders tend to deposit and form a conductive layer in the powder tube and powder hose. The entire charge generated in the gun is discharged via this layer and coating is impossible. Regular blowing out of the powder hose and powder tube is the best protection against sintering.

The installation of a swirl zone can also prevent the build-up of a conductive layer. Good results are also achieved with the use of extended nozzle pieces with increased sealing surface.

2.11 Special Powder Coating Processes

2.11.1 Powder Coating Without Guns: Electrostatic Fluidised Bed Process

In the Brennier process [43], the coating powder is not sprayed directly onto the workpiece with guns, it is applied with two surface electrodes. The size of these electrodes depends on the workpieces to be coated. Due to a modular design, different sizes can be realized at low cost.

Powder charging takes place in the range of 0 to 50 kV adjustable DC voltage, to which a 12 kV AC voltage is superimposed.. The AC voltage ensures the electrostatic charging and repulsion of the powder particles from the surface due to the constant reversal of the polarity of the electrode surface. The charged particles are then immersed in the electric field of the DC voltage built up between the electrodes and the workpiece. They follow the field lines to the grounded workpiece.

The constant attraction and repulsion of the powder particles creates a diffuse powder cloud. In small plants, the electrostatic forces are sufficient to partially move the powder. For larger volumes, the homogeneous powder cloud is achieved by blowing in air.

2.11.2 Coil Coating with Powder Coating

The use of powder coating in the field of coil coating has so far attracted little interest. Nevertheless, since 1989, individual pure powder coating plants for strip and/or sheet metal have been put into operation in Europe. Operators are mainly medium-sized companies, some of which also coat for their inhouse processing . Mainly aluminium is processed, in some plants also galvanized steel sheet, Galfan® (layer of an alloy of 5% Al/Zn) or Galvalume® (layer of 55% Al/Zn). The maximum substrate widths are between 1250 and 1700 mm, with the exception of the Alcan plant, where sheet widths of up to 2400 mm can be processed. The material thicknesses are 0.4–3.0 mm, depending on the design of the plant, and in some cases up to 5 mm for aluminium sheet.

All previous systems are of conventional design and work with electrostatic automatic spray guns with predominantly horizontal application. However, vertical application is also possible, with the metal strip running upwards through the coating zone. The current systems are equipped with up to 24 spray guns, which must achieve a uniform application over the strip width and length. The maximum throughput speed is between 8 and 25 m/min. Some horizontal lines allow the optional use of strip and sheet, but also of profiled sheets.

The heat treatment is generally carried out by a combination of infrared radiation and circulating air drying. Object temperatures of 180–250 °C are achieved, with the entire process being completed in less than 1.5 min. As a rule, the powder coatings (polyester-based) are applied in a single layer on one side of the coil, with layer thicknesses ranging from 50 to 80 µm.

The use of NIR (near infrared) drying appears promising, so that drying times of well below 1 min can be achieved [44].

2.11.2.1 Plant Engineering

Plants with coil speeds of 20 m/min apply the powder coating with corona or tribo guns, the cross-linking of the coating film takes place in infrared or convection ovens. Systems with speeds of 20 to 60 m/min also use corona or tribo guns, although more guns are required for the faster coil speed, which can lead to an obstruction in the coating process. The powder bell is the solution here. Infrared, convection or induction ovens or NIR technology can be used as curing ovens. High-speed lines with more than 60 m/min require a completely new application technology [45] (Fig. 2.85).

Structure of a Coil Coating Plant

A typical plant contains the following stations:

1. Cleaning and chemical pretreatment
2. Adhesive water drying
3. Cooling zone

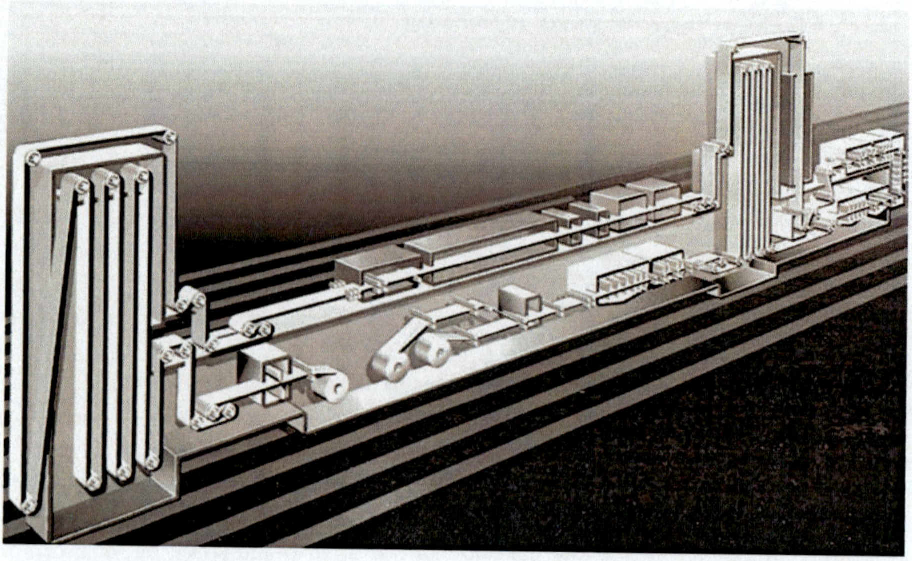

Fig. 2.85 Schematic diagram of a powder coil coating line

4. Paint application
5. curing/crosslinking
6. Cooling zone

2.11.2.2 New Process Developments

New innovative processes for coil coating with powder are the "Powder Cloud"process of Materials Sciences Corp., USA, the EMB process (electromagnetic brushing) of DSM, Netherlands, and the "TransApp" technique of the IPA Frauenhofer Institut Produktionstechnik und Automatisierung.

The Powder Cloud Process (MSC Technology) In the MSC technology, the generation and dosing of the powder cloud takes place via rotating brushes. The powder cloud with air is passed over a series of corona wires which charge the powder. The charge-to-mass ratio is very high due to the electrostatic field lines, which means that a uniform electrostatic field is generated across the entire width of the belt. The powder particles can thus penetrate the boundary layer of the fast-moving sub-static surface and deposit very uniformly [46].

The Electromagnetic Brush Process (EMB) The magnetic brush process (EMB) proposed by the Dutch company DSM Resins, derived from the principle used in copiers and laser printers, takes a completely different approach. Air is no longer required as a medium for transporting the powder particles. The basic principles are the use of generally magne-

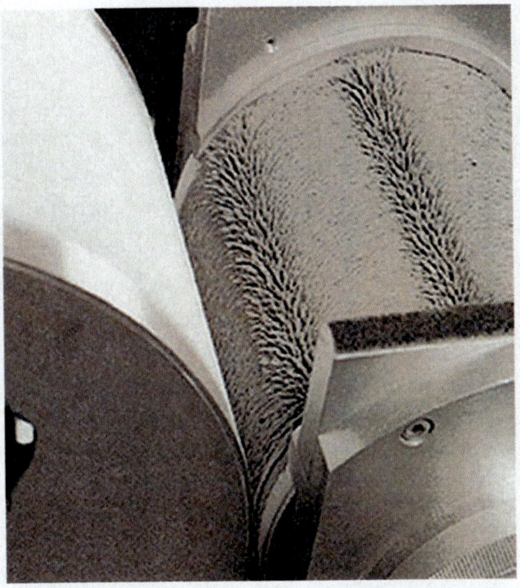

Fig. 2.86 Transfer roller and magnetic brushes [47]

tizable carrier particles and an intensive triboelectric charging of powder and carrier. The process is suitable for steel and aluminium strips and also permits high application speeds [44, 47] (Fig. 2.86).

The EHB Process (Also Known as the "Solid Block" Process) In this process, the powder coating is pressed onto the coil, which is preheated at 80 °C, in a still solid, pre-melted state. Crosslinking takes place in a tunnel oven. However, the limitations so far are the still low strip speed and an unsatisfactory quality of the film surface. This process was patented by Akzo (Courtaulds Paints) in Australia [48].

The TransApp Process In the so-called TransApp process, the powder coating is not sprayed but applied to the passing flat material via high electric field forces. The very short transfer distance ensures that the powder particles are deposited homogeneously and almost completely on the substrate. High spray losses do not occur here, making extraction booths and powder recovery systems retundant. The process is suitable for both conductive and non-conductive substrates [49].

Due to the high powder transfer rate, the achievable throughput speeds for flat substrates (strip material or sheet metal sections, e.g. blanks) are 100 m/min. These high speeds require short curing times of the powder coatings. Originally, the process was developed for high process speeds and compact systems. In the meantime, it is also suitable for the efficient, automated coating of three-dimensional serial parts. This is achieved by converting the process to a spray gap and thus to vertical application. The advantages of the fluid bed are thus retained and the process is suitable for a wide and variable range of parts [50, 51].

2.11.3 Vortex Sintering

Whirl sintering does not require charging of the powder. The workpiece is heated to the reaction temperature of the powder and immersed in the fluidized powder. The process is suitable for small parts and wire-like workpieces, such as dishwasher baskets, where large coating thicknesses are required without placing special demands on the optical coating quality [52, 53].

The layer thicknesses to be achieved depend on the preheating temperature and the dipping time. Usual values are between approximately 200 and 1000 µm. A second heat treatment of the workpieces is necessary if the heat input of the workpieces is not sufficient for complete melting or crosslinking [50].

2.11.4 Round Spray Systems in the Omega Loop

This concept is used for long, vertically hanging profiles or parts of limited width (depending on the diameter of the omega loop) arranged one above the other [54].

The characteristics of this process are:

- compact system
- high coating uniformity
- very economical operation with high parts throughput and full plant utilization
- 1 round spray head replaces up to 12 spray guns, thus you have less wear parts and less maintenance effort
- low flexibility or part geometry
- spray distance not variable
- pendulum of the hanging profiles
- not suitable for small batch sizes (high changeover costs)
- powder adjustment required

2.12 Improving the Efficiency of Electrostatic Spraying Processes

A technical solution is rarely a final solution, but usually a compromise, because many factors must be considered and weighed for the particular application. The painting process must also be considered as a holistic concept. The most important influencing variables concern:

- Processes, plants, equipment
- Application, manually or automatically
- Coating material (powder, water-based or solvent-based coatings)
- Workpiece

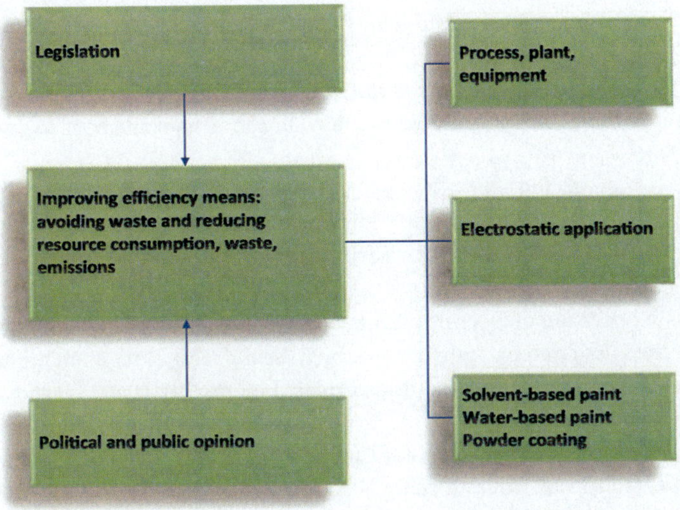

Fig. 2.87 Influences that can make an efficiency improvement necessary [55]

In the case of powder coating, the possibility of overspray recycling initially called into question the need for a high initial application efficiency, or rather judged it to be of secondary importance, since a high overall efficiency is possible. In the meantime, it has been recognized that it is absolutely necessary to already achieve a high initial application efficiency in order to achieve a high overall efficiency.

The following points can help to improve the first coat efficiency in powder coating: (Fig. 2.87)

- The thickness of the powder layer has a self-limiting effect, which is initially shown by lower application quantities per time and by the onset of backspraying. With an online control of the coating thickness, the powder application can be stopped at the optimum time. The proportion that no longer reaches the workpiece rises with increasing coating thickness, is sucked off and reduces the efficiency. Influencing variables are the height of the charge, the powder output, the distance between the spraying equipment and the workpiece, and its surface profile.
- The initial application rate depends, on the distance between the gun and the workpiece and the set gun current. With a small distance and low gun current, the maximum powder quantity to be deposited is reached faster than with a larger distance.
- The processing properties of powder coatings also have an influence on the initial application efficiency. In the case of powder, the particle size is one of them. Powder with a separated fine fraction, so-called sifted powder, leads directly to improved efficiency.

- A clear improvement is achieved with the paint-compatible design of the workpieces and hängers.
- System technology, such as gap control, application with robots, avoidance of rework or measures for a fast and efficient color change also contribute to improved efficiency [56].

2.12.1 Required Safety Installations

To ensure safe operation of electrostatic coating systems, the manufacturer and the operator are required to comply with occupational health and safety requirements. These requirements are specified in standards and rules. Figures 2.88 and 2.89 show the hazardous situations and safety criteria for electrostatic coating.

In the course of the harmonisation of European standards, several guidelines drawn up by CENELEC (European Committee for Electrotechnical Standardization) have now been given the status of national standards for electrostatic painting and powder coating systems. They are valid in all European Member States and deal mainly with the risks associated with electrostatic spraying systems. Their basis is the EU Explosion Protection Directive ATEX 2014/34/EU ("Directive 2014/34/EU of the European Parliament and of the Council of 26 February 2014 on the harmonisation of the laws of the Member States concerning equipment and protective systems intended for use in potentially explosive atmospheres (recast)").

The most important guidelines, laws and regulations to be observed in the field of electrostatic powder coating are:

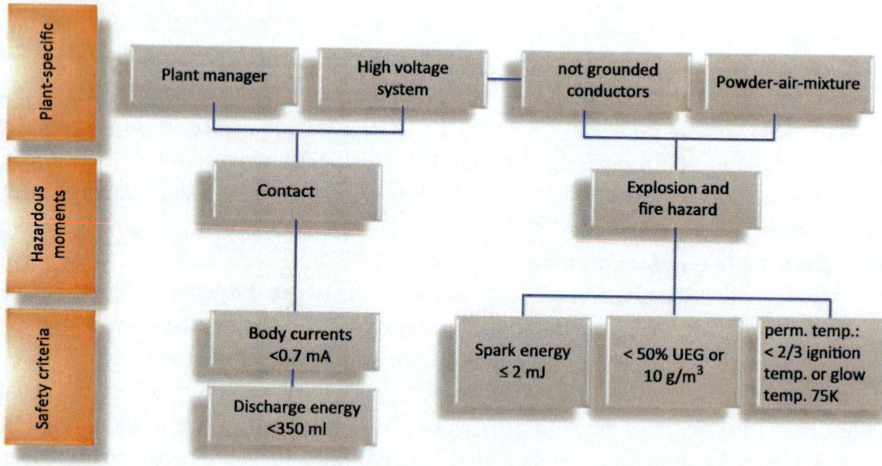

Fig. 2.88 Hazardous situations and safety criteria in electrostatic powder coating – corona charging of the powder coating [54]

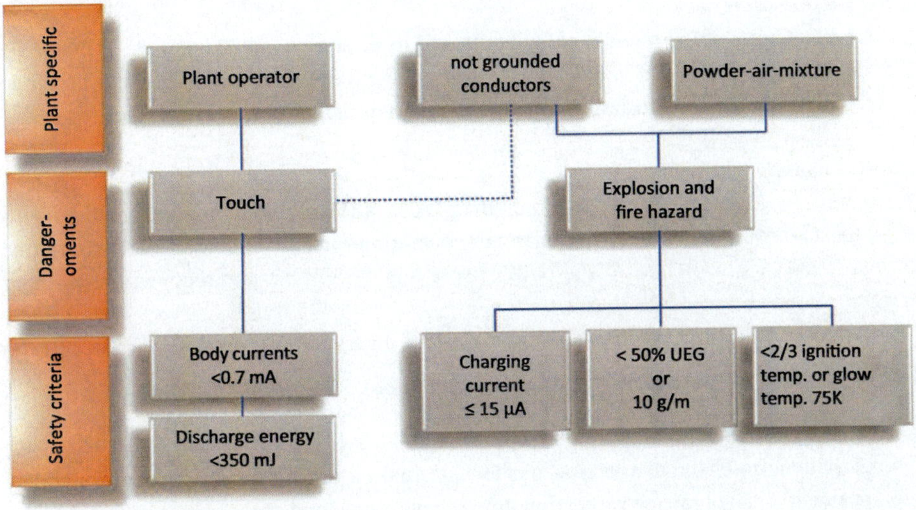

Fig. 2.89 Hazardous situations and safety criteria in electrostatic powder coating – tribocharging of the powder coating [54]

- ATEX Directive 2014/34/EU
- DIN EN 16985:2019–04 Paint booths for organic coating materials – Safety requirements.
- DIN EN 50050 Electrical apparatus for potentially explosive atmospheres – Electrostatic hand-held spraying device
- SN EN 50053 Selection, installation and use of hand-held electrostatic spraying equipment Part 2 for coating powder/energy limit 2 mJ
- DIN EN 50177 Stationary electrostatic spraying systems for flammable coating powders.
- Machinery Directive (2006/42/EC)
- Industrial Safety Regulation
- Ordinance on Hazardous Substances
- TRGS 900 "Occupational exposure limits"

DIN EN 16985:2019–04 deals with the requirements for powder coating booths and was revised in 2018/2019. The requirements of the standard include the following points for powder coating [57]:

Powder Concentration

With the exhaust air switched on, the average powder concentration must not exceed 50% of the lower explosion limit (LEL) of the powder used. If the LEL is not known, the concentration must not exceed 10 g/m3.

Construction Materials
Solid components of the powder spray booth and its equipment must be made of materials that do not support a fire. If walls made of non-conductive plastic are used, the occurrence of very high-energy electrostatic discharges (sliding brush discharges) must be avoided.

Fire Protection
Automatic powder spraying booths, including those with additional manual coating openings, must be equipped with an automatic fire detection system which, with a reaction time of less than 0.5 s, switches off the powder spraying system and the recovery system and triggers visual and acoustic alarms.

If spray guns with a discharge energy of more than 5 mJ are used, they must also be equipped with a local fire extinguishing system.

Classification of Danger Zones
The classification of the danger zones is an essential part of the safety concept for explosion protection. The ignition protection type of equipment and components installed in the powder booth depends on the concentration of flammable substances and their limitation by technical ventilation. The interior of the powder coating booth, including the open recovery system, is classified as Zone 22. The interior of closed recovery systems is to be classified as Zone 20. The exterior of the powder booth within 1 m of permanent openings shall be considered Zone 22. Doors are not considered to be permanent openings.

In the final filter, the closed area shown in purple in Fig. 2.90 is zone 20 and the area shown in green is zone 22, the non-closed area. The reason for the non-closed area is that the clean gas concentration is briefly above the LEL due to fulling during suction operation and cleaning, when the filter breaks through and dust escapes, Fig. 2.91.

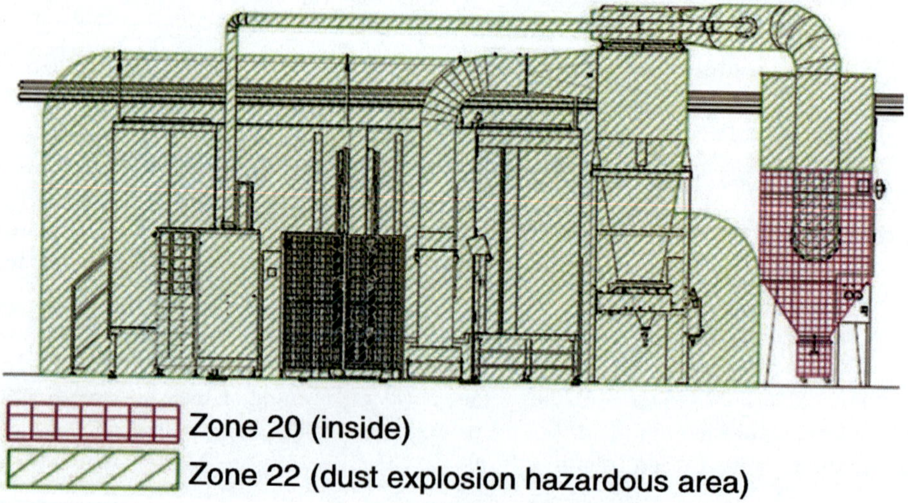

Fig. 2.90 Classification of the danger zones in the powder coating plant

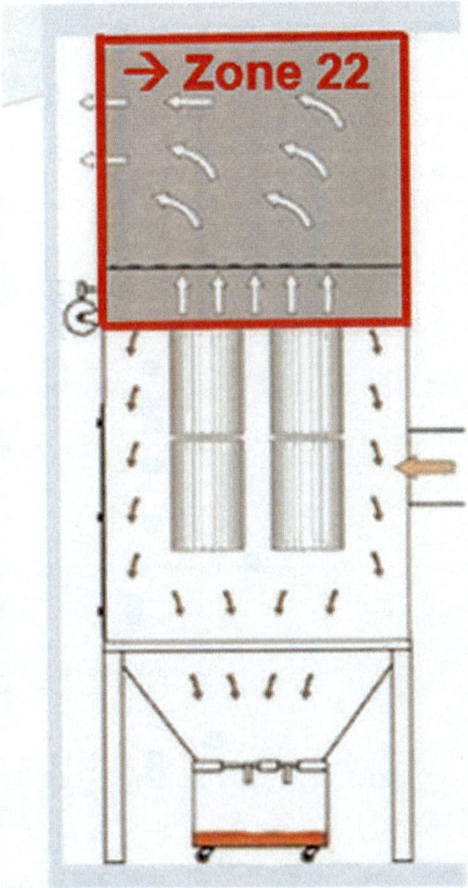

Fig. 2.91 Filter system with bag, fabric or cartridge filter

A system equipped with a sinter plate filter (Fig. 2.92) can be considered as a stagnation zone barrier according to VDI 2263, part 6.1, if the required values are achieved. Thus, an assignment of a zone is not necessary.

Workpiece Grounding/Plant Grounding

Workpieces are potential ignition sources, as they can become electrically charged during the coating process if their pick-up point on the hanger does not have sufficient earth contact. For this reason, DIN EN 50177 stipulates that the earth leakage resistance between the workpiece and the hanger must not exceed 1 MΩ or that the possible discharge energy of the workpiece is limited to a maximum of 5 mJ by special measures, for example the use of ionizers.

The measurement of the earth leakage resistance is carried out by means of an earth leakage control device installed at the inlet to the cabin, which shuts down the chain conveyor if the set value is exceeded.

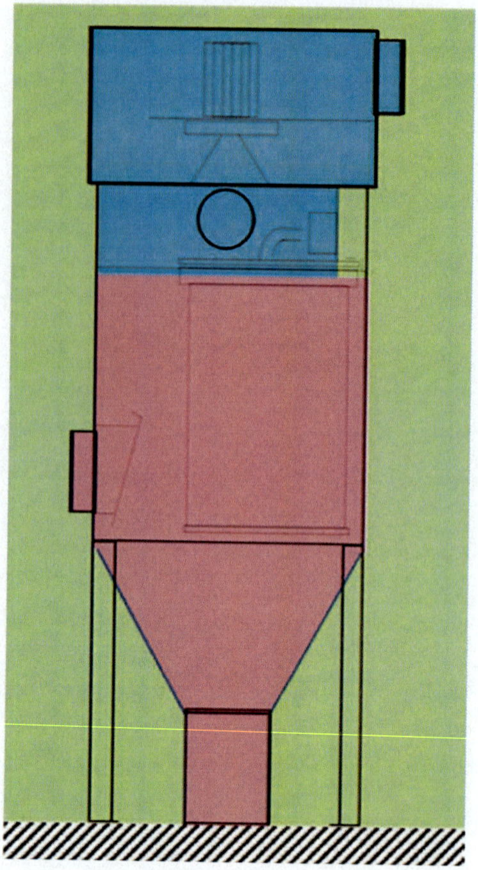

Fig. 2.92 Filter with sinter-plate filter from a manufacturer. The space marked in blue here is considered to be a clean gas space and therefore no zoning is necessary

This procedure is often problematic, since in practice the value of 1 MΩ is usually exceeded after only a few passes of the hanger through the booth. Since the alternatively approved use of grounding rods (ionizers) causes additional effort during booth cleaning, for example during a color change, and is therefore often not implemented, in these cases the workpieces are de facto to be considered as potential ignition sources.

Since the ignition of dust/air mixtures in closed containers is particularly dangerous due to the resulting explosion pressure development, closed powder recovery systems must be protected with explosion protection systems. In practice, the following options are available:

- An explosion suppression system that detects incipient dust explosions in the formation phase, i.e. within a few milliseconds, and suppresses them by introducing, for example, extinguishing powder.
- Explosion pressure relief of the recovery system, combined with explosion decoupling of the enclosure and explosion-proof design of all system components. It is important with this measure that the explosion is not only relieved to the outside, but that the pressure and flame flashback into the enclosure is prevented, because this could lead to a secondary explosion. Only one of the two measures would be ineffective on its own. However, mechanical installations in the form of pressure relief systems and quick-acting slide gates for decoupling are more of a hindrance for the colour change required in powder coating. The associated costs are also enormously high.
- A fast-acting fire suppression system to prevent the spread of flame from the booth to the recovery system. The prerequisite for this is that the recovery itself does not contain any potential ignition sources. It does not interfere in any way with the coating process and, as gaseous extinguishing agents such as CO2 or Inergen can be used, would not cause any damage even if triggered.

Personal Security

Protection against contact and exposure to high electrical voltages is achieved by arranging all live plant parts in largely enclosed coating booths. The accesses to the booths are secured in such a way that when they are opened, the high voltage is automatically switched off and the voltage carrying parts are earthed before contact is possible. In the case of manual systems, the high-voltage generator must be dimensioned in such a way that no effects harmful to persons occur as a result of approaching or touching the entire outer surface of the spray gun and supply line. The maximum permissible current is 0.7 mA = 700 μA, the maximum discharge energy of capacitive discharges must not exceed 350 mJ = 350 mWs. Since the minimum ignition energy of the spray-air mixtures is significantly lower at 2 mJ for powder coating, personal protection is included in the guarantee of the limit values for explosion protection with multiple safety. Manual equipment must be tested for ignition safety as part of a type examination. Only equipment bearing the test mark may be used. Tribo pistols are an exception. Here, charging currents <15 μA are considered equivalent to a passed ignition test in terms of safety.

2.13 Common Failures in Powder Coating and Possible Solutions

When coating with powder coating, the possibilities to change the supplied powder material are very limited. However, established routines and constant attention can prevent the occurrence of growing problems.

In most cases, the powder material supplied is adapted to a specific application area or process. Although both the coating and the equipment can vary from application area to application area, there are a number of problem areas. These are systematized here in terms of possible causes and suggestions for solutions.

For a good result and problem-free coating, it is important, to observe the supplier's specifications with regard to material and equipment.

Despite all the advances in powder coating, it is still not possible to set the correct air volume for conveying, dosing and tribological air at the push of a button.

The reliable plant operator who closely observes the powder cloud is still in demand, because only the human eye is able to determine the correct amount of air. Often observed deficiencies are:

1. Powder containers that fluidize the powder unevenly
2. Powder Feeding with too high conveying and dosing air
3. unsuitable powder hose routing
4. too long powder hoses
5. wrong hose diameter
6. too much charge

The most frequent source of error is still point 2, an excessively high air volume for feeding or dosing air. This results in the following disadvantages:

Costs	Purified compressed air is one of the most expensive energies.
Quality	Increased air volume has a negative effect on the coating.
Maintenance	Higher air velocity in the system can only cause greater wear and tear

Table 2.14 lists various problems in powder coating and describes possible solutions. They were compiled from the trouble-shooting tables of [30, 54, 58–62].

Table 2.14 Measures to be taken in the event of problems in powder coating

Phenomenon	Possible cause	Possible solution
Powder transport and application		
Powder spills out of the powder container	Fluidizing air pressure too high The proportion of fines in the powder is too high.	Fluidizing air pressure reduction Reduction of recovered powder in favor of fresh powder Consultation with powder supplier about delivery of powder with reduced fines content
No fluidization of the powder	Insufficient air pressure	Checking the compressed air supply, possibly increased pressure in the compressed air network Control of the supply line
	Clogged fluid bottom in the powder hopper	Emptying the tank and checking the membrane
Poor fluidization after some time only "Volcanoes" that blow air onto individual spots	Not enough powder in the container or container too large Solid or wet powder clogged or broken membrane	Feeding of new powder until the container is approx. 2/3 full during fluidization Increased air pressure and manual stirring in the container/control of the humidity of the compressed air (dew point <3 °C) Emptying the container Fluid bottom control
Powder separates into layers of fine and coarse particles	Too much powder in the container / too much fines in the powder	Emptying the tank to approx. 2/3 of the volume Reduction of the amount of recovery powder Control of the grain spectrum
Lumps in the powder	Too long storage or unfavorable storage conditions (temperature/humidity)	Sift powder through a 200–300 μm sieve strong fluidization with pure, dry compressed air
Powder accumulation in the hose – intermittent transport of the powder	Air velocity too low in relation to the powder quantity	Increasing the pressure of the supplementary air (if possible) on the injector hose with a smaller diameter.
	Excessive wear of the injector plugs	New injector candles, use different material if necessary
Powder output irregular	Powder air velocities too low	The ideal volume of air should be in the range of 4–5 m^3
	Fine-grained powder/powder with poor flowability	The use of a powder hopper with fluid air and vibration provides a remedy.
	Short-term, high compressed air consumption can cause pressure fluctuations	Compressed air tanks in front of the large consumption point can help
	Cross sections or radii too small for powder hose routing in injector systems	The radii must be min. 150–200 mm for injector systems.
	a poor ratio between fresh powder and recovered powder	Use fresh powder with level control
	Excessively high/low output	Adjustment of the coating parameters
	Contaminated injectors	The injectors are often neglected during daily cleaning.
Pulse	Pulsation of the powder feed is caused by an incorrect setting of the conveying air or an incorrect dosing ratio.	Check the setting of the delivery air and the dosing ratio. If the powder output is less than 80 g/min., a 9 mm powder hose should be used.

(*continued*)

Table 2.14 (continued)

Phenomenon	Possible cause	Possible solution
Low powder throughput	Low fluidization of the powder uncleaned powder supply affect transport Damaged powder hoses Worn parts low air pressure	See point poor fluidization Cleaning the hoses and Venturi tubes Check powder supply for contamination Check recovery powder before use Repair or replace the damaged parts/Avoid sharp clamps/Kink protection on the powder hoses Replacing the worn parts Checking the air supply
Powder fusion due to strong powder build-up in hoses and openings Powder sintering	Strong powder build-up/moisture and/or oils in the air supply Wrong material for the powder hose Worn Venturi tubes Powder too fine Sintering on baffle plates and nozzles due to excessive air pressure of the transport air	Cleaning or replacing the affected parts Check compressed air for moisture Use only original material Replacement of the worn parts Reduction of the amount of recovery powder Control of the grain spectrum of the fresh powder Reduction of the compressed air to 4–5 Nm3/h
	Powder	Some powder qualities have an increased tendency to sintering/fusing – Consult the manufacturer
Dust from the powder booth – insufficient ventilation	Unnecessarily large openings Clogging of the filters in the booth or at the final filter	Close openings Adapt inlet and outlet to object sizes Clean, or use new filters or filter cassettes Increased ventilation (if possible) in the booth
	Too large powder quantity injected Thermal effect due to furnance radiation	Reduction of the number of guns or reduced powder quantity per gun Partition wall between heat sources and powder booth
Separation efficiency	A poor separation efficiency is often caused by a leak in the recovery system (after-filter, cyclone, day tank, sieve, powder hopper venting, hose connections, etc.).	Checking for leaks
Level probe	The powder containers are earthed via the fluidisation hose with integral copper strand. The earthing of the Injectors are earthed in the same way via the conductive dosing air hose. This not only protects personnel, but also ensures that the electronics of the level sensor are not damaged.	always use original hoses
Badly charged powder	The high voltage generator gives too low voltage to the electrodes of the gun.	Checking cables and fuses/Checking electrodes/Increased voltage/Checking whether the voltage on the gun corresponds to the voltage read on the control panel.

(continued)

Table 2.14 (continued)

Phenomenon	Possible cause	Possible solution
Insufficient wraping and layering	Poor grounding	Control of the connections between object, carrier and earth All contact points must be free of paint and other insulating materials.
	Too large powder quantity per gun	Reduction of the quantity so that the majority of the powder passing is loaded
	Too much fine powder	Reduction of recovered powder in favor of fresh powder Consultation with powder suppliers about delivery of powder with lower fines content
	Unsuitable powder type Powder is blown past the object due to too high air speed	Clarification whether the powder is suitable for the current area of application and for the spraying equipment Reducing the air velocity and/or increasing the distance from the gun to the object
	Powder is "drawn off" by adjacent objects or hangers	Reduction of the density of objects on the hanger and/or modification of the hanger geometry.
Insufficient penetration into internal corners and recesses	Too high air velocity in the gun – the powder is blown off wrong distance gun - Object Too small powder quantity per gun Poor grounding Too wide spray pattern Voltage too high at the high-voltage generator Air velocity too high not suitable gun alignment	Reduction of the additional air Increasing or decreasing the gun-object distance Increased air pressure at the injectors control of the connection between object and hanger and between hanger and ground. All contact points must be free of paint or other insulating material. Smaller deflectors or flat jet nozzle Stress reduction to equalize the layer thickness for inside and outside corners Pressure reduction of the supplementary air, greater distance between object and gun to prevent blowing off of the powder Setting up the guns so that the powder cloud is aimed as directly as possible at complicated areas of the object
	Too much fine powder	Reduction of recovered powder by optimizing the application efficiency Powder with lower fines content
Problems to achieve a sufficient layer thickness Powder drops	Too low charge/charge effectiveness Poor grounding Air velocity too high Unfavorable humidity in the working room	Control of cables, fuses and electrodes Increase of voltage Checking the gun voltage Checking the connection between the object, conveyor and ground. All contact points must be free of paint and other insulating material. Reduction of air pressure, greater distance gun – object Avoid relative humidity below 30% and above 70% (ideal: 45–55%)

(continued)

Table 2.14 (continued)

Phenomenon	Possible cause	Possible solution
Discharges (sparking)	On hoses due to charging of the hoses as a result of friction (sparking) At the injector due to insufficient grounding Between object and hook due to insufficient grounding On the manual gun due to insufficient grounding When using metallic powders	Use original manufacturer's powder tubing Use conductive tubing Use original manufacturer pneumatic hoses Decoat hangers and hooks No rubber soles, use suitable work clothes metallic nozzles
Poor adhesion of the powder to the object	Workpieces insufficiently grounded Powder with too low resistance Grain size distribution not optimal Field strength too high	Clean hangers Consultation with the powder supplier Consultation with the powder supplier Change voltage and distance
Poor adhesion of the powder to the object	Powder is blown out of the corners and recesses due to excessive air pressure Too little penetration depth in the corners and niches due to too high tension or the use of incorrect nozzles insufficient powder quality Lifting movements which are too fast can destroy the powder cloud.	Reduce air pressure Reduce tension and use suitable nozzles Change powder, change additive, Adhere to pigmentation The lifting speed should not exceed 0.6 m/s; values between 0.3–0.5 m/s are recommended.
Backspray effects/not uniform powder layer	Unnecessarily high voltage Too small distance gun - Object Poor grounding high powder build up	Reducing the voltage Increasing the distance Control of the connection object – earth Reduction of the layer thickness or powder output
Large differences in the thickness of the layer	Unfavorable gun setup wrong distance gun/object Poor match between hanger and crosshead speed uneven air speed in the booth affect the spray pattern Unsuitable suspension technology Uneven powder transport	Control of the device and sufficient overlap Increase or decrease the distance Adjusting the speed of the hanger and/or the speed of the crosshead Consultation with equipment suppliers as far as possible uniform objects and suspension technology and correct spacing Good fluidization and at least half-filled powder container Optimization of hose diameter and length
Spraying problems after mixing with recovery powder	Change in grain size distribution	Reduction of recovery powder by optimizing the suspension technique, gun device and charging effect.

(continued)

Table 2.14 (continued)

Phenomenon	Possible cause	Possible solution
Spitters or billowy powder discharge	Accumulation of powder in guns/feed system	Reduction of recovery powder more frequent cleaning Consultation with powder supplier, delivery of powder with lower fines content
	Unsuitable air pressure or air quantity	Check the air supply. A sufficient air volume must be ensured, in order to prevent a drop in the powder output when switching on further air consumers on the equipment.
	Kinked and/or too long powder hoses	Use kink protection; shorten hoses if necessary.
	Pump, hoses or gun clogged with powder	Cleaning the parts and testing the air for moisture and/or oil
	Poor flowability of the powder	
	Humidity above 55%	Checking and, if necessary, regulating the humidity of the air
Cabin	If powder suddenly escapes from a well adjusted booth, and the pressure monitoring of the exhaust air system shows normal values, the fault is usually to be found in the booth environment.	The most common cause is drafts in the plant hall through open windows and gates (even further away from the plant). Thermally caused air currents due to high waste heat can superimpose on the cabin air.
Nozzles	To obtain optimum coating results, the choice of nozzle must be adapted to the coating requirements: The flapper generates a homogeneous powder cloud which slowly approaches the grounded workpiece. The cloud can only be controlled to a limited extent and the penetration capacity is small. It is used for flat parts and high area outputs. The flat spray nozzle produces an elliptical cross-section and can be well aligned with the object. It is used for more complex parts with recesses. Metallic powder can cause a short circuit in the gun. Metallic powder is conductive so charging of the powder must take place outside the gun. With the metallic nozzle, the electrodes are therefore outside the powder tube.	
Powder container overflows	The fluid air pressure or the air quantities are set too high. If the powder hopper is not close enough to the booth, the vent can become clogged – overpressure develops in the hopper and powder escapes.	The fluidized powder should flow only slightly, but evenly. After an extended period of use, partial clogging of the fluid plates may occur, which must be compensated for by an excessively high fluidisation air volume. In this case, the base must be replaced. The powder hopper should be as close as possible to the booth and connected to the booth with the shortest possible piece of hose. This hose must be cleaned daily.
Screening problems	The powder hopper ventilation is too weak.	Adjust Airmover more strongly or replace in case of defect
	Mesh size of the sieve too small	Replace sieve
	Incorrectly adjusted ultrasonic transducers	Adjust the intensity of the ultrasonic transmitter
	Waste bag	The waste bag of the rotary screener should have a slight negative pressure. If it is inflated too tightly, the Airmover must be higher be set.
Maintenance	The quality of the coating and the service life of the system components depend on the reliable maintenance of the system. The following maintenance work must be carried out regularly: Daily: • Blow out guns and check for sintering • Check the setting of the spraying elements • Keep gun feed pressure low	

(continued)

Table 2.14 (continued)

Phenomenon	Possible cause	Possible solution
	• Check injectors (signs of caking/grinding on the sleeve). • Blow out injectors and hoses • Clean plant, recovery and surroundings • Check the suction resistance of the recovery • Check fluid bottom Weekly: • Check grounding control • Check that the compressed air is clean, dry and free of oil • Check guns for voltage output • Check powder hoses for sintering and kinks • Check stroke movements and display • Check filter systems	
Economic efficiency	In addition to requirements for chemical and mechanical resistance, colouring and surface texture, the user's focus is also on economic aspects. When new parts are added to the portfolio, a review of the system parameters is usually beneficial. If the parts are complex, a review of the gun arrangement may also be necessary.	
Cyclone separation degree	Pinch valves close poorly. Pinch valves run too slow or too fast. Pinch valves do not open fully. Pinch valves are too small in size and thus have too small a flow rate.	Check operating pressure, set to – 2.5 bar. Adjust times according to operating instructions. Check pinch valves for material fatigue and hose inserts for correct assembly Replace pinch valve with new one
Surface/appearance		
poor uniform paint distribution	Layer too thin Too slow heating of the object Too high gun voltage Unsuitable powder coating type (too high reactivity or too coarse-grained powder)	Measure the layer with a suitable and calibrated measuring device. Adjust the layer thickness by changing the air pressure at the injectors. Check the powder curing window, if necessary. Increasing the furnace temperature, adjusting the temperature profile; if possible, IR annealing in the input zone. Reduction of the voltage or greater distance gun – object Contact with powder supplier, whether the powder is suitable for the area of application
Low gloss	Contamination with an incompatible product	Thorough cleaning of the powder booth, feed and recovery before using a new powder
Low gloss	Extreme temperature stress, e.g. during IR heating, or excessively high furnace temperatures/dwell time in the furnace. Micropores due to outgassing. Increased orange peel due to low layer thickness	Temperature reduction Check substrate for pores, moisture; check powder for moisture, possibly "outgas" workpieces or use special powders for porous substrates Increasing the layer thickness
Pinholes in the surface	Paint layer too thick moisture Base material Incompatibility with other powders Residues on the surface, carry-over from pre-treatment, oil, grease,	Reduction of the layer thickness Control of humidity in compressed air, drying of parts, control of humidity of powder Investigation of the porosity of the substrate (casting material) Clean the equipment, check the compatibility of the powders Pretreatment, final rinse, check drying Reduce coating thickness

(continued)

Table 2.14 (continued)

Phenomenon	Possible cause	Possible solution
	Spin-off products from the powder coating Wet and powder coating in one plant Overcoating of filled areas of liquid paint primer	Avoid powder and liquid coating in one plant Filler/painted part cured oven > 180 °C Since all powders are "allergic" to silicone, it must never be used in the coating area.
High gloss with matt coating	Not sufficient hardening	Increasing the oven temperature and/or extending the baking time
Paint contamination of the surface	Contamination from powder previously coated in the plant Contaminated fresh powder	Thorough cleaning of the powder booth, feed and recovery systems before inserting the new powder Consultation with Powder suppliers
Bumps/pimples in the paint surface	Sintered powder in the guns dissolves. The recovery sieve is defective. Impurities fall into the booth from the suspension or transport system. Particles/pollutants in the air from the hall are sucked into the powder booth.	Thorough cleaning of guns and hoses Avoiding too high air pressure control, Repair, replacement of the sieve routine cleaning of suspension and transport system Enclosure of the powder booth General cleanliness in the cabin environment Filtering and regulating the humidity in the room
Craters in the surface	Contamination by incompatible powder Residues on the metal surface from Insufficient pre-treatment or carry-over from pre-treatment Moisture and/or oil in the compressed air	Thorough cleaning before inserting the new powder Checking and correcting the pretreatment Checking the oil and water separators
Pores in the coating layer	Damp substrate Powder quality not suitable for the current substrate	Checking the object after pretreatment and drying
Wrong or variable color	Curing conditions outside the specified curing window. Oven atmosphere, e.g. directly heated oven, IR oven	Comparison with product data sheet, readjustment Use powder coating adjusted to oven atmosphere
	Materials/conversion layer with non-uniform inherent color. Layer thickness variations. Variation in object thickness. Unsuitable powder type	Pay attention to layer thickness; Measurement of layer thickness and regulation. Use of polyester-based systems. Consultation with powder supplier, whether powder is suitable for application area and furnace type
Droplet formation	Layer thickness too high, e.g. for wire products Powder slips to the edge. Too high residual or preheating temperature of the parts Temperature rise in the furnace too fast Powder accumulations at openings and doubles	Control of the layer thickness and necessary adjustment Increase high voltage. Allow parts to cool down longer or reduce preheating temperature Correct furnace settings Optimize workpiece suspension

(*continued*)

Table 2.14 (continued)

Phenomenon	Possible cause	Possible solution
	Varnish type not suitable for the application (too good flow)	Use more viscous coating type; consult powder supplier
Poor edge coverage	Sharp edges on the material Varnish type not suitable for the application area	Contact with supplier and processor because of changes Alternative suspension technique (sharp edges downwards) Consultation with powder suppliers
Edge structure	Too big powder cloud Too big distances between the objects	Work with targeted powder jet use gap control
Powder contracts and leaves bare spots	Powder is not charged Insufficient pre-treatment (cleaning) or drying of the substrate	See powder application Checking of pre-treatment and drying conditions (adhesive water dryer, temperature and time)
streaking/banding	With automatic coating wrong arrangement of the guns Chain speed too high In principle, there is a mathematical relationship between conveying speed, stroke speed and the number of guns moved to form and avoid stripes on the workpiece.	Determining the optimum gun arrangement or optimizing the sine curve of the automatic guns. Reduce transport speed Checking the sinusoidal curve. If the time of the double stroke of the gun is kept smaller than the time required for the conveyor to pass through the spray width with the object, no streaks occur.
Surface spotty, cloudy	Segregation for recovery powder Difference in coating thickness due to incorrect sine curve of the automatic guns High voltage not optimal Unfavorable workpiece geometry Mixing ratio fresh/recovered powder not optimal Distance gun – part not optimal for matte powders Object temperature too low during curing	Continuous or uniform addition of fresh powder adapted to actual consumption Match hoisting speed to chain speed Optimize high voltage Optimize or adapt part suspension Uniform, continuous addition of fresh powder to the recirculating powder Optimize distance gun – part (mostly increase) Increase object temperature
Surface full of holes	See craters/pinholes. High voltage too low, no closed surface, or low layer thickness	Maintain minimum coating thickness Optimize high voltage
Surface fatty	Sweating out powder additives in the furnace Powder layer thickness extremely high (>180 μm)	Optimize curing conditions Maintain target coating thickness
Surface sensitive to touch	Suspend parts while hot	Wait until part temperature < 50 °C
Cracks	See spalling embedded foreign particles in the bending edge	Clean the workpiece surface optimally, sieve the powder to remove impurities (possibly select a smaller mesh size of the sieve).

(continued)

Table 2.14 (continued)

Phenomenon	Possible cause	Possible solution
Physical properties		
Poor mechanical properties – impact strength/bendability	insufficient curing Layer too thick Paint adhesion problems due to insufficient pretreatment Paint properties Substrate has different material thickness and/or material	Measuring the temperature profile of the object and the oven. Increasing the oven temperature and/or extending the baking time. Control of the layer thickness and reduction of the powder quantity, increase of the transport speed Pretreatment control Consultation with powder manufacturer Testing the substrate
Insufficient adhesion to the substrate – flaking off	Inadequate pre-treatment under-/or over-baked embrittlement of the paint, e.g. in IR ovens Base material too thick or of different thickness	Checking the pretreatment Adjusting the curing conditions Change curing conditions/Observe powder manufacturer's curing window Use slower curing system
Insufficient corrosion resistance	Defective pretreatment not suitable for the application Not fully cured coating layer, undercrosslinking	Pretreatment control, other process Increasing the oven temperature and/or extending the baking time
Insufficient chemical resistance	Undercuring unsuitable type of paint	Increasing the oven temperature and/or extending the baking time Consultation with powder supplier
Abrasion resistance low	Lack of curing abrasive packaging coating film sensitive to scratching – too soft	Observe curing conditions or use higher temperature Choose suitable packaging Select suitable powder coating
"Picture frame effect"	Grain spectrum too coarse or not optimal High voltage too high Distance gun – object too large	Sieve the powder or clarify suitable particle size distribution with the powder manufacturer. Adjust high voltage, reduce Minimize distance Low ion gun application
Opacity/hiding power	Layer thickness too low Layer thickness varies greatly from part to part Incorrect arrangement of the individual guns for automatic coating	Observe minimum coating thickness for critical colors (red, yellow, orange) Optimize plant parameters Matching the lifting speed to the chain speed
Edge covering	Powder coating distribution too good Edge radii < 1 mm Burrs, e.g. punching burr on workpiece curing conditions unfavorable	Reduce the baking temperature or temperature rise in the melting phase of the powder. Optimize powder coating, change workpiece geometry. Deburr workpieces Measure furnace and adjust more optimal
Metallic paint properties		
No charge	Short circuit in the gun Unsuitable spraying equipment of the powder	Special nozzle for metallic coatings Consult powder supplier
Color changes Cloud formation	Uneven layer thickness	Layer thickness control If certain areas have to be coated by hand (e.g. bends in sheet metal), coat them first.
Change of the metallic effect	Dry mixed powder (segregation in the recovery)	Exclusive use of fresh powder, no recovery Use bonded powder

(continued)

Table 2.14 (continued)

Phenomenon	Possible cause	Possible solution
Textured coatings		
surface visible under the structure	Layer too thin	Increasing the layer thickness
Uneven structure	Uneven layer thickness in the edge area	Optimization of powder application to achieve the most uniform coating thickness possible Coarse-textured powder: apply sufficient coating thicknesses and coat with guns without corona rings Fine structure powder: uniform layer thickness distribution, especially in the edge area by application with guns with corona ring
Degree by degree variation in the structure	Variation of the particle size distribution	Minimization of recovery powder through optimization of powder application Exclusive use of fresh powder. More frequent and uniform addition of fresh powder
Contamination	Cleaning booth, guns. Textured powder (especially coarse textured powder) is sensitive to contamination by powder of a different color. Impurities are more likely to be seen than with smooth flowing powders.	Remedy: use only fresh powder, no recovery, new powder hoses, extra hoses for each color/color group
Overpainting		
no uniform paint distribution	Insulating effect of the first layer, so that grounding of the workpiece is limited	Reduction of the gun voltage (to 30–40 kV)
Poor adhesion between layers	first layer contamination Powder not suitable for overpainting Complete crosslinking of the first coating layer	Avoid contamination, immediate application of the second coating layer Consultation with powder suppliers For a good bond between the first and second layers, it may be advantageous to apply the first not fully cured.
Furnace problems		
Smoke/odor during the curing process	Dry powder is blown off and comes into contact with the radiators. Overheating of the coating layer	Reduction of the air speed in the entrance area of the furnace. Control of the valve and fans for sufficient ventilation of the furnace. Measuring the temperature on the object and possibly reducing the oven temperature. Control of the effect and the distance between emitters and workpiece in the IR zone.

2.13.1 The Basics of Powder Coating [35]

1. **The pistol attachment**
   Use the appropriate gun attachment for each application
2. **The hose**
   For an optimal powder conveying with injector the choice of the right hose quality as well as the hose diameter is of great importance.
3. **The hose guide**
   The powder hoses of injector systems should be as short as possible, have large bending radius and be laid without kinks.
4. **The Fluidization**
   The powder must flow in the priming container (powder hopper) like water.
5. **The fresh powder dosage**
   For a constant coating quality a continuous fresh powder dosage controlled by a level probe is necessary.
6. **The air speed**
   The air velocity in the powder hose should only be high enough to prevent powder deposits from forming.
7. **The air volume**
   The correct air volume of conveying and metering air is of decisive importance.
8. **The air**
   Use only dried and purified air.
9. **The temperature**
   Maintain a maximum temperature of 35 °C in the coating room (standard powder).
10. **The powder**
    Use powder suitable for the requirements and the charging system, e.g. for tribosystems.
11. **The delivery rules**
    Define requirements for the powder and have them confirmed by the manufacturer in the form of a test certificate or use powder materials with an independent quality label.
12. **Trials**
    Carrying out practical coating tests to determine which charging principle should be used (corona, tribo or low ion).
13. **The charging system**
    If local orange peel effects occur, use low ion charging systems.
14. **The cleanliness**
    In general, cleanliness must be ensured in the coating area. Stripped hangers must not have any loose residues (slag, blasting material, etc.) and dirt traps should be used for conveyors.

References

1. Bauch, H.: Elektrostatische Grundlagen der Pulverbeschichtung. DFO Weiterbildungslehrgang: Pulverlackiertechnik – Grundlegende Zusammenhänge für die Praxis, Berichtsband (1993)
2. Kleber, W.: Von den Experimenten der Reibungselektrizität zu hochproduktiven elektrostatischen Arbeitsverfahren. Berichte der V. Tagung „Elektrostatisches Beschichten", Hochschule für Verkehrswesen, Dresden (1975)
3. White, H.J.: Entstaubung industrieller Gase mit Elektrofiltern. VEB Deutscher Verlag für Grundstoffindustrie, Leipzig (1969)
4. Moore, A.D.: Elektrostatik, eine Einführung mit Versuchen. Chemie GmbH, Weinheim (1972)
5. Plato: Timaios, 4. Jh. v. Chr. In: Deutsche Übers. Platos sämtl. Werke, Bd. 3. L. Schneider, Berlin (1950)
6. Cajus Plinius Secundus: Historia naturalis, in 37 Büchern um 78 n. Chr. Ausg. 1524 und 1631
7. Gilbert W.: De Magnete. London Lib. II Cap. II 1600, S. 46–60
8. Descartes, R.: Principia philosophia 1644, dt. Übersetzung von A. Buchenau. F. Meiner, Hamburg (1955)
9. Guericke, O.V.: De experimento, quo praccipuae hae virtutes enumeratue per attritum in globo sulphureo excitari possunt. Phil. Trans. **7**, 5103 (Neuaufl. Leipzig 1BH1) (1672)
10. Hauksbee, F.: Several experiments on the attrition of bodics in vacuo. Phil. Trans. **24**, 2165–2175. (1704–1705)
11. Hauksbee, F.: Physico-Mechanical Experiments on Various Subjects. R. Brugis, London (franz. Übers. Paris, 1754) (1709)
12. Du Fay, C.: Premier memoire sur l'electricite, S. 31–49. Mem. Acad. Roy. Sciences. Amsterdam (1733)
13. Bose, G.M.: Die Electricität nach ihrer Entdeckung und Fortgang mit poetischer Ader. Wittenberg (1744)
14. Gray, S.: Letter concerning the electricity of water. Phil. Trans. 37, 227–230 (1731/1732)
15. Gray, S.: Experiments and observations upon the light that is produced by communicating electrical attraction to animal or inanimate bodies together with some of its most surprising effects. Phil. Trans. **39**, 16–24. (1735/1736)
16. Wollfen, C.: Allerhand nützliche Versuche, dadurch zu genauer Erkenntnis der Natur und Kunst der Weg gebähnet wird, vol. Bd. 1–3. Renger, Halle (1738)
17. Gordon, A.: Phenomona electritatis exposita. Erfordiae (1744)
18. Prinz, H.: 200 Jahre Experimentierkunst mit Reibungselektrizität. Bull. SEV. **55**, 2–21 (1964)
19. Kleber, W.: Elektrostatische Oberflächenbeschichtung, Teil 1. Meta. **54**(6), 40–43 (2000)
20. Kleber, W.: Einführung in die elektrostatische Verfahrenstechnik. Schulungsunterlagen Praxis-Forum Seminar EPS-Basiswissen, Technik + Kommunikationsverlag, Berlin
21. Bauch, H.: Auflademechanismen in Pulversprühgeräten und ihr Einfluss auf die physikalischen Prozesse bei der Schichtbildung. Schriftenreihe „Praxis-Forum", Fachbroschüre „Oberflächentechnik", Heft 23/89, Technik + Kommunikation Verlags GmbH, Berlin
22. Kleber, W.: Triboelektrische Pulveraufladung. Meta. **50**(7), 564–567 (1996)
23. Bauch, H.: Theoretische und experimentelle Untersuchungen zum Einfluss der Korngrößenverteilung und deren Änderung bei der Pulverlackierung. Schriftenreihe „Praxis-Forum", Fachbroschüre „Oberflächentechnik", Heft 25/90, Technik + Kommunikation Verlags GmbH, Berlin
24. Rohmann, H.: Methoden zur Messung der Größe von Schwebeteilchen. Z. Phys. **17**, 253–265 (1923)
25. Pauthenier, M.M., Moreau-Hanot, M.: La Charge des Particules Spheriques Dans un Champ Ionise. J. d' Physique et Radium. **3**(7), 590–613 (1932)

26. Bauch, H.: Untersuchungen zur Schichtbildung beim elektrostatischen Auftragen von Plastpulvern. Plaste Kautsch. **21**(11), 853–859 (1974)
27. Bauch, H.: Einfluss der Anlagenparameter und der Pulvereigenschaften bei der elektrostatischen Plastbeschichtung mit pistolenförmigen Sprühgeräten. Plaste Kautsch. **22**(3), 280–287 (1975)
28. Kleber, W., Bauch, H.: Was ist mit Tribo-Pulversprühpistolen? Oberfläche + JOT. **26**(1), 19–21 (1986)
29. Gustov, S.: Electrostatic Phenomena in Powder Coatings. Information Paper Nordson Corporation, USA (2017)
30. Firmenschriften zur Praxis der Pulverapplikation, ITW Gema AG, St. Gallen
31. Mendler, H.: Handbuch der Applikationspraxis. Firmenschrift J. Wagner GmbH, Markdorf (2001)
32. Topp, M.: Interne Schulungsunterlagen der Fa. Wagner, Markdorf
33. N. N.: Pulver für Pkw-Neues von der Pulverlackierung am Auto. Vortrag anlässlich des Firmenjubiläums der Fa. Eisenmann (2001)
34. Moser, J., Plenk, R.: Pulveraufbereitung und -Förderung in Perfektion. Vortrag EPS-Pulvertref, München (2002)
35. Mendler, H.: Ratschläge für Pulververarbeitung, Ausgabe April (2008)
36. Topp, M.: White Paper Ultraschallsieb, internes Papier der Fa. Wagner, Markdorf
37. Kleber, W.: Elektrostatische Oberflächenbeschichtung, Teil 4. Meta. **54**(9), 74–78 (2000)
38. DIN EN 16985:2019–04, Lackierkabinen für organische Beschichtungsstoffe, Sicherheitsanforderungen; Deutsche Fassung EN12981:2005+A1:2009
39. Stuckmann, M.: Neues Pulverbeschichtungskonzept erhöht Produktivität. JOT. **42**(5), 14–18 (2002)
40. Bildquelle JOT – Ferk – 1358
41. Adams, H.: Regelsysteme für die Zukunft. Metalloberfläche. **52**(4), 286–288 (1998)
42. Metzger, G.: Pulverausstoß optimal geregelt. JOT. **40**(2), 24–28 (2000)
43. Plonka, S.: Pulverbeschichten ohne Pistolen. Metalloberfläche. **54**(6), 46–47 (2000)
44. Meuthen, B.: Pulverbeschichten von Band und Blech. JOT. **40**(2), 16–23 (2000)
45. Mallet, K.: Powder coatings for high speed coil lines. ECJ. **2001**(12), 37–38 (2001)
46. Michaelis, R.: Pulver-Coil-Coating – Eine Alternative zur Bandbeschichtung mit Flüssiglack. JOT. **1998**(2), 14–18 (1998)
47. Binda, P., van den Berg, Jeths, R.: Electro-magnetic brush technology: a novel powder coating application process for coil coating. In: Vortrag gehalten bei der ECCA Konferenz 2003 in Malta (2003)
48. Maggiore, G.: Coils mit wärmehärtenden Pulverlacken beschichten. besser lackieren 14(9), 12 (2000)
49. Strohbeck, U.: TransApp – schnelle Beschichtung großer Flächen ohne Sprühen. JOT. **2003**(1), 43 (2003)
50. Strohbeck, U.: Pulverbeschichten. besser lackieren, Jahrbuch (2017)
51. Pulli, K. Dr: besser lackieren, 01 (2018)
52. Scholten, H.: Wirbelsintern als Alternative. Metalloberfläche. **53**(2), 42–45 (1999)
53. Kari, A.: Innovative Wirbelsintertechnik. Metalloberfläche. **52**(6), 468–469 (1998)
54. Strohbeck, U.: Elektrostatische Pulverbeschichtungsgeräte, Übersicht der Verfahren, Anlagenperipherie. Praxis Forum Seminar EPS-Basiswissen, Technik + Kommunikationsverlag, Berlin
55. Kleber, W.: Ausgereizt? Meta. **55**(9), 49–53 (2001)
56. Wellige, V.: Automatisierungstechnik in der Pulverbeschichtung. Meta. **52**(4), 290–291 (1998)
57. Liere, H.: Neue Sicherheitsanforderungen für elektrostatische Lackier- und Pulverbeschichtungsanlagen. JOT. **1999**(2), 8–12 (1999)

58. Korrigieren von Fehlern bei Beschichtungsprozessen mit Pulverlacken. Informationsbroschüre Jotun Powder Coatings, Norwegen, MIT 95-51 D 5/95
59. Tipps und Tricks zur Drylac®-Beschichtung, Anwendungstechnische Empfehlungen. Firmenschrift, Tigerwerk, Wels
60. Kleiner Leitfaden zur Behebung von Störungen beim Pulverbeschichten. Informationsschrift, Firma Morton
61. Pulverlackprobleme vermeiden. Informationsschrift, Firma Akzo Nobel Powder Coatings
62. Pulverlackhandbuch. Firmenschrift der BASF (2002)

Further Literature on Application

Bondar, W.A., et al.: Statische Elektrizität und Brandschutz. Staatsverlag der Deutschen Demokratischen Republik, Berlin (1977)

Freitag, H.: Raumexplosionen durch statische Elektrizität. Chemie, Weinheim (1955)

Gänger, B.: Der elektrische Durchschlag von Gasen. Springer, Berlin (1953)

Harper, W.R.: Contact and Frictional Electrification. Oxford Press, Oxford (1967)

Kleber, W.: Elektrostatische Oberflächenbeschichtung, Teil 3. Meta. **54**(8), 35–38 (2000)

Lunze, K.: Einführung in die Elektrotechnik. Hüthig Buch, Heidelberg (1972)

Mierdel, G.: Elektrophysik. VEB Verlag Technik, Berlin (1972)

Kleber, W.: Pulversprühpistolen mit Korona- oder Triboaufladung verwenden? Vortragsband der Seminarreihe Pulverlackierer unter sich, Tagungsband 2000, Vincentz Verlag (2000)

Lenherr, M., Bühlmann, E.T.: Pulvertransport und Förderung. Weiterbildungslehrgang „Pulverlackiertechnik-Grundlegende Zusammenhänge für die Praxis“, DFO (1993)

Liberto, N.P. (ed.): Powder Coating, the Complete Finisher's Handbook. Powder Coating Institute, Alexandria (1999)

Mendler, H.: Pulverbeschichtungsprobleme vermeiden. Industrie Lackierbetrieb, Teil 1, 1/96, S. 9–14; Teil 2, 2/96, S. 65–67; Teil 3, 3/96, S. 123–125; Teil 4, 4/96, S. 180–183

Mühlen, U.: Anlagenauslegung und Anlagenkonzepte. Weiterbildungslehrgang „Pulverlackiertechnik – Grundlegende Zusammenhänge für die Praxis“, DFO (1993)

Scott, A. (ed.): The Science of Powder Coatings – Applications, vol. Bd. 2. Sita Technology, London (1994)

Image Sources used in Chap. 2

Allgaier GmbH, Uhingen, D: 2.37 unten

DSM Coating Resins, Ceintuurbaan, Zwolle, NL: 2.86

Gema Switzerland GmbH, Mövenstrasse, St. Gallen, CH: 2.13; 2.14; 2.15; 2.16; 2.17; 2.18; 2.19; 2.21; 2.22; 2.24; 2.25; 2.26; 2.27; 2.28; 2.29; 2.31; 2.34; 2.35; 2.36; 2.37; 2.38; 2.44; 2.45; 2.46; 2.52; 2.53; 2.54; 2.55; 2.56; 2.57; 2.58; 2.61; 2.64; 2.65; 2.69; 2.73; 2.75; 2.76; 2.77; 2.78; 2.79; 2.81, 2.82, 2.84

Herding GmbH Filtertechnik, August-Borsig-Straße, Amberg, D: 2.33; 2.90; 2.91; 2.92

Industrieberatung, Hans Mendler, Abtwil CH: 2.35, 2.43, 2.48, 2.49 a,b, 2.50 a,b

Karl Bubenhofer AG, Hirschenstrasse, Gossau, CH: 2.4

Nordson Deutschland GmbH, Heinrich-Hertz-Strasse, Erkrath, D: 2.8; 2.9; 2.12; 2.23; 2.30; 2.83

J. Wagner GmbH, Otto-Lilienthalstrasse 18, Markdorf, D: 2.2; 2.3; 2.11; 2.20; 2.32; 2.39; 2.40; 2.41; 2.42; 2.47; 2.51; 2.59; 2.60, 2.62; 2.63; 2.67; 2.68 a–d; 2.69; 2.70, 2.72; 2.74; 2.80

Hangers and Conveyor Technology 3

3.1 Hangers, Product Carriers

In practice, product carriers or painting hangers “only” have the task of transporting the workpiece or the material to be painted through the system. They do not have a high priority, since investments in the overhead conveyor construction or its optimization are usually not made. However, reductions in production costs and throughput times can be achieved precisely through a sensible and targeted overhead conveyor design. Especially with robot painting systems, a good and precise suspension and positioning of the parts is necessary in order to guarantee the painting quality.

The design of the hangers and suspensions is very diverse, but should always be adapted to the conditions of the plant and the requirements of the workpieces. Adapted hangers are just as much a precondition for successful powder coating as, for example, conveyor technology, plant or powder coating.

When designing hangers, a number of principles must be observed, which are explained below [1, 2].

- Product carrier must be designed in such a way that contact or pick-up points are located outside the powder or paint application and can be completely or partially covered by the material to be painted. Blowing off excess powder on the product carrier before stoving helps to reduce material and paint stripping costs.
- The painting support can be both a goods carrier and a cover for certain parts that are not to be painted. In this case, the hanger has the function of both support and masking.
- During immersion pretreatment, attention must be paid to non-scooping material geometries in the product carrier design. The suspensions should be designed in such a way that no liquid can accumulate. Penetrating liquid must also be able to flow out again quickly and reliably.

J. Pietschmann, *Industrial powder coating*,
https://doi.org/10.1007/978-3-658-37592-8_3

- In electrostatic processes, the workpieces must be well grounded for as many painting processes as possible. The use of exchangeable suspensions, so-called sleeve suspensions, is helpful here. A paint application is carried out on the "base frame" of the construction. The suspension, whose contact points are protected and kept flexible by a soft, heat-resistant cap, can be replaced at any time. If a product carrier is not equipped to the maximum, the contact points of non-replaceable, rigid workpiece fixtures can be protected with heat-resistant caps or hose sections. For safety and quality reasons, the earthing should be checked regularly.
- Robot painting requires precise placement of the workpieces, since even the slightest deviations from the programmed path mean significant losses due to overspray and poor coating results. Screw connections are recommended in order to be able to readjust at any time. It is essential to protect the moving parts of such constructions from paint application.
- When using longitudinal and cross beams, coating of the suspension points should be avoided. The use of U-profiles or sheet metal crossbars with internal, protected suspension points offers a cost-effective suspension. The U-section acts similar to a Faraday cage, suspension hooks retain contact for a very long time as powder penetration is prevented. Flexibility is achieved through a variety of mounts. The dimensioning of the beams is based on the part spectrum. Beams with a small cross-section can be used in combination with longitudinal beams as small parts hangers. The variable division of the longitudinal beams (U-hooks as holders) makes stacking possible . The protection of not needed receptacles is done by easily attachable, high heat resistant caps. If other graduations are required after several runs, the protective and insulating caps used can be removed without great effort.
- A paint hanger should be designed to be minimal. The use of profiles, tubes or round cross-sections is a good way to avoid unnecessary powder application, unnecessary heating of material in the curing oven and high material costs.
- The loading of a goods carrier should be possible without a great deal of time. This means that the employee must be able to safely and reliably carry out the placement within the specified time frame when loading and unloading the goods. Customised suspensions are especially necessary for difficult part geometries. However, their use in practice should always be discussed in advance with the system operator and confirmed by tests.
- Employees who work on the system on a daily basis and load product carrier supports must be included in the definition of the requirements for the hangers, as practical experience is of particular importance in the subsequent implementation.

The following figures are examples of hangers and fabric receptacles in various applications (Figs. 3.1, 3.2 and 3.3).

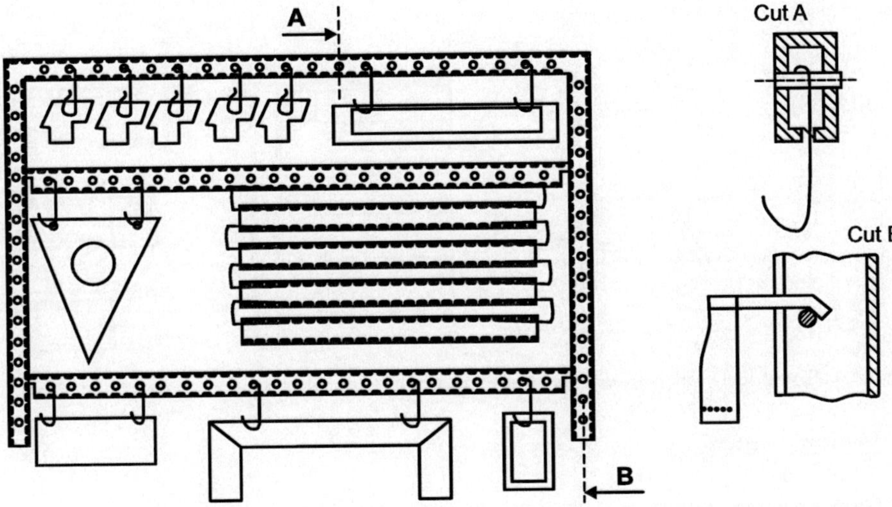

Fig. 3.1 Example of a universal hanger

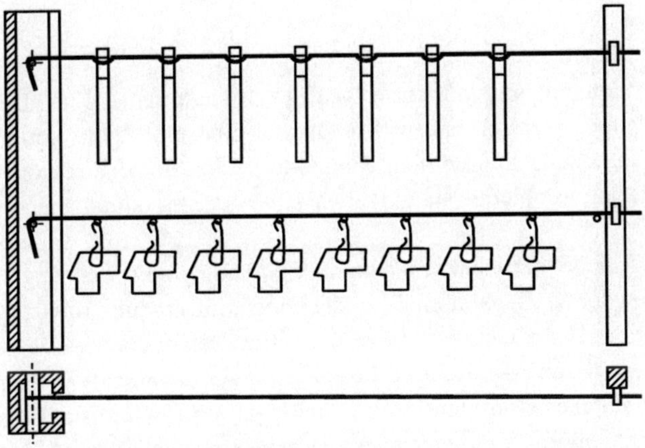

Fig. 3.2 Example of a hanger for fixing small parts

3.2 Conveying Technology

3.2.1 Requirements and Criteria

Conveyor technology is an essential component of coating plants. It offers a variety of processes and solutions and must therefore always be adapted to the requirements and possibilities of the coating systems.

Already in the planning stage, the requirements and criteria regarding the components are to clarify

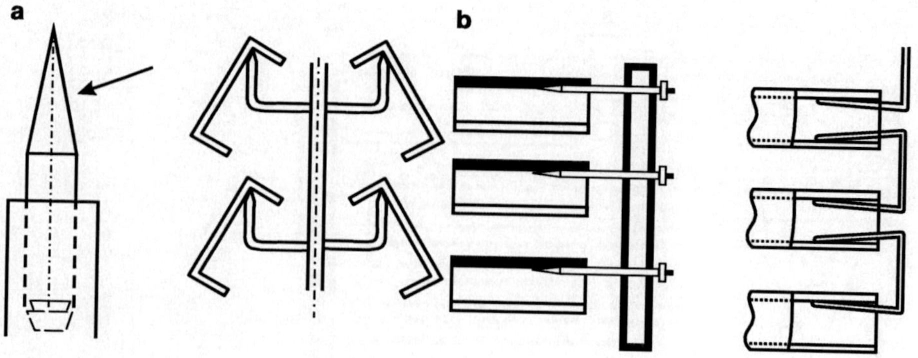

Fig. 3.3 Example for suspension of profiles

- Conveyed material
- Throughput
- Colour change and colour variations
- Goods carrier
- Environmental conditions and care
- Investment needs

The conveyor system not only influences the degree of automation, but also the basic plant design. A continuous painting line with a circular conveyor, looks completely different from a cyclic dip coating line with an automatic feeder. For this reason, conveying and process engineering problems should always be tackled simultaneously and solved together.

The field of surface treatment is a demanding application for conveyors: vapors from pretreatment, powder dusts, paint mists, high temperatures require robust systems suitable for this application. If the conveyor fails, the entire system comes to a plant stop . It is advisable not to go to the load limits of the conveyor system already in the planning stage. A somewhat more generous dimensioning usually pays off in terms of higher availability, longer service life and lower maintenance requirements.

It is essential to investigate whether the loading and unloading of workpieces can be automated or at least facilitated by auxiliary equipment. It must also be examined whether areas of surface treatment can be included in the overall conveyor concept. If necessary, for example, it may be advisable to feed workpieces directly into the production area and to remove them only in the assembly, storage, packaging or shipping area.

3.2.2 Conveyed Goods

Important here are the dimensions (length, width, height and, if applicable, diameter) and the weight of the goods to be conveyed. These specifications refer to transport units, i.e. if small parts are transported in suspension frames, the complete frame is considered to be the material to be conveyed. The properties of the material to be conveyed have a direct effect on the possible systems [3].

- Length (extension in conveying direction):
  Short conveyed goods are usually uncritical, long conveyed goods require a lot of space during curve travel due to the swivelling out beyond the conveying course (envelope curve). Since these envelope curves are directly reflected in the space requirements of the system, the use of a system with options for transverse or inclined travel is often unavoidable from lengths of 3 meters. Here one then chooses Power&Free conveyors or skid technology. The same envelope curve problem also occurs with inclined or declining sections. The rule is: the longer, the flatter the inclines must be, otherwise greater overall heights will result.
- Width (extension across the conveyor):
  The width is important for the design of the radius when cornering. The greater the width, the greater the radius and clearances must be. For technologies with the option of inclined or cross travel, the stacking distance results from the width plus the necessary clearances between the conveyors. The clearances are in turn influenced by the type of suspension.
- Height
  The required room height is directly derived from the material height. The height is also decisive for the situation at material intake and discharge. Lifting stations may have to be used..
- Weight
  Weight is a decisive criterion for the selection of conveyor systems. Due to the system, the various technologies can only be used economically for certain weight ranges. As a rule of thumb: the heavier, the more complex the applicable technology will be. Furthermore, heavy parts can only be placed on the conveyor systems with special measures. The entire spectrum of handling technology is required here.
- Capacity
  Throughput refers to the number of goods conveyed per unit of time, usually expressed in goods conveyed per hour. The conveying speed in the process areas can be calculated from the throughput, taking into account the length of the conveyed material. This speed is directly reflected in the dimensions of the pretreatment zones (exposure time), the adhesive water dryer (drying time) and the powder or paint dryer. Likewise, the length of the cooling zones depends on the throughput. Furthermore, the number of coating stations (booths) depends on the throughput and thus also the selection of a branchable conveyor system.

  The throughput determines the number of suspensions and the conveying systems that can be used (branches, buffer stations, etc.) as well as the type of drive technology.. Here too, the higher the throughput, the more complex the technology that can be used.
- Colour change and colour variations

The number of colors and the frequency of color changes results in the booth configuration (changing booths, quick cleaning, multiple booths). The booth configuration places demands on the conveyor system used, for example, branching is not possible with a conventional circular conveyor. In this case, alternating booths can be used, but throughput losses must be accepted. Again, the more variants, the more complex the technology that can be used.

- Goods carrier

  The parts to be coated must be connected to the conveyor for transport.

  Product carrier supports are always adapted to the workpieces, and the variety of designs is correspondingly large. In principle, a distinction can be made between specialized and universal product carrier supports. The universal design (use of wire hooks, ropes or chains) is preferred for large part variance or optimal part geometries, but always depends on the skill of the worker at the loading station. The specialized design is used in conjunction with difficult parts, low part variance (each variation requires its own design) or high throughputs (placement can be streamlined).

  The different conveyor systems are not all equally suitable for every goods carrier, it is true that the more complex the system, the more specialized the goods carrier.
- Environmental conditions and care

  In the area of pretreatment and painting, high demands are placed on the conveyor system with regard to the prevailing environmental conditions (such as contamination, temperature). However, as the capabilities of a conveyor system increase, so does the need for maintenance, which results from the inevitably increasing number of components. This results in requirements for the qualification of maintenance personnel and their availability.

In the future, shuttle systems or shifting technology will continue to gain importance in the field of surface technology. The difference to the chain systems is that the previously common endless conveyor chain is divided into individual controllable elements. This means that each load traverse can be controlled individually as required. Directions of movement are longitudinal and transverse, there are no bends. This gives the system an advantage over chain systems in limited spaces and where the load beams dimensions are large.

3.3 The Support Systems in Detail

3.3.1 Manual Push: Pull Conveyors

3.3.1.1 Technology

Manual push – pull conveyors are the simplest and most original conveyor technology. They consist of a guide profile and carriages that run in this track. The driving force is, the worker. (Fig. 3.4). These systems are normally suspended, if possible from the building

Fig. 3.4 Push-Pull conveyor for the coating of metal parts

structure or from a steel structure provided for this purpose. A variety of complete systems are offered.

- Path components (straight lines, arcs)
- Carriage for loads up to 4500 kg
- Switching points for branches
- Turntables
- Transfer tables.

Almost all conceivable system geometries can be assembled from the components (inclined and transverse transports). Trolley buffers are possible. The trolleys form trains together with load beams, so that parts of greater length can also be transported.

In coating plants, purely manual transfer conveyors are very seldom used; usually a combination with a drive means is used on partial sections, e.g. dryers.

- Pre-treatment
  Cycle transport: push cylinder, possibly with chain as pushing element.
  Continuous transport: Chain transfer with adjustable speed
- Dryer
  Intermittent dryer: Parallel tracks with transfer cars in front of and behind the dryer, pure manual transfer technology.
  Continuous dryer: chain transfer with adjustable speed
- Load and unload place

Lifting stations: Systems with such combinations can already require very complex control and are comparable to Power&Free systems. The advantage is that such systems can be expanded and subsequently automated; as throughput increases, lines can be equipped with drive units.

3.3.1.2 Conveyed Material

Length	Up to 12.000 mm reasonable
Wide	No limitation
Height	Difficult to push from 2000 mm
Weights	Up to 600 kg, higher weights possible but not to handle for the worker
Throughput	Small (about 5 units/h), depending on the worker
Variability	Large, almost all geometries conceivable
Goods carrier	No limitations
Care	Simply

3.3.2 Circular Conveyor

3.3.2.1 Technology

Circular conveyors consist of a guide profile and a chain that runs on or in this guide profile. The chain serves both as a driving and supporting element. Components are:

- Track components such as straight lines, curves etc.
- Chain
- Cardan chain mostly made of stamped parts, for the lower load range and runs in the guide profile
- Plug-in chain made of castings, for the upper load range, runs under the guide profile
- Drive
- Caterpillar drive, single drive, multiple drive for long length, wheel drive
- Tensioning station
- Spindle clamping station manually adjustable, no additional chain hoist, weight clamping station, spring clamping station, pneumatic clamping station
- Lubrication device is necessary for cardan joint chains, serves to protect the chain against contamination and temperature.
- Attachment parts: Goods carrier suspension, rotary hanger.

The chain only allows limited trailer loads, furthermore branching is not possible. Circular conveyors have the same speed over their entire length, and the distance between the conveyed material is always the same. Colour changes are only possible with throughput losses. In the case of changing cabins, part of the conveyor section must be cleared so that the cabins can be extended or retracted. Several booths in a row are difficult or impossible to clean during operation.

3.3.2.2 Conveyed Material

Length	Up to 3000 mm reasonable
Wide	Up to 1000 mm reasonable
Height	Difficult to hook in from 2000 mm
Weights	Up to 250 kg
Throughput	Large, practicable up to 12 m/min
Variability	Small, suitable for mass products
Goods carrier	Flexible, but not for applications requiring precision
Care	Simply

Circular conveyors are suitable as a robust, uncomplicated workpiece transport with always the same process sequence. They are only considered for plastic component painting if the workpieces are not very sensitive to dust and the drying times are not too long. The field of surface treatment of metal is the main area of application for the circular conveyor, albeit with a declining trend [4].

3.3.3 Circular Conveyors with Branching Capability

3.3.3.1 Technology

The need for fast color changes in powder coating plants generally leads to plants with two or more parallel booths. Circular conveyors could only be used to a limited extent for this plant geometry; Power&Free systems have become the state of the art in this area (Fig. 3.5) [5, 6].

Fig. 3.5 Power&Free with chain track, entering or leaving the cabin

Different systems can be found on the market that work according to similar basic principles:

A part that can be automatically separated and reassembled is inserted between the chain and the goods carrier. This part is transferred at switches from one chain to the other, branching is thus possible. There is no stacking capability as in the Power& Free technology, in this respect the system still behaves like a circular conveyor [7].

In practice, the intermediate piece is designed as a hook with guides, as a crossbeam or as chain pieces in their own guideway. All these systems have in common that the chain is used as a supporting element. This results in load ranges similar to those of circular conveyors. Up to now, these systems have only been used in conjunction with cardan chains. An important feature is that the branching sections must be of equal length, since each transferred goods carrier must encounter a gap on its return, which restricts the possible system arrangements. A simple control unit is required at the branching points, so the maintenance requirement is higher than for a pure circular conveyor.

3.3.3.2 Conveyed Material

Length	Up to 3000 mm reasonable
Wide	Up to 1000 mm reasonable
Height	Difficult to hook in from 2000 mm
Weights	Up to 120 kg
Throughput	Large, practicable up to 12 m/min.
Variability	Medium, multi-cab solutions
Goods carrier	Flexible, but not for applications requiring precision
Care	Medium

3.3.4 Power&Free Systems

3.3.4.1 Technology

A further guide profile (free track) is fitted below a circular conveyor (power track). This guide profile is designed for carrier that form a trolley train by means of connecting traverses. The The free track houses the load supporting trolleys.

The trolley train is driven by a carrier/pawl design that creates a form-fitting connection to the power chain. The pawls are designed in such a way that they release the connection to the carrier when they run onto a stacking skid. If this stacking skid is attached to the last trolley, the trolley hoists can be stowed. To do this, the skid on the last trolley and the carrier on the first trolley release the latch during stacking. If the foremost trolley train in the conveying direction is picked up again, the accumulation is released. Stopping of the foremost trolley train is performed by a stacking skid that can be moved to the correct position by using a cylinder (pneumatic or electric); this is called a stopper or lock.

In practice, two forms have developed, which differ in the arrangement of the latch:

1. moving pawls on the chain
2. moving pawls on the trolley

The advantage of moving pawls on the trolley is that only one trolley has to be removed in the event of a defect. With movable pawls on the chain, the carrier (pusher dog) is missing in the system. A risk of cycle loss is possible.

Due to the stackability, buffers or storage tanks can be realized with Power&Free.

- **Longitudinal stack:** The trolley trains are stacked one behind the other, long but narrow in space.
- **Inclined stacking:** If you construct a carriage train from four carriages and connect each pair with a shorter auxiliary crosshead, you obtain an inclined or crosswise geometry. If a stacking skid is added to the rear carriage of a pair, the carriage train can be stacked at an angle. An inclined stack consists of a Power&Free section for the front part of the trolley train, a Free section for the rear part of the trolley train as well as a normally mechanically operated entry switch and a pendulum switch at the exit. The driving is only on one side of the trolley train; thus angles up to 60° inclination are possible. For long and narrow parts, the use of inclined stacks does not result in a smaller space saving (Fig. 3.6).

Fig. 3.6 Power&Free as inclined stack 45°, buffer of sheet metal parts

- **Cross stack:** The cross stack is the 90° variant of an inclined stack. However, it is not possible to drive the trolley trains on one side because of the transverse forces that occur; two Power&Free sections are used and drive on both sides. The necessary drivers on both sides make double locks necessary on straight conveyor sections. The two chains in the cross-stack must run in parallel; this requires tensioning stations to compensate for the chain length. The cross-stack is a very efficient way of changing the direction of long parts at the ends of the system. However, the direction of movement of the payload is reversed, i.e. the parts are suddenly moving in the opposite direction (Fig. 3.7).

Another essential component are the already mentioned switches. A distinction is made between switch points at the beginning of a branch and pendulum points for merging. Pendulum points have a switch blade, which is brought into the correct position by the trolley during the passage. In the case of switch points, the switch blade determines the direction of exit; it must therefore be set before the trolley passes through. Today, switch points with electric or pneumatic actuation are mostly used for this purpose. The system control decides on the switch position on the basis of the information available to it and controls it accordingly.

In the simplest case, the information can consist of a trolley feature such as an additional trip cam, i.e. all trolleys with trip cams must pass through the feeder. For more complex tasks, identification systems are used that contain an automatically readable

Fig. 3.7 Power&Free as 90° buffer of sheet metal parts

trolley number. This trolley number is linked to the data entered when the parts are fed in (for example, all parts with colour number 4711 in branch) and then controlled accordingly. Binary systems can be used for this up to 280 °C, or ceramic QR codes up to 400 °C. In the temperature range < 100 °C, RFID tags are usually used.

The power chain used requires that the transport speed in the system is the same at all points. Different speeds may be necessary for coating, pre-treatment and transport. This is solved by using several chain circuits with different speeds, for example a process chain for pre-treatment and coating with a low adjustable speed and a transport chain with a constant, higher speed.

Lifting stations are used at infeeds and outfeeds; this simplifies the handling of heavy or bulky parts, and also makes greater conveying heights practicable. The control is usually carried out by the worker by means of a crane bulb. In this way, he can set the optimum feeding height for him. After release, the infeed and outfeed is usually carried out automatically by the system control. Lifting stations are also used for floor changes.

Power&Free technology in coating plants is a widely used process for solving complex tasks. Either for a wide range of products with different requirements or the integration of the coating line into the overall material flow of a production (Fig. 3.8).

Due to the flexibility together with the possibilities of the control technology, sorting storage, multiple passes, different types of pre-treatment etc. are possible.

The technology is proven in coating plants and has the necessary robustness for the different environmental conditions.

The arrangement of the conveyor technology above the material to be conveyed brings advantages due to low contamination of the conveyor and a greater ground clearance. However, this involves the risk of contamination of the parts to be painted. On the other hand, in the event of component failure, there is a risk of breakdown with the need for repair. This risk can be minimized by preventive maintenance, but is reflected in comparatively higher costs. The transport principle and the complexity of this type of plant make qualified personnel necessary

3.3.4.2 Conveyed Material

Length	Up to 20.000 mm
Wide	Up to 4000 mm
Height	Unlimited, but sensible from 2000 mm lift stations onwards
Weights	Up to 10.000 kg realized
Throughput	High
Variability	Very large
Goods carrier	Flexible
Care	Understanding and training required; PLC skills necessary

Fig. 3.8 Power&Free right: compact, space-saving design – left: intermediate storage unit

3.3.5 Electric Monorail System

Electric overhead conveyors, which have proven themselves as a modern and flexible conveyor system, are not considered for paint shops because the required explosion-proof design cannot be implemented in a practical manner. If electric monorail systems are used in areas upstream or downstream of the surface treatment, they are linked to the painting line conveyor by automatic transfers (Fig. 3.9).

3.3.6 Floor Conveyors

3.3.6.1 Technology

If the pick-up points of a product are located on the bottom or if accessibility from all sides without disturbing suspensions is required, the conveyor system must find its place below the conveyed product. Another criterion is the weight; a conveyor located at the bottom transfers the occurring loads more economically, a complex steel construction is not necessary.

In principle, both circular and Power&Free conveyors can be turned upside down and thus used as floor conveyors [8, 9].

Floor conveyors consist of a chain box usually cast into the floor and a plate or cardan chain guided therein. However, the use of the plate chain is more cost-effective. The link plate chain is fitted with carriers, usually of a non-separable design, into which the driving pins of the trolleys are latched. The chain drags the trolleys loaded with the payload; the characteristics correspond approximately to a circular conveyor: no branches, same speed everywhere. Driving pins are guided in the chain box. The carriages usually have two swivel castors at the front and two fixed castors at the rear. This ensures controlled curve behaviour.

The driving pins can be detached from the chain manually. The carriage can therefore be used flexibly, for example for the transport to and from production stations.

Fig. 3.9 Electric monorail through a shot blast line. Cleaning of steel sheets before surface treatment, coupling of the vehicles to a train

Floor conveyors are often used in production lines, for example as a classic assembly line (Figs. 3.10, 3.11 and 3.12). Variants with switches or stoppers are also used here, which means:

- less drive units than with skid conveyor system
- less sensors
- lower costs for control effort

In coating plants, floor conveyors should be limited to the simple design. The arrangement below the conveyed material exposes the conveyor to increasingly difficult conditions:

Fig. 3.10 The inverted Power&Free on the floor in a wheel plant, buffer on the left, entry into the powder booth on the right. To save space, the approach is at crosswise to the transport direction. In front of the booth, the goods carriers are turned in the direction of transport

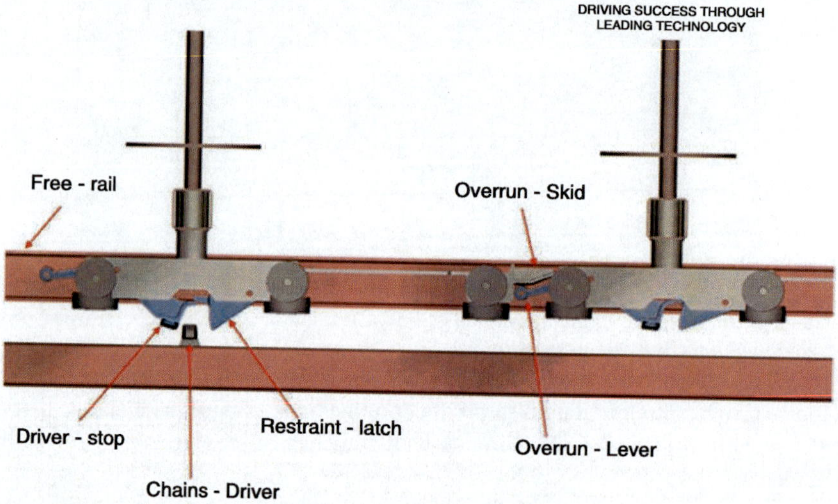

Fig. 3.11 Structure of the Power&Free rail in longitudinal section

- Pre-treatment systems: Stainless steel designs can be used here in conjunction with automatic lubrication devices and sanitary hook and pan covers. However, the compensation of the degreasing effect always remains maintenance-intensive.
- Wet paint systems: Can be used with clever chain case geometry and air guidance.
- Powder booths:with colour change, .

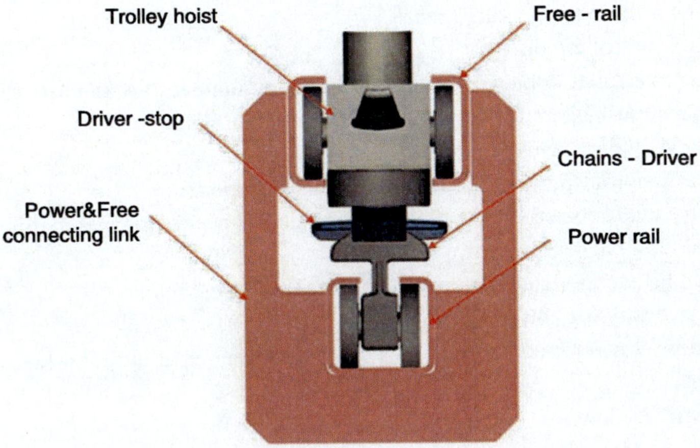

Fig. 3.12 Structure of the Power&Free rail in cross-section

3.3.6.2 Conveyed Material

Length	Up to 6000 mm reasonable
Wide	Up to 2000 mm reasonable
Height	Up to 2000 mm reasonable
Weights	Up to 4000 kg
Throughput	Very large
Variability	Small
Goods carrier	Flexible
Care	Simple but cleaning intensive

3.3.7 Skid Systems

3.3.7.1 Technology

Skid conveyor technology is another bottom-mounted system. Skid means runner. Whereas floor conveyors use trolleys as a base for the goods, a skid consists of a frame and transport skids (in some cases the frame and skids are integrated and cannot be distinguished) with no moving parts and therefore no vulnerable bearing points. Skid technology is used wherever high performance, precision or flexibility are required and the goods to be conveyed must be freely accessible from above.

Skids are transported on stationary conveyors:

- Roller conveyor
  Mostly used for longitudinal transports.
  Typical for single workstations with stop-and-go behaviour.
  Precise positioning possible through additional parts (stakeouts).

Cannot be used in the coating area
- Accumulation roller conveyor
  Variant of the roller conveyor. Skids can be accumulated; thus an cost effective buffer formation is possible in the lower load range up to 500 kg
- Suspension chain conveyor
  For transverse transport and large longitudinal transports.
  Insensitive, can be used in all process areas, such as pre-treatment, coating, dryer
- Accumulating chain conveyor
  Skids can be accumulated, even in the higher load range.
  Inexpensive way for buffering.
  Frequently seen in dryers (dwell times can be maintained at different flow rates)
- Lift tables
  Mostly with hydraulics.
  For smaller height changes.
  At workplaces for better accessibility
- Converter
  Combination of roller conveyor, lifting table and chain conveyor.
  Skid enters lengthwise on roller conveyor, is deposited on chain conveyor and exits crosswise. Used as branching (long chain conveyors, many transfer units) or as reversal of direction (two units side by side, skid then backwards)
- Turntables
  Used as branching or for direction changes without reversal of the skid conveying direction
- Lifting stations
  For level change.
  Special feature: Stroke stations in temperature-resistant design in A-dryers.

From the listed components as well as special, not mentioned set pieces, arbitrarily complex plant layouts can be developed. Compact geometries are possible due to the ability of transverse movement without extending envelopes (Fig. 3.13).

Since the components are each equipped with their own drive technology, there is a great deal of flexibility in the process, which can be used by making changes in the control software. On the other hand, skid systems require complex control technology in PLC technology. The density of components per conveyor section is higher than, for example, with Power&Free technology. Since the drive principle is based on friction and not on positive locking, skid technology is more fault-tolerant than Power&Free.

As already mentioned in the case of floor conveyors, conveyor technology located below is heavily exposed to the influences of the coating processes. This must be compensated for by increased maintenance if the same availability is to be achieved.

Fig. 3.13 Skid conveying of bumpers through the painting process

3.3.7.2 Conveyor Property

Length	Up to 7000 mm reasonable
Wide	Up to 4000 mm
Height	Up to 3000 mm
Weights	Realized up to 4000 kg
Throughput	Very large
Variability	Very large
Goods carrier	Flexible
Care	Understanding and training required, PLC (programmable logic controller) skills necessary, cleaning intensive

3.3.8 Immersion Plants

3.3.8.1 Technology

High pre-treatment and/or coating requirements necessitate the use of immersion plants. Immersion plants consist of the treatment tanks into which the parts are immersed for treatment and the transfer units for transport (Figs. 3.14 and 3.15).

Fig. 3.14 Power&Free in continious pass plant for small truck parts

Fig. 3.15 Power&Free as a dipping system for large steel parts. The dipping process is carried out by hoists

From a technical point of view, the transfer units are precisely designed overhead cranes and consist of a travelling and a lifting axle. The payload is suspended from cross-beams. These traverses have receptacles for the transfer units, which are lifted, positioned above a pool, lowered, the transfer unit moves to the next tank and the process begins again.

The sequence of the basins is determined by the chemistry used, and the treatment times. The movement times are determined by the material being conveyed.

The travel speed depends on the width of the material to be conveyed (distance between basins) and on the suspension (oscillation during acceleration). The immersion speed depends on the tendency of the parts to float. The lifting speed depends on the scooping tendency of the parts. The dripping time after lifting results from the geometry of the parts as well as from the option to tolerate media carry-over.

All movement times together result in the program length. The throughput of such a system is typically specified in the term cycle time, i.e. every x minutes a product carrier should leave the system ready treated. Since the program length is usually longer than the cycle time, several converters are provided. This results in parallel processing of the driving program. The driving programs are displayed in a time/distance diagram. The technical realization is usually done with a crane bridge running on crane rails. Depending on the span, one or two travel drives with frequency control are provided. The two lifting masts are attached to the crane bridge as guides for the lifting slides; the lifting drive or drives are also frequency controlled. The frequency control, in cooperation with displacement measuring systems (absolute encoders), allows the adjustment of the travel curves to the respective conveyed goods (swinging, floating, scooping) as well as the necessary precision in positioning. The control system works with a time table based on the travel diagram and thus triggers the necessary actions. When designing a immersion plant the available construction heightis an essential point.

As a general rule: Construction height = 3 × (conveyed goods + 700mm).

3.3.8.2 Conveyed Material

Length	Realized up to 13.000 mm
Wide	Up to 4000 mm
Height	See note above
Weights	Up to 10.000 kg
Throughput	Wide range
Variability	Drive program only
Goods carrier	Flexible
Care	Understanding and training required; PLC skills necessary

3.3.9 Automatic Feeders

Automatic feeders are program-controlled lifting and travelling devices that have been specially developed for the surface treatment of metals. They travel back and forth in a straight line over basin systems, while the workpieces – usually on multiple hangers – are transported suspended at right angles to the conveying direction. The individual plant stations are fed vertically from above in cycles, in the sequence specified by the program.

Depending on the throughput or the required cycle time, the plant is equipped with one or more automatic feeders. Programmable logic control is standard here. The transmission of the switching commands and the energy supply are usually carried out via a trailing cable system.

The advantages of such a system design result from the functional principle: vertical loading, transverse transport and compact design result in short basins and save space. This is why automatic feeders are particularly interesting for bulky workpieces and confined floor space, and also where dipping is an option from the outset. The simple program change enables a quick process changeover when changing workpieces with different requirements occur. The selectable sequence of station loading facilitates necessary expansions, since new system components can be added wherever space is available, independently of the process sequence.

Automatic feeders are often combined with other conveyor systems (circular conveyors, Power&Free, floor-mounted drag conveyors with trolleys). For example, if the loading and unloading of workpieces should be possible independently of the rhythm of the surface treatment or if, for example, the pre-treatment can take place in cycles but the coating must be continuous. In these cases, the conveyor systems are linked by automatic transfers.

3.3.10 Circular Table Conveyor

Circular table conveyors are preferably used when smaller workpieces have to be transported vertically and possibly rotating through the surface treatment area. In this case, the rail of an inverted circular conveyor or a drag chain conveyor is guided on supports at table height, and the chain is provided with special workpiece holders. Typical examples are the painting of headlight reflectors, shock absorbers, chair legs and other turned wooden parts or wheel covers made of plastic, over which there must be no moving parts because of the material's sensitivity to dirt.

Similar designs are used as so-called assembly or moving table circular conveyors in a wide variety of assembly areas. Here, the chain is usually connected to circulating platform carriages into which the conveyed material is loosely placed.

3.3.11 Cross Bar Conveyors

In the case of cross-bar conveyors, two parallel chains are connected by cross bars. The material to be conveyed is placed loosely or fixed in special workpiece holders.

In conveyor lines for the painting of flat wooden parts, cross-bar conveyors are mainly used as robust, temperature-resistant means of transport through dryers including flash-off and cooling zones. Another field of application are the so-called *belt washing machines* within the metal pre-treatment.

3.3.12 Tapes

Conveyor belts (belt conveyors) for unit loads can be made of rubber, plastic, textile or metal fabric of various qualities, depending on the intended use. Wood and steel plate belts are also common. The workpieces are placed loosely.

The main field of application for belts is transport outside the area of surface treatment. However, belts made of textile fabric can be used, for example, in belt lines for the painting of flat wooden parts, as well as brake or accelerator belts made of rubber in the area of the casting machine.

In the field of metal pretreatment, belt washers are occasionally equipped with wire mesh belts (more often with cross-bar conveyors). Wood panel belts can be advantageous, if necessary, in the painting of corpus furniture, chair and armchair frames or similar workpieces made of wood.

3.3.13 Roller Conveyors

Like belts, roller conveyors have their main area of application outside of surface treatment. They are often used, for example, for goods distribution processes in storage, order picking, packaging or shipping areas. The loosely placed workpieces must either have a smooth underside or be transported in containers or on pallets.

3.3.14 Stacking Machines

Conveyors in painting plants can be supplemented by various additional devices. As a typical example, stacking devices or automatic stacking machines are mentioned.

In the field of metal painting, sheet metal panels or blanks are often fed in stacks. The feeding station is equipped with an automatic stacking device, the removal station with a corresponding device for re-stacking the finished coated parts (lifting slides with suction cups, suction bars with driving rollers).

Within the wood varnishing process, automatic stacking machines continuously take over flat parts coming from the coating machine (curting coater) and transfer them in batches to tray trolleys so that the extremely time-consuming manual stacking is no longer necessary. Such an automatic stacker consists of a feeder belt, an intermediate basket and a multi-level transfer unit. The feeder belt serves as speed compensation between the conveyor line and the intermediate basket. The intermediate basket is fed in cycles, starting at the top in the initial position. This is followed by the closed transfer to the floors of the multi-level translator at the same level. At the same time, the new filling begins, this time in a downward cycle. After the second transfer, the multi-level transfer unit is filled, moves crosswise, lifts up and lowers onto the trolleys.

References

1. Hausmann, C.: Kosten sparen mit optimalen Lackiergehängen. JOT. **41**(12), 20–24 (2001)
2. Liberto, N.P.: Powder Coating, the Complete Finisher's Handbook. Powder Coating Institut, Alexandria (1999)
3. Moeller, E., Ritter, R.: Industrielle Lackiertechnik. Weka Fachverlag, Augsburg (1997)
4. Meyer, G.: Mit Kettenförderern Produktivität von Beschichtungsanlagen steigern. Besser lackieren. **13**(8) (2000)
5. Meyer, G.: Power + Free und Kreisförderer, Verfahrensvorteile miteinander kombinieren. Besser lackieren. **13**(8) (2000)
6. Meyer, G.: Es muss nicht immer Power + Free sein. JOT. **39**(10), 70–72 (1999)
7. Meyer, G.: Vom Kreisförderer zur Power + Free mit Kettenlaufwerk. JOT. **39**(4) (1999)
8. Meyer, G.: Power + Free steht Kopf. Fördertechnik (1996)
9. Meyer, G.: Beschichten in einem Zug. Industrie Anzeiger 51–52 (1996)

Further Reading

Koppe, M.: Gestellbau für die Pulverbeschichtung. JOT. (1996)
Götz, U.: Fördertechnik für Oberflächenanlagen. JOT 10 (1995)
Götz, U.: Fördertechnik für Oberflächenanlagen. JOT 11 (1995)
Mendler, H.: Ratschläge Pulververarbeitung. Schierholz GmbH. Bremen

Images used in Chap. 3

Caldan Conveyor A/S; Bad Hersfeld: 3.4, 3.5 (linkes Bild), 3.6, 3.7
Eisenmann GmbH, Böblingen: 3.11 bis 3.13
Mendler, Hans: 3.1 bis 3.3
RonalGroup, Härkingen CH: 3.10
Schierholz GmbH, Bremen: 3.5 (rechtes Bild), 3.8, 3.9, 3.14 und 3.15

4 Curing of Powder Coatings

The drying and curing process, i.e. the transition of a liquid or powder coating material into a mechanically solid coating, should be as rational as possible and without impairing the quality of the still sensitive surface.

The mechanism of the curing and cross-linking reaction of most powder coatings takes place in the individual reaction systems according to kinetic laws, depending on the temperature. During baking, the powder coating should be brought to the necessary curing temperature homogeneously and as quickly as possible over the layer thickness up to the substrate surface. Only in this way can the powder coating melt reach the viscosity minimum without already being considerably hindered from running by cross-linking reactions. With slower heating, the coating material already cross-links without having run sufficiently, which results in an unevenness of the surface due to a non-optimal flow of the powder. The usual drying or curing temperatures for powder coatings range between 110 and 250 °C with holding times between 5 and 30 min. The shape and material distribution (thickness) of the workpiece also play a role.

The dwell time is the time during which the part remains in the active part of the furnace; it is divided into heating time and holding time. The baking temperature and the required holding time are determined by the powder material, the heating time by the thickness of the base material and by the design of the heating zone.

Constant curing parameters via a temperature-monitored heating process lead to consistent gloss levels and prevent overburning of the coating.

4.1 Types of Dryers

Depending on the type of feed, a distinction is made between

J. Pietschmann, *Industrial powder coating*,
https://doi.org/10.1007/978-3-658-37592-8_4

- Batch dryer and
- Dryer for continuous operation.

In general, dryers with their installations should have a smooth-surface design so that the time required for cleaning work can be kept to a minimum. Dryer housings usually consist of double-walled sheet steel cassettes with insulation in between. The individual cassettes must be tightly joined together at the joints. Careful assembly using suitable sealant is an important point. Sealing compounds containing silicone should generally be avoided in the powder coating area, as silicone residues lead to defects (craters) in the paint film.

During assembly, care must be taken to ensure that there are no thermal bridges between the inner and outer skin of the dryer.. From certain lengths and temperature ranges, sufficient expansion joints must be provided to absorb length changes between the inner and outer skin of the housing. Recirculation and exhaust ducts must be designed to be airtight. Fans must be connected to the housing in such a way that no disturbing vibrations are transmitted.

4.1.1 Batch Dryer

Chamber dryers are the simplest designs for curing ovens and are charged in batches. These dryers are used for very low throughputs and/or for strongly varying curing conditions, e.g. if different curing times have to be observed due to different material thicknesses or if different curing temperatures have to be observed due to the use of different powders.

The big disadvantage of these furnaces is the batch operation. If the doors are opened for loading and unloading, the temperature in the furnace drops significantly and it takes some time until the desired temperature is reached again. However, for optimum melting and good flow, the object temperature should be reached as quickly as possible.

4.1.2 Continuous Dryer

In the case of series production, charging takes place in continuous operation – continuously or in cycles – and usually with the same conveyors that are used throughout the paint shop. In the continuous dryer, the inlet and outlet openings are on opposite sides. Figure 4.1 shows the schematic structure of a continuous dryer. A reversible dryer is one in which the conveyor line is deflected once or several times in the dryer.

Today, continuous and reversible dryers are equipped with so-called A-locks. These are zones for shielding the inlet and outlet of dryers against heat losses due to inclined upward or downward conveyor routing. The infeed and outfeed are preferably located at a level below the dryer floor, see Fig. 4.2. If the plant operates in cycles, the dryer can be equipped

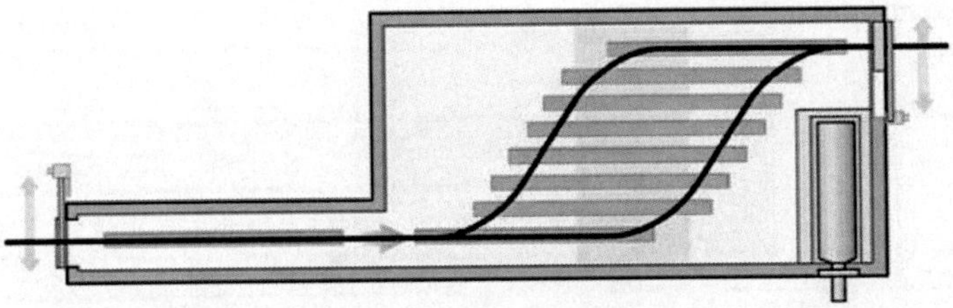

Fig. 4.1 Schematic diagram of a continuous furnace with sliding gates and Power&Free conveyors

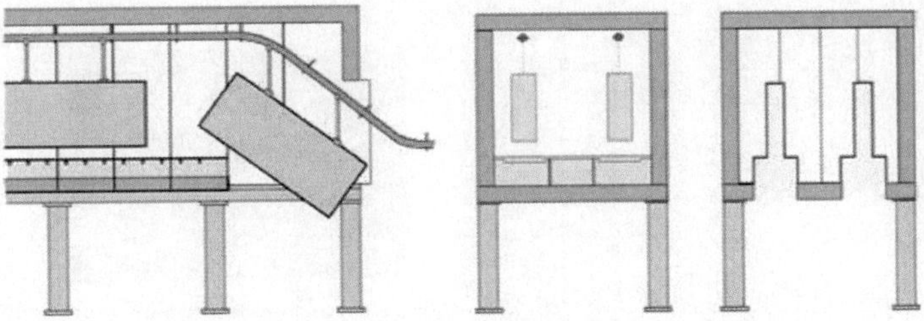

Fig. 4.2 Illustration of an A-type furnace with only partial A-type conveying system

with sliding and lifting gates to prevent heat losses. This is preferably done for larger workpiece dimensions and lower throughputs.

The furnace is extended here by the incline distance of the conveyor. This is shorter, the steeper the conveyor can be guided. The workpiece suspension must be taken into account (sufficient distance between two workpieces = 100 mm, minimum 80 mm) [1].

For reasons of space, a perfect A can often not be realized. A compromise may be that the front wall is given a cutout for the conveyor and hangers and only the workpiece itself, which requires the greater width, enters the furnace chamber from below. The heat losses in the area of the narrow cutout can be minimized by extensive masking, e.g. with elastic material.

4.1.2.1 A-Furnace for Skid Conveyor Systems

With this type of conveyor, the difference in height is not overcome by an inclined section, but with the use of a lift. The mode of operation is therefore discontinuous again, but the furnace is only extended by the length of the workpiece.

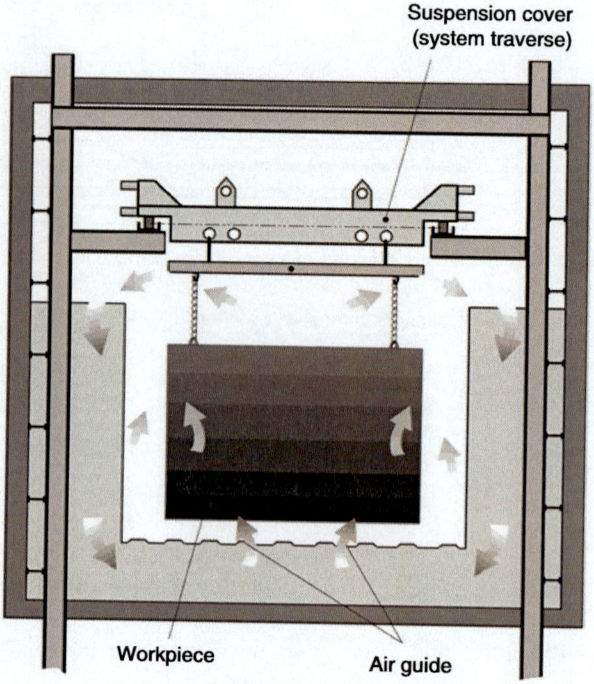

Fig. 4.3 Schematic diagram of a trough dryer

4.1.3 Special Forms

4.1.3.1 Trough Dryer

A trough dryer (Fig. 4.3) is a design in which the load is fed vertically from above in cycles. Here, flap gates prevent excessive heat loss. Trough dryers are often located in immersion basin systems with lifting and travelling gear transport.

They are used when (bulky) workpieces are transported through the plant by an automatic charging system (lifts and trolley). The furnace atmosphere is sealed by the suspended cover under which the workpiece is suspended when the furnace is in use, and by hinged or sliding covers when hangers are missing.

4.1.3.2 Combi or Block Dryers

Since the workpieces are usually chemically pretreated before powder coating, a water dryer is required in most plants in addition to the powder baking oven. If both are combined, the following savings are possible [1]:

- A common partition wall eliminates the transmission loss through an exterior wall for each furnace.
- The exhaust air of the powder furnace can usually be mixed with the adhesive water dryer and discharged from there as exhaust air. This saves an exhaust air pipe and

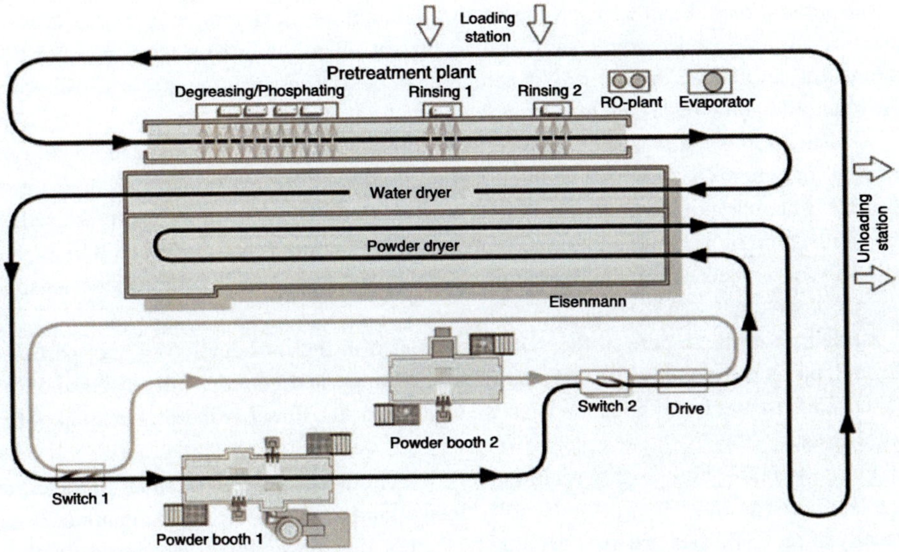

Fig. 4.4 Plant layout for powder coating with block dryer

energy is recovered according to the temperature difference between the powder oven and the adhesive water dryer.

The powder oven is usually built in a U-shape for these block dryers, so that approximately the same housing lengths often result. Figure 4.4 shows the plant layout for powder coating with a block dryer.

4.2 Drying Process

According to the type of heat transfer, a distinction is made between

- dry off/curing by convection and
- dry off/curing by radiation of various types.

4.2.1 Convection or Circulating Air Dry off

In convection or circulating air drying, a flow of warm air is directed to the workpiece, whereby an exchange of energy and material occurs at the surface. The warm air cools down and thus transfers thermal energy to the workpiece. The object temperature increases and the paint film heats up at the same time.

All known media can be used for heating circulating air dryers. Commonly used in practice are crude oil, natural gas, electric current, heat conducting oil, hot water and steam. The selection is determined by economic or operation-specific aspects as well as by the required baking or drying temperature.

A distinction is made between indirect and direct heating. In the case of indirect dryer heating, the energy is supplied to the circulating air via heat exchangers. In the case of directly heated dryers, the dryer medium is heated by admixing heating gases produced by the combustion of gas or, if necessary, fuel oil. Direct heating is more energy-efficient, but can only be considered if the degree of purity of the flue gases excludes any negative impact of the quality of the workpiece surface.

Certain hazards lie here in the yellowing or also in the introduction of soot particles through incomplete combustion. If a thermal afterburner is required for the disposal of the exhaust air stream, its combustion heat can be used for the direct or indirect heating of the paint dryer.

For particularly high quality requirements, the circulating air and fresh air of the dryer can be filtered to reliably protect the still "open" paint surface from contamination. Fans, usually of radial design, are used to circulate the hot air. Convection dryers generally operate with circulating air speeds of between 1 and 2 m/s. From time to time, despite the high energy requirements of the circulating air fans, it may be useful to increase the discharge velocity considerably. In practice, values of up to 25 m/s are selected for this purpose.

The most important advantage of circulating air dryers – and thus the explanation for their widespread use – lies in their universal applicability for a wide manufacturing program. Geometrically different parts with a similar ratio of weight to surface reach the same heating speed. Therefore, differently sized and differently shaped workpieces of similar material thickness can be dried together at the same temperature profile. Even with very differently shaped sections of large workpieces, temperature compensation takes place.

Due to the same temperature control, the risk of overburning, i.e. damage to the paint film due to overheating of certain workpiece sections, is also minimal. Due to the small difference between the circulating air and object temperatures, even operating faults with conveyor standstill do not generally lead to scrap production. Nevertheless, care must be taken not to exceed the object temperatures and holding times specified by the coating manufacturer too far, as this can lead to colour changes. In the event of malfunctions and downtimes, suitable measures must be taken to reduce the oven temperature or to drive the parts out of the oven.

4.2.2 IR Radiation Drying

Another way of supplying energy for curing coating materials is infrared radiation. The intensity of IR radiation depends on the wavelength spectrum and thus on the radiator temperature. Basically, a distinction is made between long, medium, short and ultra-short wavelength radiation [2], see Table 4.1. Figure 4.5 shows the wavelength ranges of the individual emitter types.

Table 4.1 Relationship between wavelength and temperature of IR radiation

	Wavelength in µm	Temperature in °C
Longwave	3–6	200–700
Medium wave	1.5–2.5	750–1500
Shortwave	1–1.5	1500–2700
Ultra-short wavelength	<1	3000

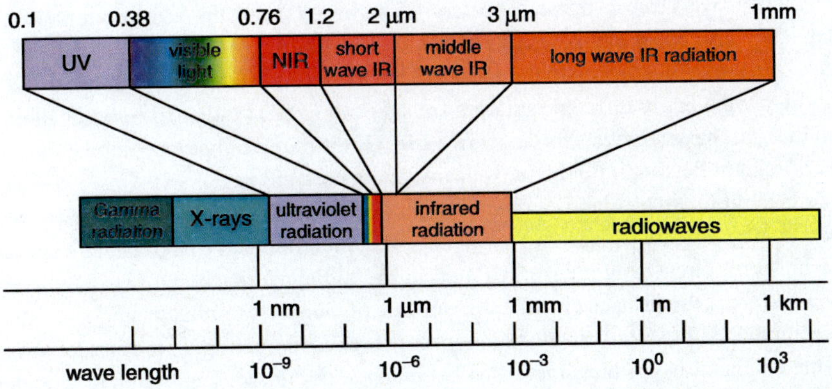

Fig. 4.5 Wavelength ranges of the different radiation sources

Instead of the wavelength, a distinction is sometimes made between light and dark radiators according to the beam wall temperature. Accordingly, the so-called *dark radiators* correspond approximately to the lower long-wave range. The radiators are channels made of black plate, in which flue gases are circulated at temperatures of 300–400 °C.. This type of radiator is used in cases where waste heat with a corresponding temperature level is available, e.g. in car body dryers with thermal exhaust air cleaning. Due to the large masses, these radiators have a relatively slow control behaviour. The convection loss is high due to the large (heat exchanger) surface and leads to considerable heating of the air.

In the medium, short and ultra-short wavelength range, electric radiators are usually used. These offer,, the advantage that a more precise control of the object temperature is possible.

IR rays are absorbed or reflected by objects depending on their nature and surface. As with light, bright, smooth surfaces reflect a larger proportion of radiation than rough and dark surfaces. Only the non-reflected part of the radiation is converted into heat. It leads to an increase in the temperature of the workpiece and a heating of the paint film from the inside. The advantage of IR radiation drying is the ability to transfer a lot of energy in a very short time. This results in a faster operational readiness of the dryers, a faster heating of the workpieces and a not inconsiderable saving of space due to shorter drying distances.

These advantages can be fully exploited if the dry material is thin-walled, flat and consistent. However, more complicated shaped workpieces with different thicknesses and radiation shadows heat up at different rates. The higher the radiation temperature, the greater the difference. Therefore, it is very easy for the coating to overheat at certain points. This can be avoided with great effort in terms of control technology or with a significant increase in the proportion of circulating air, but this negates the actual advantages of radiation drying.

Medium wave electric infrared heaters (IRM heaters) are the most common type of heater. They are robust and have a long service life. The disadvantage is a relatively slow control behaviour: Heating up to full power "from a standstill" takes approximately 1–2 min.

Short-wave electric infrared radiators are superior to IRM radiators in terms of control behaviour, but have a much shorter service life. Gas infrared radiators combine the advantages of radiant heating with the inexpensive energy source gas.

Airflow in radiant heating is an important issue because, the air in the radiant furnace is heated unintentionally. In order to avoid heat accumulation and to make the heat transfer more uniform, circulating air is circulated in the radiant furnace and exhaust air is also discharged from the furnace chamber. In the case of electric IR radiators and gas radiators, it is recommended to circulate the radiators to prevent overheating. In the case of gas emitters, the combustion products must also be removed or diluted. Ventilation is provided by separate fans or in combination with the adjacent air circulation furnace.

4.2.3 Special Procedures

In other rapid curing processes, such as UV radiation drying or electron beam curing, the radiation, in contrast to IR radiation, is not used for heating but for triggering the polymerization of the binder in so-called radiation-curing coatings.

High-frequency drying (heating by inductive or capacitive resistance of the goods in the high-frequency field) is also a special field. For the coating of metals, only inductive drying comes into consideration. It could be used for the coating of pipes, wire and packaging tapes.

4.2.3.1 Drying with "Near Infra-Red": NIR

In infrared heating, thermal energy is transmitted by means of electromagnetic waves. The infrared radiation spectrum ranges from 0.76 to 1000 μm. NIR (Near Infra Red) technology uses the infrared radiation spectrum close to visible light between 0.76 and 1.2 μm.

According to Stefan-Bolzmann's law, the energy transfer of the radiation is proportional to the difference of the fourth power of the filament and object temperature. This means that an increase in the radiator temperature leads to a disproportionately larger increase in the energy transfer. The emitted radiation intensity increases with T4 [3, 4].

In NIR technology, high-power halogen lamps are used as the actual radiation source, operating with filament surface temperatures of up to 3500 K. The lamps are then used to generate light.

Due to the high transmitted power densities, temperature-controlled heating of the object is required via control of the radiators. The low mass of the heating coil has the advantage that the radiation systems react spontaneously to current fluctuations and can be controlled extremely quickly. The heating source reaches full radiation power within a very short time and is switched off again just as quickly.

The fast response time of the new system in combination with the contactless energy transfer allows a fast regulation of the process and thus enables an automated process control.

The radiation emitted in the NIR spectral range can be directed very efficiently by means of special reflector systems. Depending on the surface contour to be heated, ellipsoidal and parabolic reflectors can be used to bundle the radiation in such a way that it can be applied homogeneously over the entire surface with extremely high radiation densities. This enables targeted heating of surfaces as well as intensity bundling on lines and points. In this way, a homogeneous energy density can be generated over a contoured surface. In addition, radiation focusing is also possible in the case of undercuts due to specially shaped reflector systems. Due to the high radiation density and the depth effect of the NIR technology, homogeneous heating of the coating to the curing temperature is achieved in just a few seconds.

4.2.3.2 Induction Heating

Inductive drying [5] takes a completely different approach. Here, the workpiece to be dried is placed in a magnetic field and heated by the eddy currents generated in the workpiece. The heat is thus generated directly in the workpiece. This means that the coating always dries from the inside out, not from the outside in as with other processes.

Induction heating is suitable for all drying processes. It can also be used to dry solvent-based coatings and powder coatings. Induction drying is said to significantly improve the adhesion between the powder coating layer and the workpiece surface. In addition, relatively fast heating – partly in the range of seconds – is possible. Even workpieces with large dimensions can be dried because, depending on the choice of frequency, the energy is only converted at the surface, precisely where you need the heating.

The induction coil used for heating is usually a ring or line inductor adapted to the workpiece. By designing the induction coils accordingly, it is also possible to heat only individual zones on a workpiece.

A requirement for the use of induction drying is that the workpiece geometries allow the current to be distributed evenly so that the same temperature is reached everywhere. Parts such as tubes, rods or bolts are ideal, for example. In the automotive industry, drive shafts, brake discs, clutch pedals or wheel bearings are also dried with this method.

Induction heating can be combined with conventional drying methods. For example, parts can be inductively preheated and cured with convection or radiation. The advantage: A temperature level slightly below the final temperature is reached very quickly, which significantly shortens the entire drying process.

4.2.3.3 Microwave Drying

With microwave drying, heating of the lacquer layer is achieved from the inside to the outside. High-frequency electromagnetic waves penetrate the coating layer and heat the substrate. This also prevents the paint film from closing first on the surface, as is the case with drying with convection heat.

The wavelength of microwaves is between 1 mm and 15 cm. They are generated in a magnetic field tube (magnetron), the frequency range is 2.45 GHz.

4.2.3.4 Thermal Reaction Drying

Another type of drying is drying with thermoreactors. The process is suitable for powder and wet coatings. Thermoreactors are catalytic infrared radiators. They emit both infrared radiation and convection heat. The radiation emitted by the thermoreactor is pure thermal radiation, which lies in the mid-infrared wave range. Since the emission spectrum ranges from 2 to 8 μm, the power can be adjusted very flexibly. These systems should also be able to greatly reduce the drying time and thus the cycle time. The energy savings should be around 50%.

4.2.4 Evaluation of Different Curing Methods

In Table 4.2, an evaluation of various curing processes was carried out with regard to their possible applications, i.e. for which workpieces or materials is the oven suitable, the space required, energy input, costs and, last but not least, the powder coating required [1].

Table 4.2 Use and evaluation of different curing systems

	Circulating air	[a]IRM + circulating air	[b]NIR	NIR + Conv. UV
Filigree workpieces	+++	++	– – –	– – –
Different workpiece thicknesses	+++	++	+	+
Preassembled products	– – –	– – –	+++	+++
Temperature sensitive parts	– – –	– – –	+++	+++
Plastics	– –	– –	+++	+++
Glass	+++	+++	+++	+++
MDF, wood-based materials	– – –	– – –	+	+++
Cast materials, porous materials	– –	– –	– – –	+++
Space requirement	– –	+	+++	++
Energy use	– –	+	++	+
System costs	1	1.3	1.5–1.8	2–2.2
Powder costs	1	1	2–3	2–3

[a]*IRM* Infrared with medium wave emitters
[b]*NIR* Near infrared

4.3 Optimization of Paint Dryers

The heat requirement of a dryer is determined by

- the material throughput,
- the curing temperature specified by the coating material,
- the necessary heating of fresh and circulating air,
- the heat losses due to air leakage at the inlet and outlet openings
- the heating of the conveyor and transmission heat losses.

The material throughput is a given quantity, the proportion of transmission heat is low and of secondary importance. All other items must be carefully examined and weighed against each other so that both an energy-saving and overall cost-effective solution can be found. Table 4.3 shows the different possibilities where energy is consumed in the furnace.

The most important factor is the curing temperature required for the coating material, because a curing temperature reduced by 10 °C reduces the energy requirement by over 7% [2]. A powder coating with a corresponding curing window must of course be available for the requirement.

In order to heat as less material as possible with no need for hardening, the conveyor, including the hangers, should be as light as possible. This request must be weighed against the demand for a robust conveyor system that is less susceptible to failure, which should have a long service life and therefore must not be too tightly designed..

The air balance in the kiln is crucial for the energy consumption and the quality of the cross-linking. Increasing the necessary minimum exhaust air volume can be helpful to reduce the discharge of hot air at the inlet and outlet openings of the dryer. However, larger exhaust air volumes require greater fresh air heating and greater sizing of the exhaust air purification system that may be required. In addition, uncontrolled air exchange at the dryer openings cannot be completely prevented, no matter how large the exhaust air volume. Therefore, it is recommended to keep the exhaust air as small as possible and to prevent hot air from escaping into the room by means of effective airlock constructions.

The most effective solution is to be found in dryers of elevated design with A-locks at the inlet and outlet. The loss of outlet heat is very low and the exhaust air volume can be

Table 4.3 Type of energy consumption in the furnace and dependence on conditions

Type of energy consumption	Dependence on
Surface losses, transmission	Surface area, heat transfer coefficient, temperature difference
Fresh air heating	Air volume, temperature difference
Lock losses	Lock area, temperature difference
Workpiece heating	Mass flow rate, temperature difference
Heating of the conveyor and the carriers	Mass flow rate, temperature difference

kept low. However, a certain air exchange in the oven is necessary in order to avoid condensation of split-off products from the powder in cooler zones of the oven. A specialist should always be consulted to adjust the air balance at the start of production. The need for good insulation and careful housing installation is self-evident.

4.4 Measuring the Curing Temperature

Optimum curing conditions are not only relevant for the quality of a coating, they also enable a reduction in energy consumption. The measurement and control of the oven temperature is therefore an important instrument in the company's own monitoring and quality assurance.

4.4.1 Basics of Temperature Measurement

Temperatures can be recorded using different measuring methods. Common techniques are thermocouples (type K, J, N and the like), resistance thermometers (PT-100, PT1000) and non-contact measurement (for example with infrared) [6, 7].

4.4.1.1 Structure of Surface Sensors

The measuring signal of a surface sensor is formed according to the following relationship:

$$T_M = T_1 + T_2$$

T_M = measured temperature
T_1 = Circulating air temperature
T_2 = Workpiece temperature

With an ideal surface sensor, the influence of the circulating air temperature $T_1 = 0$, i.e. the measuring element is completely decoupled by an insulator.

4.4.1.2 Thermocouple

A thermocouple (Fig. 4.6) consists in its basic form of two wires of different alloy. At the junction, which is either spot-welded and/or twisted, a thermoelectric voltage drops as a function of temperature. This relationship is not linear. Since this voltage is very small, the measurement signal must be extremely amplified for evaluation. The advantage of a

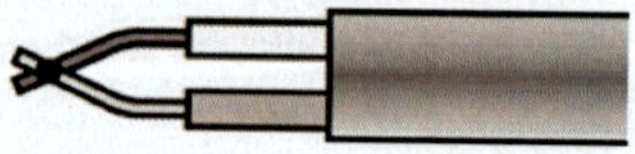

Fig. 4.6 Design of a type K thermocouple

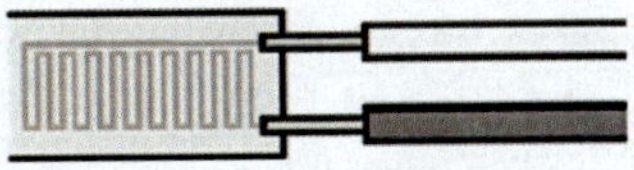

Fig. 4.7 Design of a PT-100 resistance element

thermocouple in its original form is its fast response. The disadvantage is a comparatively high measurement error due to the lack of linearity.

4.4.1.3 Resistance Element PT-100

A resistance measuring element (Fig. 4.7) consists of a ceramic plate as a carrier material with a vapour-deposited resistor. Platinum (Pt), for example, is used as the reaction material. This resistance changes as a function of temperature in a linear relationship. The advantage of this system is the high measurement accuracy and the linear relationship. A disadvantage is the theoretical inertia of such systems.

4.4.1.4 Oven Measurement

For furnace measurement, a wide range of manufacturers offer measuring devices in different price categories. For the user, the possibilities of evaluation and documentation are particularly important. The following minimum requirements for documentation should be included in the software of a measurement system:

- Measuring point plan
- Specification of the test specimen with wall thickness
- Information on the sensor type
- Lacquer data
- Evaluation of the burn-in conditions according to window or limit method
- Order system, such as file manager.

4.4.2 Application of Temperature Measurement

Temperature measurement in practice can be divided into two areas of application:

- Regular control of the stoving conditions, quality control, calibration of serial parts
- Process optimization, i.e. reduction of energy costs and cycle time

4.4.2.1 Curing Conditions

In addition to pretreatment and paint application, correct compliance with the curing conditions specified by the paint manufacturer is the third process variable for a qualitatively flawless coating. If these are not adhered to, defects will occur as listed in Table 4.4, related to the curing conditions.

The following example is intended to show how the actual temperature in a convection oven can deviate from the desired temperatures. The operator was of the opinion that

Table 4.4 Failures in the stoving of coatings and their causes

Error	Cause	Failure detection
Lack of adhesion	Curing time too short; temperature too low cycle time error	Oven measurement; cross cut testing
Too high gloss level	Curing time too short; temperature too low cycle time error	Oven measurement; gloss level measurement
Gloss level too low	Curing time too long; temperature too high cycle time error	Oven measurement; gloss level measurement
Severe yellowing	Curing time too long; temperature too high cycle time error	Oven measurement;colour measurement
Insufficiently cross-linked paint film → poor weather resistance	Curing time too short; temperature too low cycle time error	Oven measurement; gloss and colour measurement cross-linking test

everything was in order, as the display on the control cabinet for the desired setpoint of 210 °C indicated an actual value of between 208 °C and 214 °C.

However, a temperature measurement in the furnace then revealed high deviations, as shown in the measurement report, Fig. 4.8. The following causes for the deviations could be determined:

- the temperature display was not calibrated (error + 3 °C, offset -0.8 °C),
- the only air sensor in the dryer was very far from the warm air inlet, i.e. approximately at the coldest point of the dryer,
- the airflow was poor, through two air inlets in the first third of the dryer.

This resulted in high energy costs, the powder coating showed over-burning and a lack of gloss, and in addition, depending on the colour shade, there were colour differences due to yellowing.

As a solution to the problem, the air sensor was relocated and calibrated and the air ducting was changed accordingly.

Another example shows the condition of an oven of a vertical coating plant (Fig. 4.9). The main problem in this plant was the too low air temperature in the inlet area of the dryer. As a result, the object temperature at the lower end of the profile only reached the required 170 °C for 12:25 min, while this temperature is exceeded at the upper end for 29:45 min. At the lower end, the powder coating is not sufficiently cured.

As a remedy, in addition to the new setting of all supply air grilles, two gas emitters were installed on both sides in the lower inlet area. As a result, the distance between the upper and lower profile ends above the required temperature limit of 170 °C is

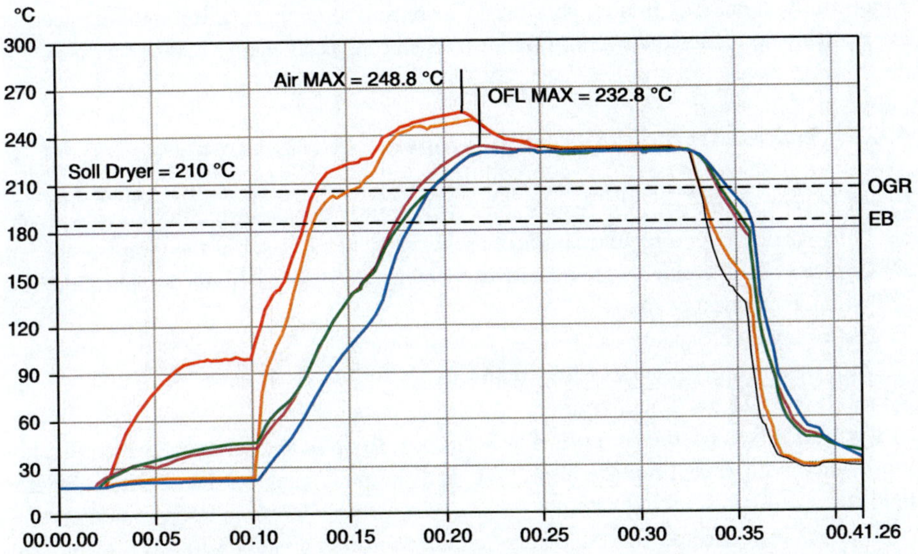

Fig. 4.8 Result of the temperature measurement of a convection oven. With curing temperatures of 180 °C object temperature and a desired circulating air temperature of 210 °C, it becomes clear that the circulating air temperature is much too high

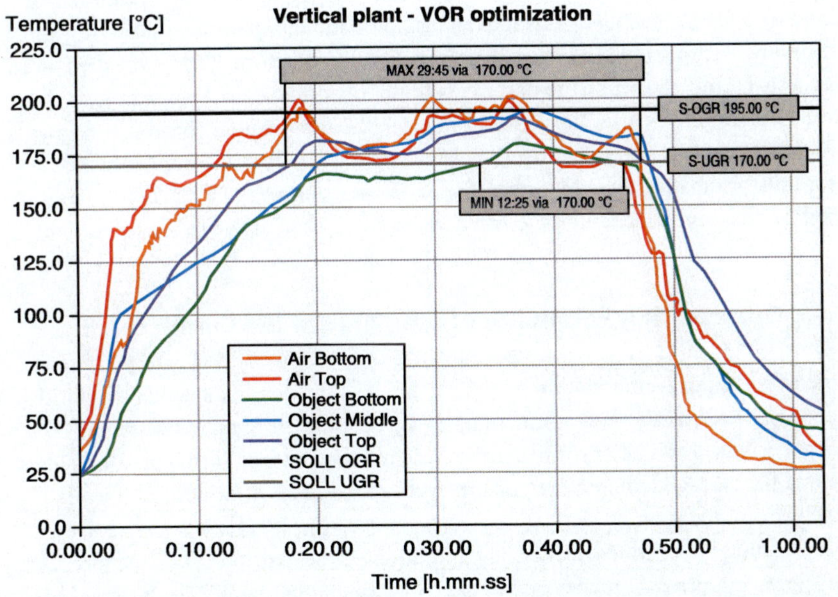

Fig. 4.9 Temperature profile of the furnace of a vertical plant before optimization, requirement: 170 °C/20 min object temperature

significantly smaller. It is now possible to increase the cycle time/throughput speed or, alternatively, to reduce the overall furnace temperature as an energy-saving measure.

4.4.3 Process Optimization with Temperature Measurement

Errors in connection with the temperature measurement or control can be avoided very easily by carrying out a regular check of the furnace and a check of the temperature sensors. Since PT-100 elements are usually used, every user can carry out a calibration himself with the simplest of means.

The following procedure can be used for calibration:

One measuring resistor each with R1 = 100 Ω (Ohm) = 0.00 °C and R2 = 180 Ω = 211.34 °C are required.

First, the offset, i.e. the zero offset, is corrected. To do this, disconnect the temperature sensor on the dryer and replace it with R1, and set the display on the control cabinet to 0.00 °C.

Then R2 is connected and the display on the control cabinet is set to 211.34 °C. This procedure can be carried out within a few minutes for each temperature measuring point. The complete measuring chain, i.e. line error and error at the amplifier in the control cabinet, is thus adjusted.

Another important criterion is the cleaning of the sensors. Just as a layer of dust and paint residues is deposited on the dryer walls over the years, the temperature sensor is also covered by a layer. This layer acts like an insulator and dampens the measuring signal, i.e. the measured temperature is lower than the actual temperature. This gives the controller a lower actual temperature, so the dryer gets hotter over the years. Depending on the degree of contamination, it is usually sufficient to clean the temperature sensors once a year.

With large dryers, a single air sensor for control is too few and not process-safe. If this sensor fails or the display is incorrect, the entire plant comes to a standstill. At least three sensors are recommended, with the mean value being used as the reference variable [7, 8].

4.4.4 Optimisation Possibilities in the Area of the Curing Oven

Before any practical optimization of a dryer can be considered, the actual conditions must be recorded. At least the measurement of the circulating air temperature in the area of the product carriers is necessary. Measuring the air temperatures in the area of the cooling zone and the surface temperatures of the dryer in the area of the roof and walls is useful for further optimization.

An essential factor for correct curing conditions is the distribution of the circulating air. If large temperature differences occur, the dwell time must finally be designed according to the worst conditions. Furthermore, a certain correspondence must be established between the dryer temperature displayed on the control cabinet and the actual temperature in the area of the product carriers. Here, depending on the placement of the temperature sensors, very large differences can occur.

Like the coating material itself, the product carriers with the required hangers must also be heated with each furnace pass. In this case, it is almost impossible to modify existing equipment. When purchasing new equipment, however, it is essential to pay attention to the wall thicknesses of both the traverses or product carrier and the hangers.

Another important point is the insulation. With an existing dryer, re-insulation is usually difficult and costly. In the case of a new acquisition, attention should therefore be paid to the outside temperatures. Above all, good insulation in the roof area is important.

Separating the conveyor technology can also lead to energy savings. If the parts are transferred to an oven chain directly at the oven, the chain and spindle do not cool down and there is no need to heat up the goods carrier.

Exhaust air and cooling zones must also be included in the consideration. The amount of energy introduced to cure the paint must be extracted again in the area of the cooling zone. The waste heat can be used to preheat the fresh air. Furthermore, a cycle-controlled cooling zone is useful. The exhaust should only be activated when the cooling zone is really occupied and should especially be switched off when the dryer doors are open. Otherwise, recirculated air will be sucked out of the dryer into the cooling zone, especially in the upper area of the doors.

References

1. Eberhard, J.: Trockner, Energieübertragungsarten und Energiebilanzen, DFO Weiterbildungslehrgang „Pulverlackiertechnik – Grundlegende Zusammenhänge für die Praxis“. Paderborn (1993)
2. Möller, E., Ritter, R.: Industrielle Lackiertechnik. Weka Fachverlag, Augsburg (1997)
3. Bär, K.: Sekundenschnelle Aushärtung von Pulverlacken. JOT. **38**(2), 26–29 (1998)
4. Bär, K.: Die NIR-Technologie – Anwendungsbeispiele zur Kostensenkung bei der Pulverlackierung. DFO Tag. Pulverlack-Prax. **9** (2000)
5. N. N.: Neue Trocknungsverfahren im Kommen. JOT. **3** (2000)
6. Sander, C.: Kosten reduzieren mit der richtigen Einbrenntemperatur. JOT. **41**(6), 50–55 (2001)
7. Sander, C.: Qualität steigern und Kosten reduzieren mit der richtigen Einbrenntemperatur. EPSTagung – Der Pulvertreff 2 (2001)
8. Scott, A. (ed.): The Science of Powder Coatings – Applications, vol. Bd. 2. Sita Technology, London (1994)

Images used in Chap. 4

Eisenmann GmbH, Böblingen: 4.1 bis 4.4
Axalta Powder Coatings, früher Du Pont Pulverlack GmbH: 4.5, 4.6
Solu Tech GmbH: 4.6 bis 4.9

Surface Pretreatment of Metals

5

5.1 Cleaning and Pre-Treatment

A thorough pre-treatment is always necessary as a basis for a flawless paint finish.

This includes that:

- removal of surface contaminants such as dirt, grease, oil, corrosion products,
- activation of the surface for coating and
- application of adhesion-promoting and/or corrosion-protecting layers, the so-called conversion layers.

Inadequate surface pre-treatment leads to impaired coating adhesion and reduced corrosion protection. Cleaning can be carried out with solvents or, more common in industry, as a wet chemical pre-treatment with aqueous degreasing with appropriate additives.

The procedure chosen is based on:

- the material of the substrate,
- the surface contaminants present,
- the regulations of the legislator with regard to the choice of cleaning agent,
- the required degree of purity and
- the pre-treatment process itself.

After correct degreasing, a metallic clean surface is obtained, which already allows painting, but is not sufficient according to today's quality requirements. Therefore, the cleaning is followed by the layer-forming pre-treatment. The layers formed on the surface of the workpiece cause an enlargement of the surface due to a certain roughness of the formed layer, and offer a better corrosion protection. Special functional groups are also applied for

J. Pietschmann, *Industrial powder coating*,
https://doi.org/10.1007/978-3-658-37592-8_5

paint bonding. The efficieny of the corrosion protection is significantly influenced by the chemical nature of the layers.

Another important criterion for pre-treatment is rinsing. In all chemical processes, one or more rinsing processes with water, process water or deionised water must be switched between the individual treatment steps in order to avoid carry-over of bath contents. Particular importance should be attached to the final rinse. Here, rinsing should be carried out with deionised water in order to avoid drying of electrolyte residues or rinsing water (usually "contaminated" with lime and salts). This drying or contamination on the substrate can later lead to problems in the adhesion of the paint.

5.2 Surface Condition Requirements

5.2.1 Degree of Purity

In metal surface treatment, a distinction is generally made between the physically clean and the chemically clean surface. The physically clean surface in the sense of the absence of solid dirt and solid impurities is generally not suitable for use as an adhesive base for coatings.

Chemical purity is the absence of all impurities. This is often confused with the ability of the surface to be properly wetted by water.

In rinsing, the absence of a water-repellent effect is often considered to indicate the absence of coarse impurities such as oil and grease. However, effective wetting depends on the surface tension of the contaminants as well as the thickness of the water film and may be masked by evaporation of the water and presence of wetting agents on the metal surface.

Water wetting by simple immersion in deionized water is not sufficient to assess the cleaning process or the cleanliness of the metal surface.

Due to pre-machining and pre-treatment, a component surface consists of various layers which differ both in their composition and in their structure. First of all, there is a deformed boundary layer on the base material, followed by a reaction layer consisting mainly of metal oxides. Additives, for example from lubricants, can also be incorporated into the reaction layer. The reaction layer passes into a sorption or adsorption layer, which consists of the elements carbon and oxygen. Carbon and oxygen-containing layers form on all metal surfaces as soon as they come into contact with the atmosphere. Finally, the surface is covered by a contamination layer. A large proportion of the contamination layer is made up of machining oils, greases and emulsions as well as metal chips or dust. Figure 5.1 shows the schematic structure of a component surface with the various layers covering the metal from manufacturing.

On cold-rolled sheet, emulsion residues are found in the order of 100–300 mg/m^2 surface area and rolling abrasion of 50–200 mg iron/m^2 surface area. Deep-drawn parts show oil deposits of 100–200 mg/m^2 surface, corrosion oil protected even up to 2 g/m^2 [1].

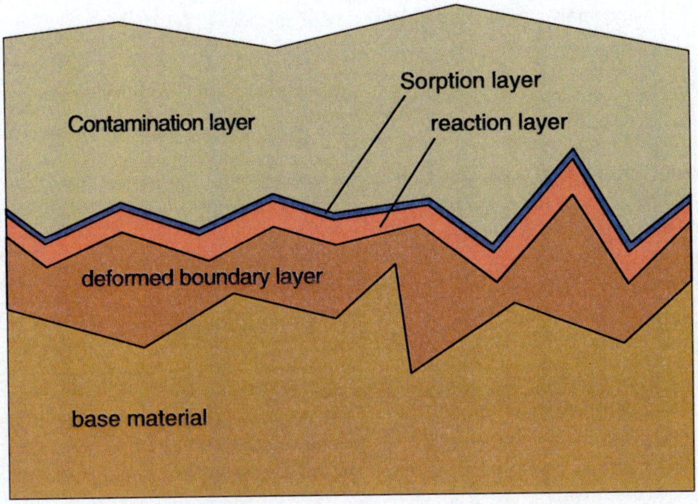

Fig. 5.1 Layer of an technical metal surface

On the other hand, there are requirements for the cleanliness of the surface, according to which only residual oil or metal abrasion of a maximum of 5–10 mg/m^2 is tolerated after cleaning, especially in the case of a following coating with inorganic or organic layers.

In order to obtain a surface finish of the workpiece that is suitable for the followingprocesses, it must be pretreated correctly. In order to assess the success of a cleaning process, one wants to check and evaluate the degree of cleanliness of the surface. Unfortunately, there are only a few processes that show a satisfactory result.

Due to the different nature of the component surface and due to the numerous types of impurities, the selection of the appropriate test is always made taking into account the entire treatment process. Single measurement techniques that include all aspects of the characterization of a surface do not exist. A combination of different methods makes an analysis with regard to several impurities possible. Criteria for selecting a test can be: cost, duration, effort, non-destructive or not, online testing possible or not, component dimensions, type of information provided (quantification, characteristic values).

A large number of different methods are described in the literature, with which the cleaning effect or the cleanliness of the surface can be assessed.

Table 5.1 gives an overview of different possibilities for testing the purity of surfaces.

Clean metal surfaces have high surface energies and can be wetted with water (hydrophilic). If there are greases and oils on the surface, water droplets form. Their size and number can be a measure of cleanliness. This property is used as a basis for the water wetting tests and spray mist tests. Dye can be added to the test solutions for clearer visualization (nigrosine test, grease red test) [2].

Table 5.1 Tests and analyses for testing the purity of surfaces [2]

Quick test for visible dirt	
Visual inspection	Assessment of the uniformity/spottiness of a surface
Rapid tests for fatty layers	
Water drainage test (wetting test), spray mist test, nigrosine test, grease red test, test inks, edge angle measurement	Visualization of the change in wettability of the metal surface
UV test	Visualization of oils fluorescing in UV radiation
Rapid tests for pigment soiling	
Wipe test, "scotch tape test", detachment method	Visualization of particle dirt
Detection of passive layers	
Copper sulphate test – Prussian blue test	Passive layers prevent copper coating in copper sulphate solution and bruising in hexacyanoferrate(III) solution ($K_3Fe(CN)_6$)
Electrochemical processes	
Anodic polarization measurement	A very low anodic current is imposed on the system. The voltage-time characteristics are evaluated
Other methods	
Weighing method: Removing the dirt and measuring	Determination of the difference in weight or the amount of dirt in the solution
Carbon determination by oxidation	Burning of the hydrocarbon chains and measurement of the C-content
Instrumental surface analysis	
IR (infrared), IR-RA (infrared reflectance absorption) spectroscopy	Detection of mineral oil hydrocarbons
Detachment procedure according to DEV H18, Detachment procedure according to DIN 38409, part 18 *(withdrawn)*	Dissolving the contaminants and quantification by means of IR
GDOS and RFA (glow discharge spectroscopy and X-ray fluorescence analysis)	Element depth analysis of the layers
ESCA (electron spectroscopy for chemical analysis), XPS (X-ray photoelectron spectroscopy) SEM (scanning electron microscopy)	Qualitative and quantitative analysis of thin films
AFM (atomic force microscopy)	Surface topography

The best known is the so-called water break test. It is used in two variants. One is used to determine and compare the cleaning effect of different processes or for control within the production.

In the water spray test, sample sheets to be compared are placed next to each other and sprayed with a fine water mist. Individual water droplets form in oil-covered areas, while a coherent film of water forms in the clean areas. In addition to the qualitative statement, this can be quantified by determining the oil-covered surface. In this way, different cleaning methods and the cleaning ability of a method for certain contaminants can be compared. If the average of five tests is evaluated, sufficiently reproducible results are obtained.

Especially for the control of cleaning in production, the atomizer test proposed by the American Electroplaters Soc. [3] has proven its worth. The test specimen or workpiece is immersed in a hydrochloric acid solution containing 1 wt% HCl for one min, rinsed in clean running water and dried in air without the use of compressed air dryers. Distilled water is then sprayed onto the surface to be tested from a distance of 60 cm for approximately 30–45 s. The distilled water is then sprayed onto the surface to be tested from a distance of 60 cm. It is important that the spray or atomizer nozzle produces a fine mist. Clean surfaces are covered with a continuous film of water after this treatment, dirty ones with small, irregularly distributed water droplets.

The standards mainly describe tests for the cleanliness of steel surfaces. The ISO standards [4–10] explain methods for determining oil and grease residues, water-soluble sulphates, iron ions, soluble salts, moisture and chlorides [11]. describes the determination of surface roughness after blasting and [12] generally the surface treatment methods for steel substrates before coating.

Test ink methods use the spreading tendency of liquids on degreased surfaces [13]. The test inks are water-ethanol mixtures or also solutions of different surfactants with different surface tensions. According to DIN 53364 [14] or ASTM D 2578-84 [15] different test series are defined. If they stand on the surface after application without contracting, the surface tension of the workpiece is the same or higher. The test allows a relatively rough statement about the surface tension of the tested materials, but only on smooth surfaces. A cast surface or grinding grooves lead to wrong results.

Another method is to measure the contact angle. The contact angle is the angle that the edge of a liqid droplet makes with the surface of the component. If the contact angle is greater than 90°, then the droplet runs off the substrate and this is a sign of a greasy surface. If the angle is between 90° and 0°, the surface can be wetted by the droplet. The highest degree of purity is achieved at a contact angle of 0°. The contact angle can be measured with a microscope, although this method is very complex.

Rapid tests for pigment soiling include the wipe test, tesa film test and the peel test. In the wipe test, the surface is wiped with a white cloth and the blackening is used as a measure to assess the cleanliness.

A similar procedure is used for the tesa film test. For this purpose, a transparent adhesive tape is stuck onto the component. After removal, mainly solid contaminants (metal dust, abrasion, chips) adhere to the adhesive tape. If this adhesive tape is stuck onto a white paper, the individual impurities can be made visible.

In the detachment method, the sample or workpiece is weighed before and after the contamination is removed, or the amount of contamination introduced into the solution is determined.

Passive layers on steel can be detected with the copper sulphate or Prussian blue test. An oxide layer can also be made visible on aluminium with the so-called drop tests. This is based on porosity tests of oxide layers. If the test solution is in direct contact with the bare metal, chemical reactions cause colour changes.

One of the electrochemical methods is anodic polarization. With this method, typical voltage-time curves occur depending on the type and amount of contamination and the activity of the metal surface. The metal surfaces are galvanostatically loaded with a very low anodic current (μA range). Oxide or passive layers show a completely different electrochemical behavior than a clean, bare metal surface. The principle of anodic polarization is used as the basis for the CT 100 cleaning test. With this method it is possible to evaluate the condition of a metallic surface. Inorganic and organic impurities, oxides, passive layers and conversion layers can be detected [13].

5.2.2 Porosity and Voids

Pores and cavities in the zones near the surface of the workpiece have a negative effect on the following coating. Corrosion resistance and adhesion are affected by enclosed impurities or efflorescence of liquid residues. The blooming out of liquid residues during the baking process also leads to disturbances (craters, bubbles) in the paint film.

Mechanical processing, such as grinding, polishing or blasting, can in some cases close pores. However, pores and cavities can also be opened up by chemical pre-treatment with corresponding metal removal by pickling or passivating the surface.

Electrolyte residues cannot be removed from pores or can only be removed insufficiently, depending on pore size and depth. Higher temperatures ($T > 25$ °C) and correspondingly long times ($t > 2$ min) during rinsing have a beneficial effect on the result. Additional movement of the parts or circulation of the sink is also beneficial.

5.2.3 Contamination: Dirt on the Surface

5.2.3.1 Anti-Corrosion Oils, Waxes and Lubricants

These products are applied to the metal surface to protect against corrosion during handling, transport and storage and to facilitate non-cutting and cutting shaping. They can be based on mineral oil, esters, corrosion inhibitors, additives, antioxidants, pigments and surfactants and are present in a film thickness of about 1–6 g/m^2. For proper surface treatment, it must be possible to remove the oil film completely from the surface. If the oil film remains in contact with the metal surface at a higher temperature for a longer period of time, it may age and become difficult or impossible to remove completely (resin deposits).

5.2.3.2 Particles

Particles can be: chips, fluff or fibres, pigments, metal abrasion from brushing or grinding, dust, residues of abrasives or blasting media.

The smaller the particles, the more difficult they are to remove. Particles <100 μm are problematic for a process-safe cleaning.

5.3 Mechanical Pre-Treatment

The aim of mechanical surface preparation is to remove impurities by removing them and at the same time creating a surface condition that enables good adhesion of coatings. There is usually also a simultaneous removal of the substrate surface. The mechanical methods are primarily suitable for removing solid contaminants. This primarily involves scale, corrosion products and also old coatings.

A special feature of mechanical preparation is that not only the impurities are removed, but also a new active surface is created. In the process, a new surface profile, i.e. a changed roughness, is imprinted on the substrate. In addition, the surface may be hardened or notch effects may occur. The type of mechanical surface treatment to be used depends on factors such as the properties of the contaminants, the type of cleaning object, the cleaning objective and local and economic conditions. The requirements of environmental protection and occupational safety must also be taken into account.

It should be noted that blasting, grinding and polishing can remove foreign oxide layers, greases and oils as well as release agents from the surface and also remove or cover defects, but on the other hand new foreign layers and defects can also be created by residues of the polishing or auxiliary agents. In many cases, the mechanical surface treatment "displaces" foreign layers and defects below the surface, and problems with adhesion and corrosion protection only occur after some time. Mechanical surface preparation is reported in detail in [16, 17] and especially in [18].

5.3.1 Grinding and Brushing

5.3.1.1 Grinding

Grinding and brushing are processes in which the existing surface substance is removed and a new surface is created. In terms of process technology, grinding is considered to be a machining operation with geometrically undefined cutting edges. The process of grinding is shown schematically in Fig. 5.2. The pressure of an abrasive grain with its cutting edge and the movement between the abrasive grain and the surface produces a chip and an abrasive groove. The deformations and dislocations caused by grinding in the metallic

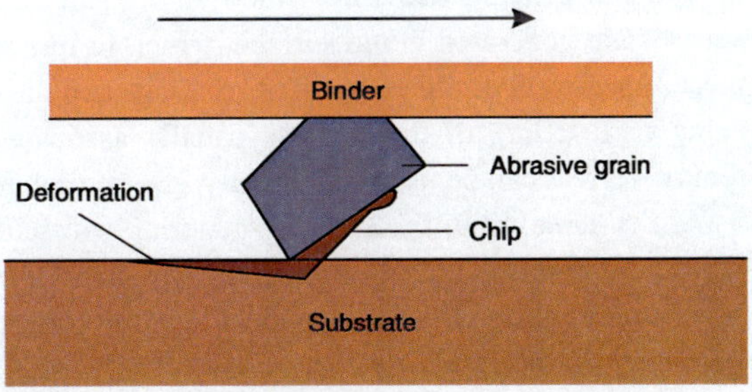

Fig. 5.2 Figure of shematic grinding process

substrate can reach almost a millimeter deep. The contaminants are pushed in front of the chip or they flake off the surface. Hard material grains are used in grinding and, when the resistance from the bulging chip becomes too great, brittle fracturing of the abrasive grain occurs, giving it a new sharp cutting edge.

5.3.1.2 Brushes

Brushing is closely related to grinding. The effectiveness here is based on the use of geometrically determined cutting edges, namely the right-angled cut wire ends of the brush. If the wire ends are worn, new sharp cutting edges can be brought into engagement by reversing the direction of rotation.

Therefore, only the wire ends are effective when brushing, i.e. no better grinding effect can be achieved by increasing the pressure.

Related to grinding is polishing. Hard edged polishes like diamond dust essentially act like an abrasive material. Softer polishing agents, tend to depress into the surface. However, polishing is rarely considered for cleaning purposes, as it usually leaves residues of the polishing material on the surface.

5.3.2 Blasting

Blasting is the most effective and widely used method of mechanical cleaning and preparation.

During blasting, granular solids (mineral, metallic or organic), but also liquids and even vapours or frozen gases reach the surface to be cleaned at high speed. This is done by mechanical, pneumatic or hydraulic means. The process is described in the standards [19–22].

Blasting with solid abrasive can change the purity, the surface shape (roughness) and the properties of the layer close to the surface, for example by hardening. In addition, there is an activation of the surface.

5.3.2.1 Basics

During blasting, various interactions occur between the blasting medium and the surface. The impact energy of the blasting grain primarily causes plastic deformations on the surface of the blasting material, which also result in hardening. In addition, compressive stresses are generated in the surface, which in turn can lead to deformation of the workpiece. As a result of the plastic deformations and the grinding effect, there is also a roughening or levelling of the surface, usually associated with a removal of material. This removal effect can be used specifically, e.g. for deburring [23].

At the same time, the blasting material also undergoes changes. The impact energy plastically deforms the blasting grain, which in turn causes it to solidify. Stresses also occur in the grain, which can lead to the grain cracking. Finally, abrasion rounds off the originally sharp grain edges.

The mode of operation of blasting is comparable to that of grinding. The surface is abraded and roughened by the blasting grains. The power increases with the number of impacting particles up to a maximum. However, in the case of even larger quantities, obstructions occur due to rebounding particles, for example when blasting vertical to the surface. The yet angle is of great importance for the result.

When the abrasive strikes vertically, there is a predominant hammering effect. This effect is particularly high with spherical abrasives. The hammering effect determines on the one hand the roughness of the surface and on the other hand its hardening. With a small blasting angle, there is more of an abrasive effect, which is particularly noticeable with an angular grain. Under these conditions the difference to grinding is the smallest.

These two borderline cases already make it clear that the blasting medium is of particular importance. Metal particles made of chilled cast iron, steel, aluminium or zinc are used as solid blasting media; these are often wire cuttings. Mineral blasting media are also widely used. Originally, sand was used for this purpose, but it produces dust that often causes silicosis among the workers and is therefore no longer permitted today. Corundum, slag or glass beads are considered alternatives. A third class of blasting media are plastic beads, but also particles of nutshells, apricot kernels or corn meal.

How blasting is to be carried out in detail, which blasting media and blasting conditions are to be used, depends on the purpose for which this type of surface treatment has been selected. After all, blasting is mainly used for heavily contaminated surfaces. This may involve derusting or descaling or the removal of old coatings.

The right choice of abrasive is crucial to the success of blasting. A blasting medium type and grain size must be found that is suitable for the task in every respect. The suitability depends on the blasting system. For centrifugal blasting, only metallic blasting media, preferably in spherical form, are suitable. For dry compressed air blasting there are hardly any restrictions regarding the use of blasting media. For blasting with the addition of water, only mineral blasting media can be considered, as metallic blasting media corrode. Furthermore, the decision has to be made whether the abrasive should be used only once or whether it is better to keep it in circulation. The material of the parts to be blasted must also be taken into account in the selection.

The blasting material should usually be harder than the surface and the possibility of contact corrosion must be taken into account. This applies, for example, to stainless steel when blasting with steel grit. In addition, there are the interactions of the grain on impact with the surface, the initial contamination and the required cleanliness after blasting.

5.4 Liquid Cleaning Procedures

Aqueous cleaning is described in detail by Molz [1] and Grün [24]. Common cleaning processes are based on treatment with aqueous detergent solutions (alkaline, neutral or acidic) or with organic solvents (halogenated hydrocarbons, high boiling gasoline fractions or polar solvents).

The aqueous systems are used in immersion, spray, brush, ultrasonic or electrolytic cleaning systems or corresponding combinations, while the organic cleaners are mainly used in immersion systems with and without ultrasound as well as in vapour degreasing systems.

Today, cleaners and processes are not only evaluated with regard to their cleaning effect. Other important considerations are the associated environmental impact and possible health hazards; among other reasons, this is one of the reasons why organic cleaning agents hardly play a role in many applications or are implemented in special plant technology (closed systems).

5.4.1 Cleaning Parameters

Every cleaning process is based on the interaction of chemistry, mechanics, temperature and time. The individual parameters must be coordinated with each other, whereby each one decisively influences the cleaning effect.

5.4.1.1 Chemistry

The choice of cleaner depends on:

- the type of material: steel/iron, non-ferrous metals,
- the type of contamination: oils, greases, rolling abrasion, pigments, oxides,
- the cleaning requirements: accepted residual coatings,
- the requirements for the surface with regard to the subsequent processes: Activation, passivation, metal removal,
- the application procedures for the cleaner solution,
- the requirements of the local authorities.

5.4.1.2 Mechanics

The greater the mechanical support, the better the cleaning effect. On the other hand, a lower system mechanics, e.g. with a simple immersion cleaning, can be compensated with bath movement by pumping over, higher cleaner concentration and extended cleaning times. The mechanics that can be used is essentially determined by the workpiece geometry, the type of contamination and the treatment time available.

Spray cleaning is mainly used where large-area parts are to be cleaned which are uniformly covered by the spray jets. With increasing spray pressure, the cleaning effect is increased on the one hand, but on the other hand there is also the problem of higher foam formation. Regular maintenance and cleaning of the lines and nozzles prevents nozzle blockage and reduced line cross-sections. In practice the spraying pressures are generally between 1 and 6 bar, in high pressure systems also up to 15 bar.

Simply **dipping** the workpiece into the cleaning solution achieves the least cleaning effect. A relative movement between the two, such as pumping the solution around, significantly improves the cleaning effect.

Ultrasonic cleaning is an efficient form of immersion cleaning and is particularly effective in removing firmly adhering contamination. Due to the ultrasonic support, materials with porous and very rough surfaces can also be cleaned well.

In **electrolytic cleaning**, the mechanical effect of gas evolution and the current-induced forces come together, the latter repelling electrically charged particles such as iron abrasion when the polarity is appropriate.

Brush cleaning is mainly encountered in strip treatment and is preferably used here to remove pigments and roller abrasion. Although brushes are very maintenance-intensive, they deliver an excellent cleaning result.

5.4.1.3 Temperature

An increase in temperature generally causes a significant improvement in the cleaning effect, with a few very special exceptions. The minimum working temperature depends, among other things, on the type of greasing, as it should always be above the melting point of the grease to be removed.

A temperature increase of +10 °C halves the cleaning time, a temperature decrease of −10 °C doubles the time.

5.4.1.4 Time

The treatment time depends on the type and amount of contamination and the overall production process (specifies the number of cleaned workpieces per time unit) and depends above all on the type and design of the cleaning system.

5.4.2 Purification Mechanism in Aqueous Solutions

The cleaning process with aqueous cleaning agents is mainly chemical in nature and takes place in several steps.

The cleaning process of a metal surface contaminated with non-emulsifiable oil/grease, pigments, dust and chips starts with the surfactants diffusing from the aqueous solution into the oil/water interface. As soon as the wrapping energy is reached, the oil contracts into droplets and rolls off the surface (roll-up effect), see Fig. 5.3, taking with it the other contaminants dissolved and dispersed in the oil. After this coarse cleaning, oil and dirt components are still present on the surface, but now in very low concentrations of a few molecular layers.

A comparable final state is also achieved in this first sub-step of the cleaning process if an emulsifiable oil is present on the surface. In this case, however, the main amount of oil is not washed away by rolling off the oil droplets, but via an emulsion formation.

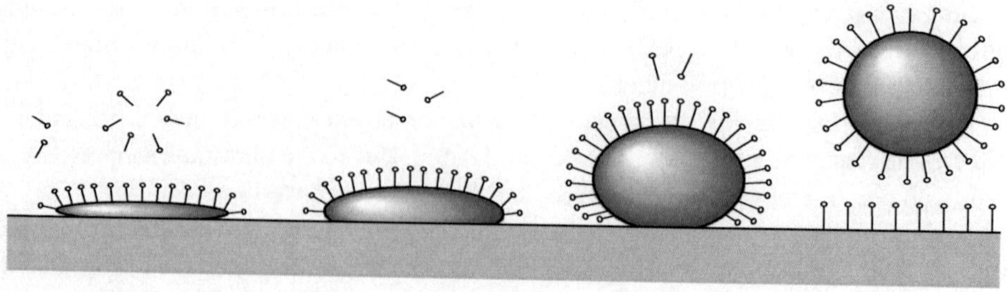

Fig. 5.3 Schematic figure of the displacement phases of an oily soiling by a wetting agent-containing cleaner solution

In the second step, the remaining thin oil layer is removed by displacement processes using surfactants and builders. In parallel, pigments are dispersed and washed off the surface. Since these processes follow adsorption and desorption mechanisms, a relatively long time is required until equilibrium is reached. An existing inorganic builder of the cleaning product works especially in this process in a synergetic way with the surfactants. It increases the speed of equilibration (cleaning speed) very much and shifts the position of equilibrium to decidedly lower oil occupancies.

After this second reaction step, the surface is hydrophilic and is wetted by the aqueous solution. The other ingredients of the cleaning product can now perform which leads to an activation or passivation of the substrate. In the case of amphoteric aluminium, oxidic top layers and insoluble metal soaps are detached not only in the acidic but also in the alkaline solutions. This is immediately followed by an attack of the base metal and hydrogen is formed. In both cases the surface is activated.

This mechanism can be used to explain and understand many observations in practice:

a. When cleaning in aqueous solutions, there is a lower temperature limit below which the temperature must not fall.. The reason for this is the roll-up effect of the first reaction step. The oil/grease can only contract into droplets if it is liquid. The melting point of the oil/grease on the surface determines the minimum degreasing temperature.
b. The thickness of the oiling is insignificant for the cleaning speed and intensity. Thin as well as thick oil layers are washed off the surface comparably fast with the rolling oil droplets.
c. Oiling that has gummed up due to improper storage of the workpieces is particularly difficult to remove. The resinification has increased the melting point of the oil, so that a higher cleaning temperature is required. If the melting point is above the boiling point of water, degreasing is no longer possible. Furthermore, due to the resination reactions, polar functional groups are incorporated into the molecule, through which the molecule

is more strongly bound to the metal. The adsorption and desorption mechanisms in the second step are negatively influenced.

d. A non-emulsifiable oiling is preferable to an emulsifiable one according to cleaning criteria. The roll-up effect makes it easier and faster to remove from the surface and the washed-off oil " comes up to the electrolyte-surface" in the cleaning solution and can be easily removed from it. The result is a long service life of the cleaning bath.

5.4.2.1 Structure of Aqueous Cleaners

A typical cleaner consists of a framework of inorganic salts, the so-called builders, surfactants and, if necessary, additives such as organic complexing agents, antifoaming agents, anti-corrosion additives, inhibitors, etc.

5.4.2.2 Builders and Complexing Agents

Alkalis

The alkalis sodium and potassium hydroxide cause high pH values of 10.3–13 in the usual application concentrations. They saponify the ester components of the grease to be removed and ensure conductivity in electrolytic cleaning baths. The cleaning effect is increased by the forming soaps . These strongly alkaline products are used in practice for cleaning iron, steel and magnesium, as they attack non-ferrous metals – copper and brass – as well as aluminium, zinc, tin and lead.

By selecting a suitable temperature, dwell time and/or an additional inhibitor, the metal removal can be controlled.

Silicates

The silicates have a good dispersing and emulsifying effect and are characterized by a high dirt carrying capacity, i.e. they prevent rapid re-contamination of the cleaned parts. In practice, cleaning solutions containing silicates have considerably longer service lives than silicate-free solutions. During cleaning, a thin silicate film is adsorbed on the metal surface, which on the one hand reduces the corrosion of aluminium, zinc and brass in alkaline solutions, and on the other hand can impair the coating of steel (e.g. phosphating). Under no circumstances should they be used for cleaning in anodising lines, as the silicate acts as a very strong densification poison even in trace amounts. Silicate coatings must be reliably removed from the metal surface in a hydrofluoric acid dip process.

The strength of the silicate deposit and the inhibition can be controlled by the composition of the cleaner, more precisely by the SiO_2/Na_2O ratio.

Phosphates

The ortho-, pyro- and polyphosphates used in the technical cleaning products are among the most effective dispersants for inorganic dirt particles and enhance the effect of the surfactants. Due to their complexing properties, they take over the water softening in solution and also partially the removal of metal oxides and metal hydroxides from the substrate surface. Despite these excellent cleaning properties, phosphates have been substituted in many products today for ecological reasons.

Borates and Carbonates

Borates and carbonates are used as builders in mildly alkaline products. Alone or in combination with phosphates they stabilize the pH-value of the solutions in use. Through this buffering effect, a defined removal is set for aluminium surfaces. In addition to this control via the pH value, the attack on the metal is also inhibited by borates (conditioners).

Surfactants

Surfactants are surface-active compounds that migrate from a solution to its interface, accumulate there and reduce the interfacial tension. Surfaces that were originally non-wettable thus become water-wettable. Surfactants basically consist of a hydrophilic, i.e. water-soluble, and a hydrophobic, water-insoluble part of the molecule.

Depending on the charge of the hydrophilic part of the molecule, a distinction is made between anionic, cationic and non-ionic surfactants. The following criteria have to be considered for the selection:

- Degreasing capacity
- Foaming behaviour
- Electrolyte compatibility
- Adsorption and desorption on the surface
- Stability
- Emulsifying and demulsifying behaviour
- Biodegradability and toxicity

Anionic Surfactants

The best known representatives are the soaps – carboxylates, alkylaryl sulphonates and alkyl sulphates. Today, alkyl phosphates and alkyl ether phosphates are also gaining in importance. Anionic surfactants tend to form strong foam and have a good emulsifying capacity and thus a strong binding to oil. They are mainly used in immersion degreasing baths. Due to their good emulsifying capacity, practically no oil separation takes place on the bath surface.

Nonionic Surfactants

Both low-foaming and foaming surfactants as well as demulsifying and emulsifying surfactants are possible. The most important representatives are alkyl, alkylamine and fatty acid ethoxylates, fatty acid alkanolamides as well as polyhydroxy compounds (such as alkyl polyglycosides) or combinations thereof.

Non-ionic surfactants are the most common surface-active compounds in industrial cleaners. Since there are also low-foaming nonionic surfactants with good cleaning properties, this group of surfactants in particular is suitable for use in spray cleaners.

Cationic Surfactants

This group of surfactants is relatively rarely used in industrial cleaners because their cleaning effect is not very good. They are found in some neutral cleaners where their defoaming, demulsifying and antibacterial effects are desired.

Products such as quaternary ammonium compounds have a bactericidal effect and a high foaming capacity. Due to their strong adsorption capacity on hard surfaces, they are rarely used in metal cleaning, as they are difficult to remove in the rinsing zones and can thus have a negative influence on subsequent process steps. Cationic surfactants react with anionic surfactants to form inactive, poorly soluble compounds. This effect is used to regulate the foam of anionic surfactants and to break down emulsions (waste water treatment).

The majority of industrial cleaners contain not only a single surfactant, but a combination of different surfactants. In the case of fatty alcohols, for example, addition products with 5–6 moles of ethylene oxide serve as emulsifiers, with 8–12 moles of ethylene oxide as wetting and washing agents, and with 12–15 moles of ethylene oxide as dispersing and leveling agents. Thus, if the same cleaning solution is to emulsify, wash and disperse, then combinations of these different surfactants must be used. The optimum system cannot be found empirically by theoretical considerations, but only by extensive series of tests.

Complexing Agent

In order to avoid undesirable precipitation of hardness components and the associated staining, as well as to achieve activation of metal surfaces during cleaning, other complexing agents are used in addition to polyphosphates. The most important representatives of this class are alkanolamines, gluconates, polycarboxylic acids, polyoxycarboxylic acids and phosphonates.

Inhibitors

Pickling inhibitors are surface-active organic or inorganic substances. Organic inhibitors usually have electron-rich groups that interact with the metal surface. Due to electrostatic interactions, they coat the metal surface with an atomic layer and thus reduce the pickling attack. The strength of this interaction depends on the polarity of the metal surface, the nature of the inhibitor and the pickling medium. The polarity of the metal oxide layer is different from that of the bare metal, which leads to a much weaker occupation of the oxide layer and thus to its almost undiminished removal. Inhibitors can reduce both metal dissolution at active centres (anodic corrosion protection) and hydrogen evolution or oxygen reduction (cathodic corrosion protection).

5.4.3 Types of Cleaner

According to the respective ingredients and the pH value depending on them, a distinction is made between strongly alkaline, weakly alkaline, neutral, weakly acidic and strongly acidic products. Table 5.2 shows the special areas of application of the various cleaners.

The most effective cleaners are the **alkaline** to **strong alkaline** ones, which are particularly suitable for heavy cleaning tasks, such as the removal of aged oils and greases as well as pigment contamination. They are also used for fine cleaning before finishing processes, where metalic clean surfaces are required.

The **mildly alkaline** or **neutral cleaners** are intended for intermediate and final cleaning in process lines where mainly light staining is removed from non-cutting or machined parts and temporary corrosion protection is required after cleaning.

The weak acid products are based on a combination of acid phosphates and surfactants. In addition to cleaning, a phosphate layer is simultaneously created on iron surfaces (iron phosphating), which can be a basis for subsequent painting.

The strongly acidic products contain surfactants and mineral acids or strong organic acids. Metal attack is reduced by the addition of so-called pickling inhibitors. The main applications of this product class are the pickling of surfaces, the removal of oxides and scale as well as the removal of solid deposits in installations and piping systems.

Acidic cleaners/pickle degreasers consist of mineral acids and surfactants. They are used when no strong grease/oil layers are to be removed, when oxide layers as well as cast and rolled skins are to be removed and/or there is a risk of increased attack on the grain boundaries by alkaline products. Phosphoric acid and sulphuric acid are used as the basis for acid cleaners for aluminium. Hydrochloric acid is ruled out because of the risk of pitting corrosion even after intensive rinsing. The strongly oxidising nitric acid is also not

Table 5.2 List of the different types of cleaning and their areas of application

Class	pH value	Ingredients	Application
Strongly alkaline	11.5–13	Alkali; silicates; phosphates; complexing agents; surfactants	Steel; heavy staining; high cleaning requirements
Strongly alkaline	10.5–11.5	Alkali; silicates; phosphates; complexing agents; surfactants	Steel; light metals: Heavy staining; high cleaning requirements
Weakly alkaline	8–10	Phosphates; borates; carbonates; complexing agents; surfactants	Light metals; copper; zinc; weak contamination; high cleaning requirement
Neutral	7–8.5	Surfactants; corrosion inhibitors; phosphates; solubilizers	Sensitive surfaces; weak contamination; temporary corrosion protection
Weakly acidic	4–5	Acid salts; surfactants	Steel, plastics; alkali-sensitive workpieces; cleaning and Fe-phosphating
Acid	<1.5	Acids; inhibitors	Metals; acid dipping/pickling; pickling/degreasing

used, as the common wetting agent combinations are not resistant here. The phosphoric and sulphuric acids dissolve the natural oxide layer quickly and then attack the metal only moderately, although the hydrogen development is still sufficient to remove the pigment dirt. The attack on the aluminium can additionally be controlled by the addition of inhibitors.

5.4.4 Pickling

Contaminants that cannot be removed from the metal surface by the usual cleaning processes, e.g. oxide layers, are subjected to a pickling treatment. Table 5.3 gives an overview of the pickling processes used as part of the pretreatment before painting [25, 26]. Corroded iron surfaces can be treated most effectively in aqueous solutions of sulphuric acid or hydrochloric acid.

Table 5.3 Technically common used pickling processes for metal surfaces

Aqueous solutions of sulphuric acid	Concentration	10–15%
	Temperature	40–60 °C
	Pickle time	1–15 min
	Inhibitor content	0.1%
	Application for steel, aluminium (without inhibitor) and magnesium	
Aqueous solutions of hydrochloric acid	Concentration	10–15%
	Temperature	20–25 °C
	Pickle time	1–15 min
	Inhibitor content	0.1%
	Application for steel only	
Aqueous solutions of phosphoric acid	Concentration	2–15
	Temperature	20–60 °C
	Pickle time	5–20 min for steel, aluminium and magnesium
		1–5 min for zinc
Aqueous solutions of sodium hydroxide	Concentration	5–20
	Temperature	40–70 °C
	Pickle time	1–5 min for aluminium
Aqueous solutions of nitric acid	Concentration	10–15%
	Temperature	40–60 °C
	Pickle time	1–5 min for aluminium and magnesium

Due to the amphoteric nature of aluminium, both acids and alkalis are used for pickling. In practice, acid pickling based on nitric acid, hydrofluoric acid, hydrochloric acid, sulphuric acid and phosphoric acid as well as their mixtures are used. Since the bath tanks, piping and pumps for acid pickling must be made of acid-resistant material, the service life of the baths is low and thus the chemical costs are high, and waste water problems frequently occur, alkaline solutions are preferably used for pickling. Alloying components insoluble in alkalis, such as copper, silicon, zinc, must be removed in an additional acidic dipping step. Freshly pickled workpieces are very susceptible to corrosion and must be carefully rinsed and immediately provided with a conversion coating, e.g. phosphating or a Cr-free process. In the case of complexly designed workpieces, especially those which are spot-welded, have folds or other capillary cavities, sulphuric or hydrochloric acid may generally not be used for pickling . The quantities of acid remaining even with optimum rinsing would cause the formation of reaction products, which would lead to blooming and breakthroughs in the lacquer film on the finished part after only a short time. If the pickling of such complex shaped workpieces cannot be avoided, only the use of phosphoric acid pickling processes can be considered. Pickling with aqueous phosphoric acid solutions, however, requires more time and chemicals. Here, a thin layer consisting of iron phosphate is formed on the surface at the same time.

Galvanized parts do not need to be pickled. If pickling is nevertheless necessary, only phosphoric acid is used and the concentration is as low as possible (e.g. 2%).

5.4.5 Acid Dipping

The metal surface is activated in a acid dipping process. Thin surface films (oxides, carbonates, silicates, sulphides) are removed from the metal surface in order to give it good adhesion for surface treatment (conversion treatment, electroplating or painting). Iron, steel and zinc can be treated with dilute solutions of hydrochloric or sulphuric acid. Aluminium in diluted solutions of sulphuric or nitric acid.

5.4.6 Rinsing

Rinsing is the dilution of dissolved contamination (dirt, cleaner, acid, pigments, chips, etc.) on a cleaned surface.

Sufficient rinsing must be carried out between the individual treatment steps. As rinsing is a diffusion-controlled process, not only the amount of water, its purity and the temperature play a decisive role, but also the rinsing duration is an important factor. The type of rinsing technique, immersion or spraying, also influences the result.

5.4.7 Water Quality

If we assume that a cleaning bath contains a maximum of 5% cleaner and 95% water, then the water quality plays a significant role. Not all ingredients of the water are useful for cleaning and rinsing in industrial parts cleaning.

Possible ingredients in tap water and their consequences:

- Calcium, magnesium → higher salinity, lime stains
- Chloride, sulphate → higher salinity
- Chloride → Corrosion
- Sodium, potassium, nitrate, carbonate → higher salinity
- Different organics → Stains, foam

The water hardness or the total salt content plays a major role here. The total salt content of the baths is increased due to the continious dosing of tap water, evaporation and drag-out losses. Water hardness leads to deposition of lime, lime soaps, phosphates, silicates, surfactants and corrosion inhibitors. The consequences of this are:

- Mud in the tank and on the parts
- Stains after drying
- Reduction of the cleaning effect due to deposition of the builders and surfactants
- Corrosion attack on the material
- Insufficient cleaning
- Reduced corrosion protection

The chloride and sulphate contents, which are also increased due to salting, lead to corrosion of the parts and the installation or a reduction in the corrosion protection applied to the workpieces. The problem can only be solved by using fully demineralised water (deionised water).

In practice, two main processes are used to produce deionised water: Reverse osmosis plants and mixed bed ion exchange systems (anion and cation exchangers).

5.4.8 Electrolyte Maintenance

The task of bath maintenance is the removal of impurities from the baths and appropriate analytical methods for their monitoring.

There are a large number of factors influencing the service life of cleaning, rinsing and corrosion protection baths.

- Introduction of impurities from the processing components
- Oil, grease, pigments, chips, foreign ions, preservatives, organic sulphur and phosphorus compounds, surfactants, silicones, alkalinity, amines, complexing agents

- drag-out of bath solution, loss of cleaner and water
- higher salt concentration due to evaporation losses and subsequent dosing of tap water, carry-over, subsequent dosing of 1-component cleaners
- Chemical change of the bath
- Saponification of oils and fats, hydrolysis of components of processing aids, precipitation of lime, lime soaps, sulphides and phosphates, carbonation, dissolution of chips
- Biological alteration of the bath
- Surfactant and organic degradation (oil, fat) by microbiology, fungal, algal, bacterial growth, increasing bacterial count, increasing dead mass of microbiology

However, the main cause of poor cleaning results and short service lives is the type and quantity of oils, greases, cooling lubricants and particles introduced. Various systems are in use to remove these contaminants from the bath, such as filtration, oil separators, separators, membrane filtration (micro- to ultrafiltration) and evaporators. An overview of the most common impurities, their avoidance and removal from the bath as well as the corresponding analysis methods is shown in Table 5.4.

Table 5.4 Impurities in the pre-treatment and their prevention

Type	Avoid	Remove	Analysis method
Builder	Extend drip off time Avoid drying of residues	Ion exchanger Evaporator	
Surfactants	Extend drip off time Avoid drying of residues	Activated carbon filter Evaporator	COD value
Coolant Mineral and synthetic oils, greases and cooling lubricants	Continuous removal from the cleaning bath	Skimmer, oil separator, separators, centrifuge, membrane filtration, adsorption filtration, activated carbon filter	Analysis according to DIN 38409 COD value
Particle Chips, pigments, dust, metal abrasion	Continuous removal from the cleaning bath	Filters, magnetic separators, sedimentation tanks, membrane filtration, adsorption filtration	Filter bath solution Determination in accordance with VDA 19
Foreign ions	In practice not possible	Ion exchanger, evaporator, Reverse osmosis (depending on ion)	Chem. Analyses Detection reactions
Water contents	Use of deionised water	Ion exchanger, evaporator, Reverse osmosis (depending on ion type)	See foreign ions
Bacteria	In practice not possible, important is to avoid bacterial growth Short distances from VE plant to VBH; regular checks, preventive measures such as UV lamp, ozone dosing, etc.	UV lamp, ozone dosing activated carbon filter, Ultrafiltration to remove dead biological material	Bacteria test with appropriate test strips

5.5 Phosphating Process

Phosphating with its manifold possibilities will only be presented here in a very abbreviated form. Since there is extensive literature and experience of the manufacturers and many different phosphating baths are used, reference is also made to the literature [27].

The term phosphating refers to the production of water-insoluble metal phosphate crystals on metals. The formation is always associated with a pickling reaction of the base metal. Of technical and economic importance are the phosphating processes based on aqueous solutions, which can be applied by spraying or dipping in the appropriate solutions.

Depending on whether all components of the layer or only a part of it are supplied from the treatment solution, a distinction is made between layer-forming and non-layer-forming phosphating.

It can be formed very thin, or layers of a few microns thick. Most of them consist of zinc phosphate. The solution itself is slightly acidic.

The main components of a Zn phosphating process are:

- Phosphoric acid
- Heavy metal cations: Zinc Zn^{2+}, Manganese Mn^{2+}, Nickel Ni^{2+}, Iron Fe^{2+} or Fe^{3+}, Copper Cu^{2+}, Cobalt Co^{2+}.
- Alkaline earth cations: Calcium Ca^{2+}, Magnesium Mg^{2+}
- Accelerators: nitrates NaNO3, $Ni(NO_3)_2$, sodium nitrite $NaNO_2$, sodium chlorate $NaClO_3$, hydrogen peroxide H_2O_2, organic nitro compounds/hydroxylamine NH_2OH and hydroxylammonium sulfate $(NH_2OH)_2 \times H_2SO_4$
- Fluorides HF, NaF and hydrogen fluorides $NaHF_2$, NH_4HF_2
- Complex fluorides, SiF_6^{2-}, BF_4^-
- Organic carboxylic acids (tartaric acid, citric acid)

The main components of an Fe-phosphating process are:

- Phosphoric acid
- Alkali phosphate, Na/ KH_2PO_4 or
- Ammonium phosphate, $NH_4H_2PO_4$
- Accelerators, e.g. nitrates ($NaNO_3$), sodium nitrite ($NaNO_2$), sodium chlorate ($NaClO_3$), sodium bromate ($NaBrO_3$), organic nitro compounds, molybdates (Na_2MoO_4)
- Fluorides (HF, NaF)
- Surface active substances

Table 5.5 shows the different types of phosphating and their suitability as a paint primer on the metal surfaces steel, galvanised steel and aluminium.

Table 5.5 Suitability of phosphatings as paint adhesion primers for various metal surfaces

Procedure type	Metal to be treated		
	Steel	Galvanized steel	Aluminium
Fe phosphating	•	•	*
Zn phosphating	•	•	•
Thick film phosphating	•	•	•
Mn modified thick film phosphate coating	•	•	•
Mn-containing Zn-phosphating	•	–	–
Zinc calcium phosphating	•	o	–

Explanations: • suitable; o conditionally suitable; – not suitable; * cleaning only

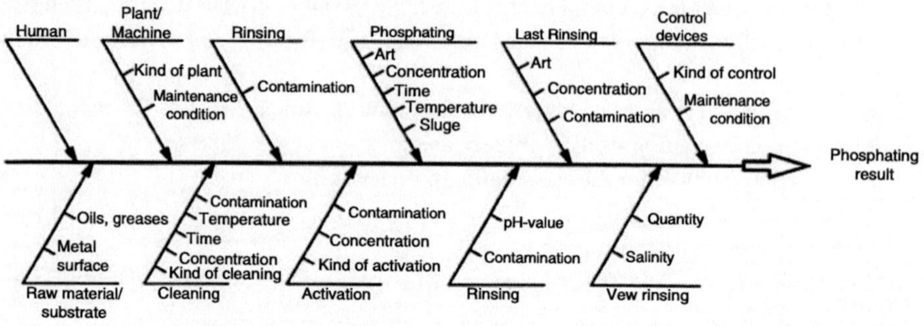

Fig. 5.4 Factors influencing the phosphating result

The result and quality of phosphating depend on a large number of details and influencing factors, such as the starting material, the cleaning, the rinsing process through to the testing and control units. Figure 5.4 shows the various influencing factors on the way to the phosphating result.

5.5.1 Layer-Forming Phosphatisation

The layer-forming phosphating is carried out from diluted, aqueous solutions which contain all the essential components required to build up the phosphate layer. Of the numerous phosphating systems, the following have attained technical importance as pretreatment prior to painting: zinc phosphate (Znph), zinc iron phosphate (ZnFeph) and zinc manganese phosphate (ZnMnph).

5.5.1.1 Reactions during Layer Formation

A prerequisite for the formation of layers during phosphating is the pickling or corrosion reaction, i.e. the attack of the metal surface by H+ ions.

On steel and iron	$Fe + 2H^+ \rightarrow Fe^{++} + H_2$
On zinc (e.g. galvanised steel)	$Zn + 2H^+ \rightarrow Zn^{++} + H_2$

On aluminum:

1. Dissolving the oxide skin	$Al_2O_3 + 6H^+ \rightarrow 2Al^{3+} + 3H_2O$
2. Pickling attack of the metallic Al	$Al + 3H^+ \rightarrow Al^{3+} + 1½\ H_2$
3. Complexation of the Al3+ by fluoride	$Al^{3+} + 3F^- \rightarrow AlF_3$
4. Separation of the complex as cryolite	$AlF_3 + 2F^- + 3Na^+ \rightarrow Na_3AlF_6 \downarrow$ ↓ Cryolite (sludge)

The layer formation can be represented by the following reactions:

a) on steel and iron

$$3Zn^{++} + 2H_2PO_4^- + 4H_2O \rightarrow Zn_3(PO_4)_2 \times 4H_2O \downarrow + 4H^+$$
$$2Zn^{++} + Fe^{++} + 2H_2PO_4^- + 4H_2O \rightarrow Zn_2Fe(PO_4)_2 \times 4H_2O \downarrow + 4H^+$$
$$Zn^{++} + 2Mn^{++} + 2H_2PO_4^- + 4H_2O \rightarrow Mn_2Zn(PO_4)_2 \times 4H_2O \downarrow + 4H^+$$

b) on zinc and aluminium

$$3Zn^{++} + 2H_2PO_4^- + 4H_2O \rightarrow Zn_3(PO_4)_2 \times 4H_2O \downarrow + 4H^+$$
$$2Mn^{++} + Zn^{++} + 2H_2PO_4^- + 4H_2O \rightarrow Mn_2Zn(PO_4)_2 \times 4H_2O \downarrow + 4H^+$$

The so-called accelerators are usually oxidizing agents. They have two tasks in a phosphating process:

1. The oxidizing agents accelerate the pickling reaction by removing the mechanical blockage of the metal surface by hydrogen bubbles, which are oxidized to water:

$$H_2(\uparrow) + \langle O \rangle \rightarrow H_2O$$

2. The metal dissolved in the pickling reaction (e.g. iron in the phosphating of steel) is oxidised, in this case to trivalent iron, which then reacts with phosphate ions to form iron phosphate, which is difficult to dissolve, and separation (sludge formation).

$$2Fe^{++} + 2H^{+} + \langle O \rangle \rightarrow 2Fe^{3+} + H_2O$$
$$2Fe^{3+} + 2H_3PO_4 \rightarrow 2FePO_4 \downarrow + 6H^{+}$$

The zinc phosphate layers produced on the surface of steel, zinc and also aluminium are firmly interlinked to the substrate metal. This type of bonding between the substrate and the coating deposited on it, known as epitaxy (preferred orientation of the Znph crystals to the substrate crystal lattice), is considered to be one of the main causes of the corrosion-protective properties of phosphate coatings.

The phosphate coating must have a uniform fine crystalline structure over the entire surface of the workpiece, with a uniformly low weight per unit area, so that optimum conditions exist for the following organic coating.

This can be achieved by:

- Selection of a suitable process adapted to the specifics of the organic coating used as well as the metallic material
- Compliance with the specified process conditions in each individual section of the pretreatment (electrolyte temperature, concentrations of the components, treatment times, spray pressures or flow conditions).

Figures 5.5, 5.6, 5.7, 5.8, 5.9, 5.10, 5.11 and 5.12 show SEM images of a zinc phosphate coating on various metals prepared by spraying and dipping.

Fig. 5.5 Substrate: Cold rolled steel, GB MBS 37; process: Spraying, layer weight 2.3 g/m²

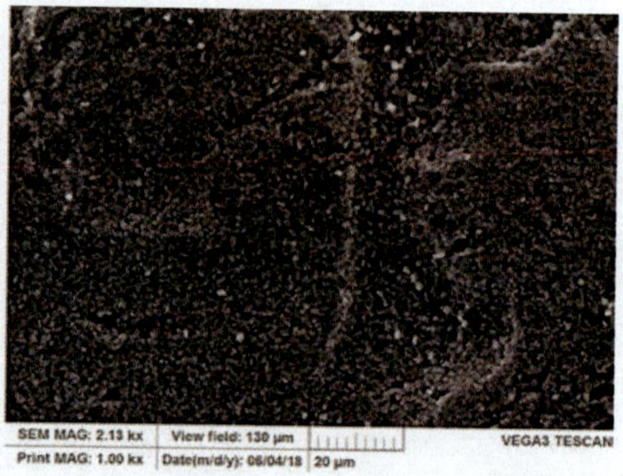

Fig. 5.6 Substrate: Cold rolled steel, GB MBS 37; process: Dipping, layer weight 1.9 g/m²

Fig. 5.7 Substrate: Hot-dip galvanised steel; GB MBZ 1, process: Spraying; coating weight 3.6 g/m^2

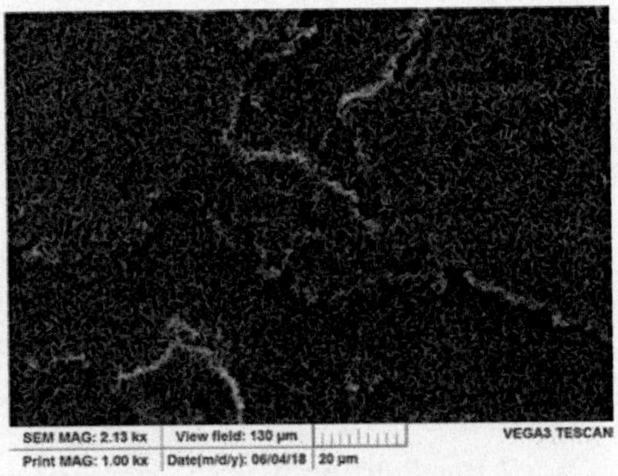

Fig. 5.8 Substrate: Hot-dip galvanised steel; GB MBZ 1, process: Dipping Layer weight 2.0 g/m^2

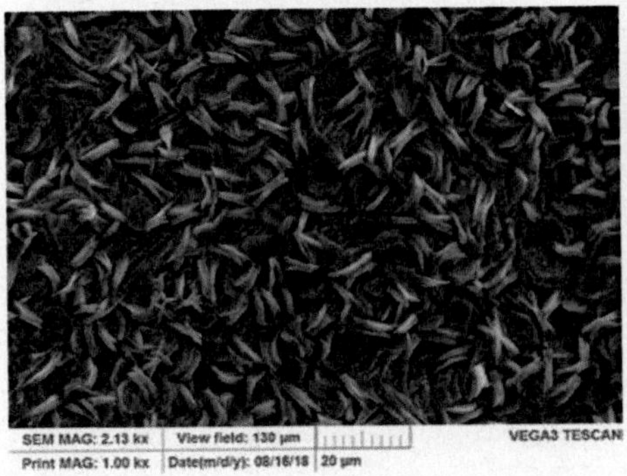

Fig. 5.9 Substrate: electrogalvanised steel, GB MBZE 13, process: Spraying; coating weight 3.9 g/m^2

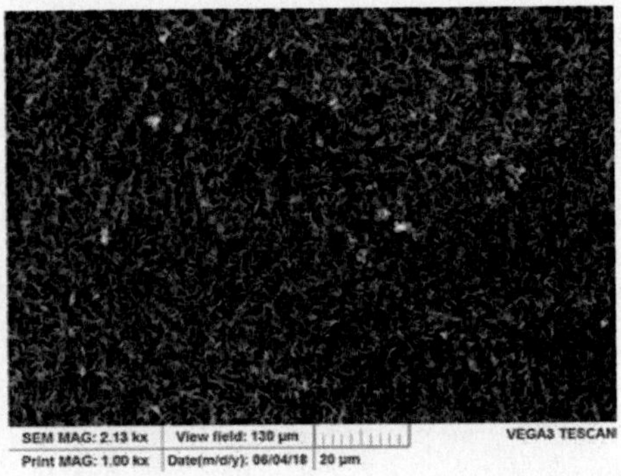

Fig. 5.10 Substrate: electrogalvanised steel, GB MBZE 13, process: Dipping; coating weight 2.2 g/m^2

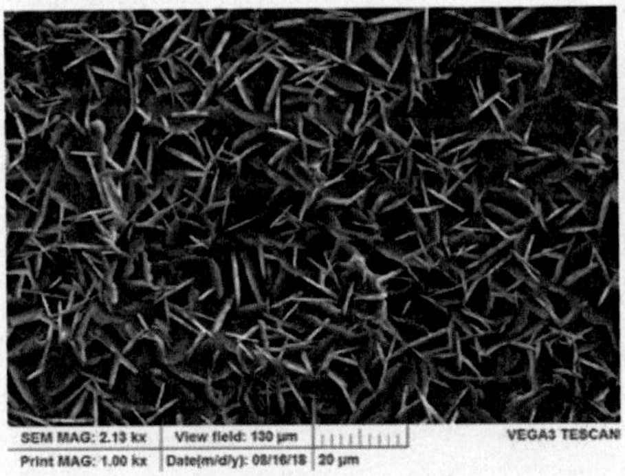

Fig. 5.11 Substrate: Aluminium; GB AA6014/2; spray process; coating weight 2.3 g/m^2

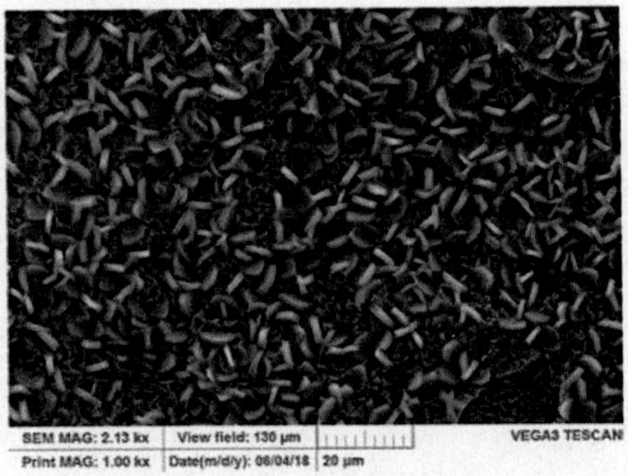

Fig. 5.12 Substrate: aluminium; GB AA6014/2; process: Dipping; layer weight 3.1 g/m^2

5.5.1.2 Activation of Layer Formation

Activation of the metal surface prior to phosphating plays an important role in achieving fine crystalline phosphate coatings. By treating the workpieces with special aqueous solutions containing titanium phosphate, a particularly favourable condition for phosphating is created on the metal surface. The number of places where crystallization nuclei can form is thus increased. As a result, crystallization begins in the form of a much denser network during subsequent phosphating. It is also achieved that a closed phosphate layer is formed more quickly, whereby the crystal size achieved in the final state and thus also the thickness of the coating remain small, see Fig. 5.13.

5.5.1.3 Passivating After-Treatment of the Zinc phosphate Coating

The anti-corrosive properties of the coating may be strongly influenced by the type and manner of post-treatment. In any case, it is important to remove all water-soluble substances from the metal surface by sufficiently rinsing the phosphated workpieces with water.

In order to avoid drying of rinsing water or carry-over from the baths, fully demineralised water should be used in the final rinsing stage.

Rinsing the phosphated surfaces with Cr-free aqueous solutions can significantly improve the corrosion protection of the painted parts. These systems are mostly based on titanium and/or zirconium compounds.

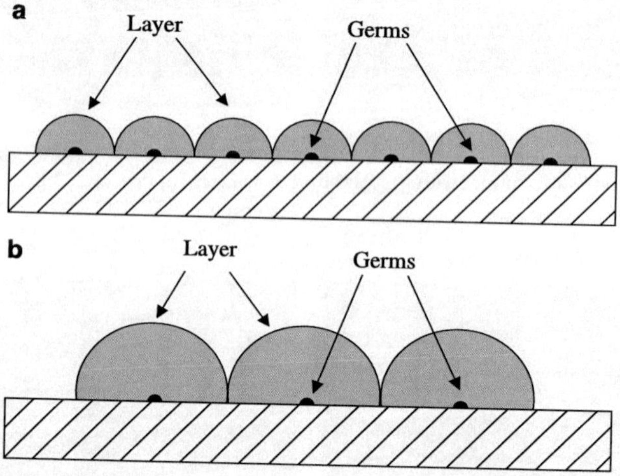

Fig. 5.13 Influence of the crystallisation nuclei on the metal surface on the thickness and roughness of the phosphate layer formed. (**a**) Metal surface with many nuclei = rapid formation of a fine crystalline, thin layer. (**b**) Metal surface with few nuclei = slow formation of a coarsely crystalline, thick layer

5.5.2 Iron Phosphating

The term "non-layer-forming" phosphating, which is frequently used in linguistic usage and in the literature, is misleading in that top layers are also formed in these processes. The coatings produced are very thin in comparison with the layer-forming Znph processes and generally have a weight per unit area of significantly <1 g/m². Since the coating type on steel consists of iron phosphate, these processes are also very often referred to as "iron phosphating".

The difference to the layer-forming processes is that only the anion, the PO_4^{3-} ion, is contained in the electrolyte solutions used. The cation is supplied by the metal to be treated itself.

An improved corrosion protection compared to the "simple" iron phosphatings are the so-called thick layer iron phosphatings [28].

5.5.2.1 Reaction at the Surface

The main components of the treatment solution for "non-layer-forming" phosphating are alkali or ammonium dihydrogen phosphates. The pH value of these solutions is in the range of 3.5–6.0.

Here, too, a distinction is made between two reaction steps:

1. the pickling and corrosion reaction
2. the film formation reaction

Pickling reaction, e.g. on steel:

$$Fe + 4NaH_2PO_4 \rightarrow Fe(H_2PO_4)_2 + 2Na_2HPO_4 + H_2$$

Film forming reaction:

$$2Fe(H_2PO_4)_2 + 2Na_2HPO_4 + \langle O \rangle \rightarrow 2FePO_4 \downarrow + 4NaH_2PO_4 + H_2O$$

$$2Fe(H_2PO_4)_2 + 4Na_2HPO_4 + \langle O \rangle + 5H_2O \rightarrow Fe_2O_3 \times H_2O \downarrow + 8NaH_2PO_4$$

In addition to the layer types mentioned here, other phases can be detected. These are, for example, iron(II) phosphate and iron(II) oxide. Thin zinc phosphate coatings are obtained on zinc surfaces, whereas no coating is formed on aluminium.

5.5.2.2 Composition of Alkali Phosphate Electrolytes

The treatment solutions for non-layer forming phosphating generally contain 2–20 g/l alkali or ammonium dihydrogen phosphate. Phosphoric acid or dialkali or diammonium phosphate is used to adjust the pH value. In addition to the phosphates, the solutions also contain oxidizing agents to accelerate the layer formation. The most common oxidizing agents are chlorates, bromates, nitrates, molybdates or nitrites. The acidic alkali phosphate solutions are used together with surface-active substances (surfactants). These solutions have a good degreasing effect. The surfactant concentration is between 0.1 and 1 g/l.

5.5.3 Iron Thick Film Phosphating

Iron phosphating is frequently used as a pre-treatment before coating steel and galvanized steel due to its simple application and comparatively low-cost process technology. Phosphating can be carried out in 1-chamber washing plants up to multi-zone continuous plants. Due to the higher coating weight, 0.4–1.0 g/m², a higher corrosion protection is achieved than with the classical iron phosphatings (coating weight 0.1–0.4 g/m²) [28].

The advantages of this pre-treatment can be summarized as follows:

- Very good corrosion protection properties on steel; with additional sealing of the layer, e.g. a chromium-free passivation, this can be further improved.
- Simple bath control via pH value, concentration and conductance.
- Degreasing and iron phosphating can be combined in one process stage. A separate alkaline pre-degreasing like in zinc phosphating is not mandatory.

In order to benefit from the advantages of this pretreatment with regard to corrosion protection, a final rinsing step with deionised water, conductivity <50 µS/cm, must be carried out. If higher corrosion protection values are required, it is also necessary to passivate the iron phosphate layer formed with chromium-free passivations (Table 5.6).

Table 5.6 Layer weights of the different phosphatings

Procedure	Layer weight [g/m²]
Fe phosphating	0.4–1.0
Zn phosphating	2–6
Thick film phosphating	6–40
Zn-Mn phosphating	2–4
Zn-ca phosphating	2–5
Mn phosphating	4–20

5.5.4 Methods for the Characterisation of Phosphate Layers

The characterization of phosphate coatings may be required for a variety of reasons. In practice, it is mostly used for production control, whereby the layer weight is determined as the characteristic feature. In practice, the coating thickness measurement is used in some cases, but this involves considerable inaccuracies..

The following criteria are used to characterize the phosphate layer:

- Uniformity of the formed layer
- Complete coverage of the metal surface with phosphate crystals
- Layer weight in g/m^2 – Crystal size
- Phase content
- Carbon accumulation in the steel surface
- Paint adhesion and corrosion protection behaviour of the phosphate coatings

5.5.4.1 Uniformity of the Phosphate Layer/Coverage of the Metal Surface

Unevenly phosphated surfaces are usually obtained when the pre-treatment before phosphating is not correct performed. Insufficient cleaning or partial activation of the surface by wiping can lead to non-uniformity. Layers that are not completely closed are the result of poor bath management.

The uniformity of the formed layers as well as the complete coverage of the metal surface with phosphate crystals is determined by visual assessment. The irregularities of the phosphate layers may still be noticeable later on the painted surface. Poorly phosphated surfaces downgrade the corrosion properties of the painted workpiece.

5.5.4.2 The Mass Per Unit Area (Layer Weight)

The quantitative determination of the layers deposited on the metal surface is best done by determining the layer weight in grams (g) of phosphate layer per m^2 of metal surface·

The phosphate layer is chemically removed from a sample of phosphated steel sheet and the resulting weight loss is determined by weighing the sample before and after removal. The weight of the phosphate layer influences the corrosion behaviour of the painted surface. When using conventional coatings (e.g. water-based dip coatings), it has been shown that the infiltration of the coating film in the salt spray test decreases with increasing coating weight of the phosphate layer. When electrodeposition coatings are used, this tendency cannot be clearly proven. While the corrosion resistance increases with increasing film weight, the paint adhesion decreases. This is due to the brittleness of the phosphate layer, which is more noticeable with thicker layers. When strongly deformed, the relatively brittle phosphate layer breaks, and the coating on top cracks and flakes off. The coating weight is generally in the range of 1–40 g/m^2, see Table 5.6. in order to obtain good corrosion resistance and paint adhesion. Higher coating weights >20 g/m² are generally coarsely crystalline and unsuitable for painting.

5.5.4.3 Crystal Size and Phase Composition

The layer formation and crystal size of a phosphate layer is determined by scanning electron microscopy (SEM).

The investigation of the phase composition provides information about the adhesive bond between metal and phosphate layer and can be used to explain the different corrosion protection of the various layers. For the investigation of zinc phosphate layers, the X-ray diffraction method (X-ray diffractometry) has established itself as a successful procedure.

With low zinc phosphating processes, phosphate layers with phosphophyllite contents ($Zn_2Fe(PO_4)_2 - 4H_2O$) of 60 to over 90% are achieved on steel by spraying and between 80 and 100% by dipping. The remainder in each case is hopeite ($Zn_3(PO_4)_2 - 4H_2O$).

The further development of the low-zinc phosphating processes led to the manganese-modified systems (trication phosphating processes), which deposit manganese zinc phosphate ($Mn_2Zn(PO_4)_2 - 4H_2O$) in addition to the phases mentioned above. This further improved the quality of the metal/phosphate coating/organic coating composite.

5.5.4.4 Carbon Accumulation in the Steel Surface

Carbon accumulation in the steel surface, e.g. from the decomposition of rolling oil residues during annealing in a closed coil, greatly impairs the corrosion protection of applied phosphate coatings in conjunction with a paint finish. One of the methods used to determine the surface carbon is the Ford power wash test. The principle of the method is that the steel sheet to be tested is subjected to a standardised cleaning method. The surface is then rubbed with a glass fleece moistened in diluted hydrochloric acid until no more visible dirt is transferred from the steel surface.

The collected glass fleeces are burnt and the surface carbon is inferred from the CO_2 formed. Experience has shown that, for example, good salt spray resistance is obtained on phosphated and painted steel sheets when the surface carbon is <4 mg/m². Poorer, but still acceptable protection is observed at 4–6 mg/m² C; values above 6 mg/m² C generally lead to unsatisfactory salt spray resistance.

5.5.4.5 Application Tests

The most important tests on phosphate coatings with regard to their corrosion protection behaviour under paint and as an adhesion primer are the application tests (corrosionand technological tests of painted workpieces). There are a large number of tests of various kinds. Various corrosion tests and adhesion tests are carried out both for production control and for checking development work, see Chap. 7.

5.5.5 Defects and Failure Prevention During Phosphating

Errors can occur during phosphating if the determining process parameters are not kept within the target range. In addition to increased chemical consumption and sludge accumulation, this results in poorly formed phosphate layers and less corrosion protection and paint adhesion.

The following parameters, make a decisive contribution to the smooth running of the phosphating process and the achievement of the desired quality level:

- Type and condition of the surface to be phosphated
- The organic-chemical impurities present on the metal surface (anti-corrosion oil, lubricants, metal abrasion)
- Type, condition and application conditions of the surface treatment process steps (cleaning, rinsing, activating, phosphating, rinsing, post-rinsing)
- Design and maintenance condition of the surface treatment plant
- Activation

The establishment and application of checklists, automatic bath analysis and statistical process control have proven to be effective tools for quality assurance.

5.5.5.1 Requirements for the Metal Surface Before Phosphating

Cold rolled steel strip or steel sheet in deep drawing quality is one of the most important materials for the application of phosphate coatings before painting (DIN EN 10130). For visible surfaces, surface type 05 from this standard should be selected, as only this is free from mechanical surface defects, e.g. scratches and scars, and the appearance of a paint finish is therefore not affected.. The surface finish usually used is quality “m”, which corresponds to a uniformly matt surface with a centre roughness Ra of 0.6–1.9 μm. Smoother surfaces often require more aggressive phosphating conditions to obtain flawless phosphate films, while rougher ones reduce the gloss of the paint film. In addition, practice has shown that the phosphatability of steel sheet, i.e. the rate of formation of the phosphate layer and its crystal size, can be influenced by the condition of the steel surface. Accumulations of carbon in the steel surface can reduce the corrosion protection in phosphated and painted condition. The surface carbon should be below 6 mg/m^2, preferably below 4 mg/m^2.

Measures to avoid defects include roughness measurement, determination of the surface carbon and sample phosphating with visual assessment of the phosphate layer.

The phosphating of hot-dip and electrolytically galvanized steel usually does not cause any problems as long as only oiled and non-passivated material is used. Thicker zinc oxide films, which may occur on hot-dip galvanized steel strip, require appropriate adjustment of the alkaline cleaner and/or the use of a fluoride-containing phosphating system.

Metal castings have a casting skin that requires more aggressive processes for proper phosphating.

5.5.5.2 Phosphating Process Sequence

The chemical surface treatment is divided into the steps cleaning, rinsing, activating (if necessary), phosphating/rinsing, passivating rinsing with fully demineralized water, drying (if necessary).

Each of these stages has its own potential for mistakes, as illustrated below.

5.5.5.3 Alkaline Cleaning

The aqueous alkaline cleaning baths are prepared at a concentration of between 3 and 30 g/l in water, the lower values being for spraying and the higher for dipping. The concentration is determined, for example, by titrating a bath sample with acid against bromocresol green. Fresh cleaner concentrate in liquid or solid form must be added to the bath to compensate consumption during operation. A drop in the concentration reduces the cleaning performance and thus the quality of the produced phosphate coatings (uniformity, degree of coverage, adhesion and, derived from this, paint flow, paint adhesion and resistance to paint infiltration).

Although keeping the cleaner concentration constant is a necessary requirement for a good treatment result, but it is not sufficient as it does not allow any statement to be made about the extent of the impurities absorbed and the possible deterioration of the cleaning performance caused by this. For this reason, when a plant is commissioned, a visual assessment of the cleaning result (wetting of the cleaned surface with water, appearance of the phosphate layers) is used to determine the rhythm at which the cleaner bath must be partially or completely drained and restarted.

To maintain the cleaning bath and extend its service life, processes are available to remove the impurities that have been absorbed. These include filtration, ultrafiltration, flotation, centrifugation and settling or floating.

The bath temperature and the treatment time also have a high influence on the cleaning performance. A reduction in both parameters leads to a degradation in quality, whereby a reduction in temperature in the spraying process can also increase the formation of foam. An temperature reduction already occurs, for example, when evaporation losses are compensated by the supply of larger quantities of cold water. Excessive foaming during spraying can also develop if the circulation pump sucks in air because the liquid level is too low or if the impurities introduced contain strong foaming surfactants. Special attention should be paid to this when selecting the oil.

5.5.5.4 Rinsing

The rinse after cleaning should be run in such a way that the rinse water in the last rinse bath before a subsequent treatment zone (activation or phosphating) does not exceed a certain alkali content resulting from the cleaner input. A typical maximum value is 0.5 ml of 0.1 M hydrochloric acid for neutralising a 10 ml rinse bath sample against bromocresol green. Higher alkali contents can lead to an impairment of the function of the following treatment bath and must therefore be avoided.

5.5.5.5 Activation

The pH value of this bath must not fall below 8, otherwise the activation effect will decrease abruptly.. The titanium phosphate colloid of the activation also shows a gradual loss of quality in the course of the service life, which is additionally influenced by the content of hardness formers in the water and the application temperature. The titanium concentration in the activation bath can be determined quantitatively. However, the presence of the target amount of titanium does not guarantee that the activation effect will occur to the required extent. This is mainly dependent on the colloidal chemical condition of the system. The most reliable method for determining the activation effect is therefore the phosphating test and the evaluation of the layer formation. A common method of operation is to reject the treatment bath continuously or from time to time partially or fully.

5.5.5.6 Phosphatiing Bath

a) Zinc content

Low-zinc phosphating processes provide phosphate coatings with a high proportion of phosphophyllite ($Zn_2Fe(PO_4)_2 \cdot 4H_2O$) on steel, even when sprayed, which is the basic conditions for the good application properties in conjunction with paints. The amount of phosphophyllite in the phosphate layer is largely determined by the zinc concentration in the phosphating bath. In order to achieve a stable, good quality of the phosphate layers, the zinc concentration must be kept in the range of about 0.5–1.0 g/l.. At lower contents, qualitatively poorer mixed layers of crystalline phosphates and iron oxide iron phosphate are formed, which are immediately noticeable by their iridescent colour when visually assessed. Higher zinc contents provide visually correct coatings, although their quality is clearly inferior in terms of paint adhesion and resistance to infiltration. Optimal work with this type of process requires frequent control of the zinc content of the baths.

b) Free acid FA

Below FA = 0.7, the bath is too oversaturated, so that more zinc phosphate precipitates as bath sludge, while above about FA = 1 the phosphating rate drops and increasingly non-covering, qualitatively insufficient phosphate layers are formed.Accelerator

c) Accelerator

The content of nitrite accelerator has a high influence on the layer formation and the phosphating rate. It is usually given in accelerator points, which are equal to the ml consumption of 0.02 M $KMnO_4$ for the titration of a 100 ml bath sample. The coverage of the phosphate layer increases with increasing spray time and accelerator content. In practice, however, the accelerator content should be kept as low as possible, as its overdosage leads to increased chemical consumption, increased sludge formation and, in extreme cases, passivation.

d) Basis weight

In phosphating technology, the course of the coating weight/treatment time curve is often used to characterize a phosphating system or to describe the phosphating

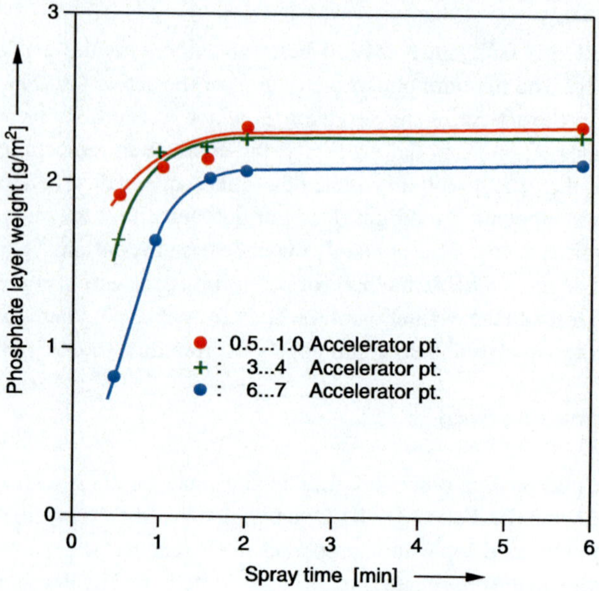

Fig. 5.14 Layer weight/time curves of a low zinc phosphating process on steel with spray application [26]

behavior of metal substrates. In Fig. 5.14, the coating weights on steel for different accelerator contents are shown against the spray time. The reactivity of the steel surfaces and the condition also have an effect on the film formation. The final coating weight and the minimum phosphating time (treatment time until a visually covering phosphate layer is formed) are significantly higher for steel A than for steel B, Fig. 5.15.

5.5.5.7 Treatment Temperature

Of the physical bath parameters, the treatment temperature in particular has a high influence on the phosphating behaviour. On the one hand, the phosphating bath is characterized by a temperature-dependent equilibrium of the free acid to the total acid, which means that every major temperature change requires an adjustment of this ratio. The higher (lower) the bath temperature, the greater (smaller) the acid ratio must be to keep the system in the optimum range. Secondly, the temperature still practices an influence on the reaction rate of the phosphating process, where a reduction in temperature equivalent to a decrease in reaction speed.

The optimum temperature range for a perfect coating formation is between 50 and 65 °C. Today, there are also phosphate coatings that form opaque layers at 40 °C already. Below the recommended temperature range, phosphate coatings with a worse coverage are increasingly obtained. With higher tempertures, beyond 65 °C coatings with good coverage can still be obtained.

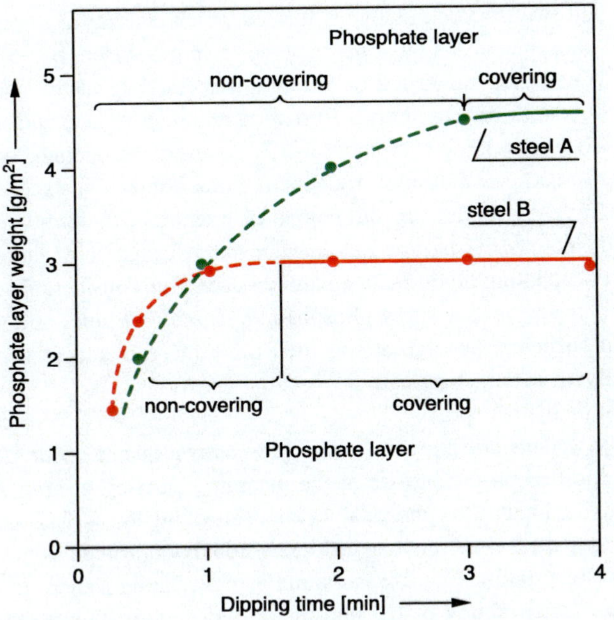

Fig. 5.15 Layer weight/time curves for two different grades of body steel sheet in a low zinc immersion phosphating bath [26]

If the recommended temperature range is exceeded, the electrolyte forms more sludge and deposits, consumes more chemicals and causes increased expenditure for system cleaning.

a) Bath poisons

Phosphate layer formation can be considerably impaired by bath poisons. A typical example is the aluminium cation (Al^{3+}), which in a spray phosphating bath already at 3 mg/l causes only tarnishing instead of a crystalline, covering layer. The harmful effect can only be eliminated by special measures, which include in particular the co-use of fluoride in the phosphating bath. When treating galvanised steel, the critical limit for Al^{3+} in fluoride-free baths is >10 mg/l Al.

In addition to the parameters already mentioned, the total acid, other cations (except zinc), phosphate, other oxidants (except nitrite), activators (fluoride, complex fluoride) and, if necessary, additives for influencing the layer and sludge should be mentioned as additional influencing variables for the phosphating process. There are dependencies in the concentration in the treatment bath and the effect on the phosphating process, the phosphate layer quality and the bath behavior.

b) Sluge and crust formation

During phosphatisation, a sludge is formed which consists to a considerable extent of iron(III) phosphate. This must be removed from the system by sedimentation (settling tank, lamella separator) and/or filtration. Excessive levels of sludge dispersed in the bath will lead to increased crust formation and interference with phosphatisation. For this reason, desludging equipment should be sized and operated so that the sludge content in the bath, as measured by Imhoff funnel, does not exceed approximately 10 ml/l (0.25–1 g/l solids).. In the course of operation, the cross-sections of spray nozzles and pipelines can narrow so much due to the formation of sludge crusts that the amount of phosphating liquid pumped around decreases considerably and is no longer sufficient to ensure proper spray phosphating. Therefore, pipes and nozzles must be checked for sufficient passage at specified time intervals and, if necessary, cleaned mechanically or, better, chemically.

c) Pre-passivation by overspray

In spraying plants, the phenomenon of pre-passivation can occur, if droplets or thin films are deposited on the surface of the incoming parts by pre-misting phosphating solution into the inlet of the phosphating zone, attacking the surface and, in the case of steel, leading to the formation of an iridescent iron oxide-iron phosphate layer. At these points, proper crystalline film formation in the phosphating zone is no longer possible, resulting in a deterioration of the phosphate film quality (flow disturbances, poorer paint adhesion and reduced corrosion protection).

5.5.5.8 Passive Post Rinsing

In order to optimize the phosphate layer quality, a passivating post-rinse is usually used after the last rinse. Today, mainly Cr-free passivations based on zirconium are used. The active ingredient content in the post-rinse bath must be determined analytically and, if necessary, supplemented by adding concentrate. Underconcentration and contamination of the bath by carry-over of electrolyte from the preceding process stages must be avoided for quality reasons. An effective measure for this is to run the bath with continuous overflow and to add fresh electrolyte.

5.5.5.9 Rinsing with Demineralised Water

To remove residual water-soluble salts from the phosphated surface, a final rinse with fully demineralised water takes place. The quantity and quality of this water must be such that the liquid running off the treated workpieces has an electrical conductivity <50μS/m (20 °C).

If a powder or wet paint is applied, the residual salts on the surface can have a negative effect on the adhesion and resistance of the paint films.

The possibility of water droplets from the transport system or from flanges and folds running onto the partially or completely dried phosphate coating must be excluded by suitable measures. Otherwise, disturbances in the deposited paint film (overlays, runners, etc.) may occur during e-coating or, in the case of powder or solvent-based paints, reduced humidity resistance (formation of bubbles in the paint film).

5.5.5.10 Control of Phosphate Layers

Properties that can be used to control and characterize the phosphate layers include:

- Basis weight (g/m^2)
- Crystal dimensions (SEM)
- Crystal density (SEM)
- Coverage
- Appearance (colour, uniformity, crystallinity)
- Roughness
- Chemical composition
- Crystal phases

The weight per unit area is a simple quantity to determine, which depends on the properties of the metal substrate and on the chemical surface treatment. Very low basis weights can be caused by coverage gaps, and cause a reduction of the corrosion protection.

Too high layer weights mean increased chemical consumption and lower paint adhesion under bending stress. This also applies to the crystal dimensions.

The crystal density is a parameter derived from the crystal dimensions. Incomplete coverage of the surface, i.e. high porosity, means poorer corrosion protection. A simple but effective coating control is done visually. The coating must be uniformly gray, fine crystalline and covering, the bare metal must not show through. In the zinc phosphate processes usually used as paint pretreatment, the macro-roughness of the phosphated surface is mainly determined by the roughness of the sheet metal substrate, to which the micro-roughness of the phosphate coating is added. Reduced solubility of the phosphate layer in caustic solutions is normally associated with better corrosion protection in conjunction with paint. Higher contents of iron (5–8%) in the layer mean quality improvement due to the presence of phosphophyllite.

Coatings with 4–8% Mn are formed in manganese-modified low-zinc phosphating baths.In contrast, the nickel content of the layer, even at bath concentrations, is comparatively low at about 0.5–1.3%. X-ray diffractometry allows the determination of phosphophyllite and hopeite on the metal surface. The manganese modification of the low-zinc phosphating process results in the formation of layers which, compared to the manganese-free layer, only lose water of crystallization at higher temperatures.

5.6 Chromating

The term "chromating" covers the chemical or electrochemical processes in which metal surfaces are brought into contact with solutions containing chromic acid by dipping, spraying or brushing, so that inorganic top layers are formed. These layers impart a degree of corrosion protection and improve the adhesion of organic coatings.

Since the coatings are formed by a chemical reaction between the metal and the treatment solution, in a sense "converting" the metal surface, chromate conversion coatings are referred to in English and American literature as "conversion coatings". Solutions that contain only trivalent chromium instead of hexavalent chromium are by definition not counted among the chromating processes.

Especially with aluminium and zinc, yellow and green chromating were very well suited as adhesion promoter and corrosion protection layer before painting with very good corrosion properties.

The chromium trioxide contained in yellow and green chromates falls under REACH Annex XIV, which means that its use is generally prohibited after September 21, 2017, unless an authorization or exemption has been granted. Today, the classic chromates are no longer important in the area of pre-treatment before painting.

As of September 21, 2017, the following substances will no longer be manufactured or articles produced with them [29]:

- Ammonium dichromates (CAS: 7789-09-5)
- Chromic acid (CAS: 7738-94-5)
- Dichromic acid (CAS: 13530-68-2)
- Chromium trioxide (CAS: 1330-82-0)
- Potassium chromate (CAS: 7789-00-6)
- Potassium dichromate (CAS: 7778-50-9)
- Sodium chromate (CAS: 7775-11-3)
- Sodium dichromate (CAS: 10588-01-9, 7789-12-0)

Exceptions to these rules are:

a. Authorisation has been requested from the manufacturer/user. The substance can be used until the decision is made and beyond if the decision is positive (the authorisation is not generally valid, only participating companies can use it).
b. Products containing this substance can be imported from outside the EU, not the regulated substance itself.

Exemptions have been requested, by the electroplating industry and the aeronautical industry.

5.6.1 Green Chromating: CrIII

Green chromate layers are obtained in solutions of chromic acid, fluorides and phosphoric acid. The protective layers that form consist mainly of phosphates of chromium and aluminium. They are free of toxic Cr^{6+}compounds in the layer. They form a good adhesion primer as well as a reliable diffusion barrier for water vapor permeable organic coatings and their application loads [26, 30–35]. The layers are formed within a few seconds up to 5 min at temperatures up to 50 °C. The greenish iridescent color of the coating is due to the $CrPO_4$ formed. No-rinse processes in roller application operate completely without waste water [36–41].

5.6.2 Yellow Chromate Layers: CrVI

Yellow chromates are produced in acidic solutions containing chromic acid, activating complex salts and fluorides or fluoride complexes [30, 33, 34, 42]. Hydrogen gas is formed by the aluminium dissolution. This increases the pH in the diffusion layer and causes precipitation of Al(OOH) or Cr(OOH). Depending on the reaction conditions and time, a colourless to golden yellow layer is formed. The color is due to the incorporation of additional Cr^{6+} ions into the top layer. The layers form a good adhesion primer as well as a reliable diffusion barrier for subsequent water vapor permeable organic coatings and their application loads [30–35, 43]. Even at damaged areas of the coating and at cut edges, infiltration of organic coatings is prevented in the unprotected area [32]. The coatings are produced by dipping, spraying or no-rinse processes [38–41].

5.7 Pre-Treatment of Ferrous Materials

The pretreatment of carbon steels can be carried out according to different procedures, but certain common guidelines can be established in each case. The first stage of treatment is usually pre-cleaning with solvents, emulsions or hot cleaners. Rusted or scaled workpieces are then pickled, thoroughly rinsed, and dried. Iron and carbon steels are usually pickled in hydrochloric or sulfuric acid. Hydrochloric acid is usually preferred because of its faster reaction and the possibility of working at room temperature. The pickling temperature should not exceed 35 °C to prevent the escape of hydrogen chloride mist. Pickling times range from 5 to 15 min, depending on the degree of scaling. High-strength steel (tensile strength greater than 1200 N/mm^2) containing more than 0.35% carbon, as well as case-hardened steel, require special attention with respect to hydrogen embrittlement [44]. To avoid embrittlement due to excessively long pickling times, such steels are often descaled after degreasing by sandblasting, brushing or alkaline agents. Soak cleaning must be neither too long nor too short. If it is too long, the adhesion of the coatings suffers.

5.8 Zinc and Galvanised Surfaces

Zinc or galvanized surfaces can be treated in both acidic and alkaline solutions. A typical workflow for galvanized surfaces is as follows:

1. Mild alkaline cleanser or acidic (aqueous solutions of phosphoric acid)
2. Rinse
3. pickling, mostly in aqueous solutions of phosphoric acid
4. Rinse
5. Layer-forming phosphating or Cr-free processes
6. Rinse (deionized water),
7. Passivation (for phosphating)
8. Rinse
9. Rinse (deionized water)
10. Drying

The number of individual cleaning steps depends here, as with all other materials, on the contamination of the surfaces, the result to be achieved and the requirements placed on the workpiece. If chrome-free rinsing solutions are used in phosphating process, a preceding rinse in DI-water is necessary.

5.9 Aluminium Alloys

5.9.1 Properties of Aluminium as a Material

From a chemical point of view, aluminium is characterised by three properties that plays a role during pre-treatment:

- The base character of aluminium (standard potential against hydrogen: −1.66 V). This allows the dissolution of the metal (oxidation) in acidic solutions. The resulting increase in the pH value in the diffusion layer near the surface leads to the separation of poorly soluble compounds and thus to the build-up of conversion layers.
- Aluminium behaves amphoterically.
  Metal dissolution at the surface is achieved not only by treatment with acidic but also with alkaline solutions. This behaviour plays an important role in pickling cleaners that prepare surface of aluminium workpieces for the following conversion treatment.
- Formation of a closed, firmly adhering oxide layer.
  This enables the process of anodization or pre-anodization. Furthermore, this is used to build up mixed oxides, as they are present in various conversion layers. Also the adsorption layers, which are used in some Cr-free processes, are only possible because they can be bound to this oxide layer towards the metal.

The different alloy compositions of aluminium materials are also important for pre-treatment. In the automotive sector, the naturally hard magnesium-containing alloys (AA 5xxx) and the age-hardenable magnesium- and silicon-containing alloys (AA 6xxx) play

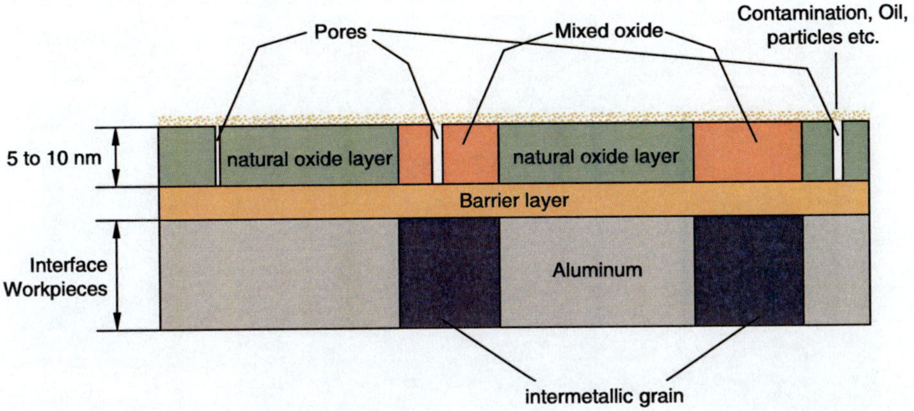

Fig. 5.16 Schematic structure of an untreated aluminium surface

a major role in Europe. In the architectural sector, the sheet alloys AlMg1, AlMg3 and the extrusion alloys AlMgSi0.5 are mainly used.

In addition to the alloy composition, the manufacturing process of the workpiece plays a major role in the pretreatment; depending on whether it is a rolled strip, an extruded profile or a cast workpiece. The surface layer is quite different in terms of thickness, number and type of defects and inhomogeneities, as well as in terms of composition. A schematic cross-section of a chemically untreated technical aluminium surface is shown in Fig. 5.16. As the figure shows, the aluminum surface is by no means uniformly composed, but has a complex structure.

5.9.2 Pre-Treatment of Aluminium Alloys

The process of pre-treatment of aluminium parts prior to paint coating is divided into the following process steps:

1. Cleaning
2. Rinse
3. Pickling or pickling degreasing
4. Rinse
5. Acid dip
6. Rinse
7. Conversion layering
8. Rinsing
9. Drying

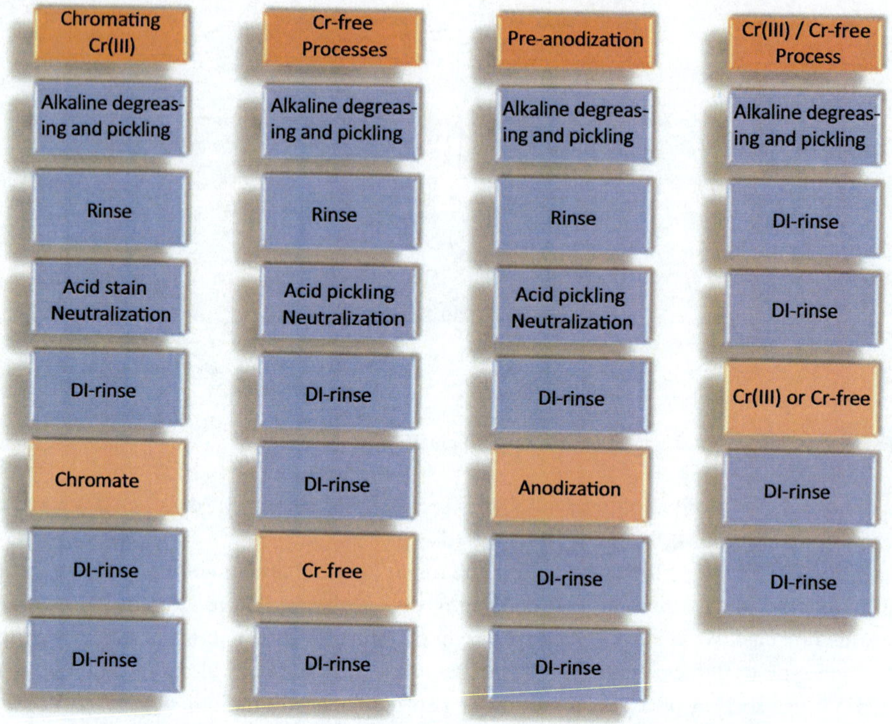

Fig. 5.17 Steps in the pretreatment of aluminium materials. The number of rinsing steps between the active baths and the water quality influence the result and the resistance of the paint. In general, for good paint adhesion and corrosion resistance, the last rinse should always be carried out with deionised water

Figure 5.17 shows the individual typical pretreatment steps for cleaning and conversion treatment of aluminium materials.

5.9.2.1 Cleaning (Degreasing)

The surface must be clean, free from contamination by firmly adhering materials of all kinds, e.g. adhesive residues, greases and oils, metal chips, grinding dust, corrosion products. Cleaning can be carried out in aqueous systems, with additions of surfactants in combination with an acidic or mild alkaline builder system.

These aqueous systems extend the bath service life (waste avoidance) and enable cleaning at low temperatures (energy savings). The so-called pickling degreasers, they contain the process steps cleaning and pickling in one bath, but have only a limited service life due to the accumulation of dissolved aluminium in the electrolyte.

5.9.2.2 Pickling

Natural and preformed oxide layers must be removed to ensure a "bright" surface for reaction with the conversion coating forming chemicals. Alkaline or acidic solutions can be used as pickling agents. Alkaline pickling solutions, e.g. 30–50 g/l NaOH, at temperatures of 20–60 °C are predominantly used for aluminium materials. The pickling attack is controlled by the concentration of free NaOH, the temperature and the treatment time. The dissolving aluminium is stabilised by the addition of auxiliary substances (e.g. sodium gluconate).

Commercially available pickles contain alkaline salts with wetting agents and inhibitors that guarantee complete removal of the oxide skin while protecting the base material.

Advantages of alkaline pickling are the high removal of the base metal and the good additional degreasing effect by saponification of oils and greases. The disadvantage is that certain elements, such as copper, magnesium and silicon, remain on the surface and make an acid dip necessary.

Acidic pickling usually consists of mixed acids based on nitric and hydrofluoric acid or ammonium bifluoride-sulphuric acid. The advantages here are the simultaneous removal of silicon and copper with less material removal than with alkaline pickling. Disadvantage is, due to the lower material removal, no removal of indented dirt.

5.9.2.3 The Acid Dipping Process

During pickling, certain alloying constituents of the aluminium are exposed and enriched on the surface. These components are hardly soluble in alkaline solutions (Cu, Mg, Si or their mixed crystals).

The task of decapping is to remove these substances in order to create a defect-free surface for the following conversion treatment.

Accumulations of foreign metals, such as magnesium, should be removed in acid solutions at room temperature (e.g. in nitric acid, sulphuric acid). Effectiveness is determined by the type of acid, concentration and treatment time. Silicon and silicate formed during pickling require the addition of fluorides or hydrofluoric acid.

The chemical composition of these acid solutions depends essentially on the type of aluminium alloy and is usually based on nitric, sulphuric or phosphoric acid.

5.9.2.4 Conversion Treatment

During the conversion treatment, a microcrystalline or microrough layer is built up which is the adhesion base for the coating layers and at the same time should have a good corrosion protection effect – especially against layer infiltration. It should act as a barrier to moisture and corrosion-promoting substances such as chlorides and sulphates.

5.9.2.5 The Drying Process

After rinsing, adsorptively bound water adheres to the parts. Poorly dripping water can be contained in the cavities and grooves of the workpiece. This water must be removed by gentle drying so that the water does not evaporate during the baking of the lacquer layers. If drying is carried out too quickly at too high temperatures, the stored water can break up the fine crystal bond of the conversion layer. This leads to adhesion problems.

5.9.3 Pre-Treatment of Aluminium Casting Alloys

After buffering of the castings, a mechanical surface treatment can be carried out preferably by grinding or blasting to remove release agent residues or casting skins. In this case, suitable abrasives or blasting media must be selected to prevent impurities from being pressed into the aluminium surface. A suitable abrasive is, for example, corundum, whereas silicon carbide causes problems due to its high hardness and poor removability. The mechanical effect of the frequently used vibratory grinding processes is often not sufficient to reliably remove cast skins. Independently of this, the adhesion of a subsequent coating is influenced by the choice of vibratory finishing agents (chips and compounds). For coating pretreatment, vibratory finishing processes are only applicable to a limited extent, so that blasting processes are rather recommended. Ceramic media, such as corundum or pure blast furnace slag, are particularly recommended as blasting media. Glass beads or aluminium granules can also be used in principle, but lead to only slight material removal. Foreign metals and plastics are not suitable, as they cause adhesion problems due to the penetration of flakes into the workpiece surface. When using steel grit as a blasting medium, for example, it is absolutely necessary to remove the pressed-in abrasion by a pickling treatment in nitric acid. Remaining iron particles would form the seeds for pitting corrosion.

Degreasing of the workpiece is necessary prior to blasting, as blasting processes have a compacting effect which would drive grease residues into the workpiece surface. Mould release agents and piston lubricants are particularly problematic as they cause burnt-in oil carbon deposits on the workpiece. Heavy oil carbon deposits cannot be removed even by blasting, but must be pickled using a highly concentrated acid mixture. Even the most sparing use of mould release agents and piston lubricants must never contain silicones or graphite, which make the casting skin water-repellent and therefore make subsequent chemical pretreatment and coating impossible [45].

5.9.3.1 Degreasing

Chemical cleaning of cast materials is generally carried out using the same processes and electrolytes as for sheet metal or extruded alloys. In the case of processes which attack the surface (pickling), care must be taken, particularly with alloys with a high Si content, to ensure that no "pickle bast" is formed which is hardly to remove.

5.9.3.2 Pickling and Acid Dipping

Alkaline pickling for targeted roughening of the surface after degreasing and cleaning is not recommended for the surface treatment of aluminium castings.

Practically all aluminium casting alloys contain Si, Mg or Cu, which cause dark, insoluble residues (pickle bast) during alkaline pickling. To remove this "pickle bast", a subsequent acid pickling is then indispensable. In particular, the surface accumulations of metasilicate and elemental silicon on AlSi materials formed during alkaline pickling (but also in acidic pickling solutions with a too low fluoride content) must be removed with a concentrated acidic fluoride-containing pickle. Besides silicon, less critical are other residues of the alkaline pickling attack. These can be removed in a diluted acidic, fluoride-containing decap bath.

5.9.3.3 Conversion Layers

The known Cr-free processes based on zirconium/titanium, silanes or SAM (self assembling molecules) and pre-anodisation are used as corrosion protection and adhesion layers for cast alloys.

5.9.4 Conversion Coatings for Aluminium Surfaces

The aim of pretreatment is to ensure adhesion between the base material and the coating (powder or liquid coating) and the corrosion protection effect of the system.

The Cr-free processes, pre-anodising and briefly chromating are described below.

The following Table 5.7 lists the various processes for the production of conversion coatings and explains their properties and areas of application.

5.9.4.1 Yellow Chromating

For the conversion treatment of aluminium surfaces, the yellow chromating process uses an acidic chromate solution with a pH value between 1 and 2, which may additionally contain bound and free fluoride ions and accelerators. The protective layer usually obtained during the treatment, with a coating weight for painting of 0.4 to max. 1.0 g/m^2, consists of aluminium/chromium(III) mixed oxides which additionally have the accelerators and soluble chromate compounds trapped. These chromates can be redissolved at a later stage, thus providing the coating with a depot of corrosion protection, which is important in case the coating is damaged down to the metal. The clod structure created by dehydration during drying provides many anchorage points for an applied coating, resulting in good adhesion for subsequent painting.

Reaction sequence and layer composition for yellow chromating on aluminium:

Pickling reaction:

$$2Al + 6H^{+} \rightarrow 2Al^{3+} + 6(H)$$

The hydrogen does not escape, but is oxidized by the chromic acid and oxide hydrates are formed.

Yellow chromate:

Table 5.7 Processes for the pretreatment of aluminium surfaces

Procedure	Layered structure	Mass per unit area, g/m²	Application techniques	Application area
Green chromate	Phosphates of chromium III, hydrated oxides of chromium III and aluminium	0.2–5	Spraying, dipping, painting (with a brush), roller application	Architecture, food sector, cans, barrels, jugs
Phosphate	Phosphates of aluminium and base metals	1.5–10	Spraying dipping	General industry, automotive, food sector, cans
NO-RINSE chromate coating	Chromium III in organic and inorganic matrixes	0.1–1	Roller application without rinsing process	Coil coating architecture
Zircon/titanium process	Speration of TiO_2/ZrO_2, mainly on the intermetallic phases, partly in organic matrix	Approximately 30 mg/m² Ti or Zr	Dipping, spraying roller application	Construction/architecture, food sector, automotive
Sam	Organic molecules with functional groups	Monomolecular layer	dippingSpraying	Automotive (wheels), food sector, coil coating
Silan	Si-O bond to the metal and functional groups	50–100 nm	Spraying dipping	Automotive General industry
Cr-III process	Hydrated oxides of Cr-III and aluminium	Approx. 400 nm	Dipping spraying	Construction/architecture General industry
Pre-anodization	Al oxide, oxide hydrate	1–6 µm	Dipping	Construction
Nano ceramics	Oxides and nanoparticles	20–50 nm	Dipping	General industry

$$2CrO_3 + 3H_2 \rightarrow Cr_2O_3 \times \text{aq.} + 3H_2O$$
$$2Al^{3+} + 6OH^- \rightarrow Al_2O_3 \times \text{aq.} + 3H_2O$$

Figure 5.18 shows the surface of a conversion layer produced from a yellow chromate coating. The surface analysis of the layer components using EDX analysis, Fig. 5.19 shows the main components chromium, oxygen and aluminium.

Figures 5.20 and 5.21 show the conversion layer of a green chromate coating and its components.

Fig. 5.18 SEM image of a layer produced from a yellow chromate coating, substrate AlMg1, layer weight 0.8 g/m2

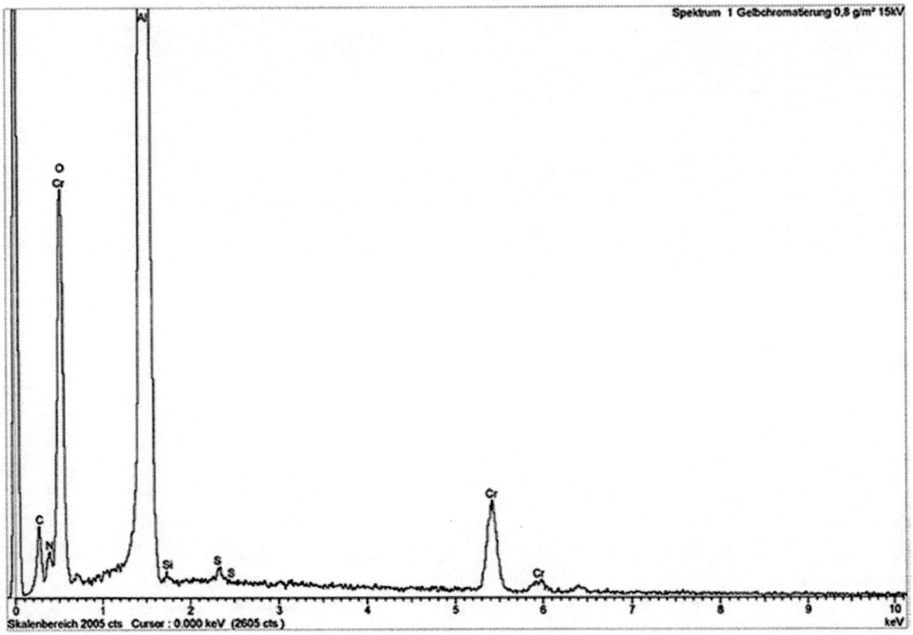

Fig. 5.19 EDX spectrum of a yellow chromate layer on AlMg1

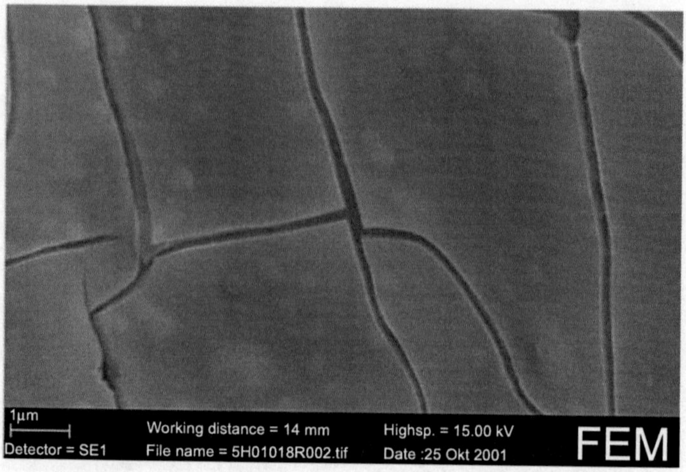

Fig. 5.20 SEM image of a layer produced from a green chromate coating, substrate AlMg1, layer weight 1.0 g/m2

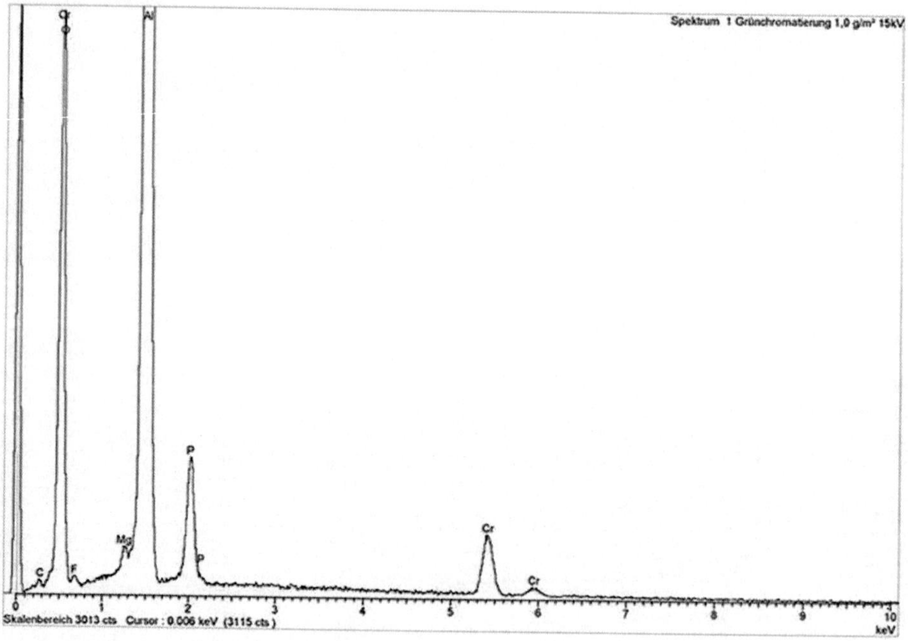

Fig. 5.21 EDX spectrum of a green chromate layer on AlMg1

5.9.4.2 Green Chromating

Green chromate solutions contain phosphoric acid and usually none of the accelerators used in yellow chromating.

Green chromates show a lower corrosion protection than the yellow chromates without a following coating. They are more sensitive to critical alloy compositions than the yellow chromates. In combination with paint coatings, there is practically no difference in corrosion protection.

Reaction sequence and layer composition for green chromating on aluminium:

Pickling reaction:

$$2Al + 6H^{+} + \rightarrow 2Al^{3+} + 6(H)$$

Green chromate coating:

$$Cr_2O_7^{3-} + 8H^{+} + 6(H) \rightarrow Cr^{3+} + Cr(OH)_3 + 4H_2O + 3H^{+}$$

$$Cr^{3+} + PO_4^{3-} \rightarrow CrPO_4$$

$$2Al^{3+} + 6H_2O \rightarrow Al_2O_3 3H_2O + 6H^{+}$$

Chromium phosphate layer:

$CrPO_4x6H_2O$	76–88%
$Al_2O_33H_2O$	12–23%

The reaction of the chemicals in the chromating solution with the aluminium surface produces fine crystals that are difficult to dissolve, which grow firmly together with the surface and form a dense microrough layer. In green chromating, a crystalline chromium(III) aluminium phosphate oxide hydrate layer is formed.

5.9.4.3 Chromium (III) Passivation

The reaction mechanism in the deposition of the passivation layer from solutions containing Cr^{3+} can be divided into two parts:

1. the activation of the metal surface
2. the deposition of the passivation layer

The activation of the aluminium surface is also carried out by fluorides in this process. The layer is a mixed oxide of Cr and aluminium oxides or hydroxides.

Properties of the Cr(III) layers:

- Cr(III) oxides are very stable
- good corrosion protection and adhesion primer for coatings

- to be used for aluminum and galvanized surfaces. For optimum corrosion protection, it is recommended to optimize the passivation to the respective base material.a certain corrosion protection is also given without painting
- coloured layer, depending on the base material and alloy a yellowish shade can be achieved
- layer weight is determined in the same way as for yellow chromating
- disadvantage is a very narrow process window with regard to the pH value
- the process is sensitive to carry-over, hydroxide carry-over leads to precipitation of $Cr(OH)_3$, phosphate carry-over leads to precipitation of $CrPO_4$

Today, formulations based on Cr(III) salts and organic acids (oxalic acid) or based on Cr(III) salts and mineral acids are generally used.

New generations of Cr(III) pretreatments are also said to have self-healing properties, through the incorporation of chemically inert nanoparticles. The migration of these particles into cracks or weak points/damage in the conversion film causes the self-healing effect. According to the manufacturers, the nanoparticles form a quasi top coat on the chromate film during dipping. The thickness of the layer is about 400 nm. Optimum bath temperature is T = 30 °C, pH range 1.8–2.2, layer formation takes place within 30 s. At pH values <1.8 film dissolution occurs, at higher pH values >2.2 the nanoparticles separates before film formation occurs.

Cr(III) passivations are suitable not only for aluminium alloys but also for zinc and galvanised surfaces.

5.9.5 Chromium-Free Processes

Due to the ban on the use of chromium (VI) salts and acids, Cr-free processes are used in painting today, with a few exceptions (Table 5.8).

Table 5.8 Chromium-free processes and their possible applications

Aluminium	Zinc	Steel
Anodization	Phosphatization	Phosphatization
Zr/Ti	Cerium	Silane
Cerium	Silane	Nano ceramic layers
Phosphatization	Nano ceramic layers	
SAM		
Silane		
Nano ceramic layers		

5.9.5.1 Titanium/Zirconium Process

The use of transition metal compounds – especially zirconium and titanium – for the pretreatment of aluminium surfaces has been known for a long time. Acidic solutions are used in which zirconium or titanium or both are dissolved in complex form and which contain further inorganic and/or organic additives. Fluoride is usually used as a complexing agent. A three-step mechanism is assumed for the build-up of the conversion layer in the treatment bath, which begins with a pickling reaction in which aluminium oxide and metal are dissolved. The released aluminum ions react with both water (hydrolysis) and the transition metal complex (recomplexation). Finally, hydrated oxide layers of the aluminium and transition metal separate on the surface, which may also contain complexing agents and other solution components. No redox reaction takes place, the oxidation state of the Ti/Zr is +4 in the solution as well as in the layer. The layer formation starts preferentially in the area of the intermetallic phases.

The layer formation takes place according to the following principle:

1. Hydrogen (H_2) is formed at the intermetallic phases (in the case of aluminium).
2. There is an increase in the pH value, whereby the solubility of the oxide decreases.
3. Oxide separation occurs, first and predominantly on the intermetallic phases.

Figure 5.22 shows the basic structure of such a colorless layer.

The layer thickness or the layer weight is usually not determined directly, but via the titanium or zirconium coatings measured on the surface by X-ray fluorescence. They are in the range of up to 30 mg/m^2. A requirement for pretreatment with Ti/Zr compounds is a high water quality directly before and, in the case of rinsing processes, also after the conversion treatment (deionized water).

Properties of the Zr/Ti layers:

- the layer formation takes place in seconds
- good paint adhesion and corrosion resistance, but depending on the aluminium alloy and the paint system (paint type powder or liquid)

Fig. 5.22 Schematic layer structure for titanium/zirconium process [46]

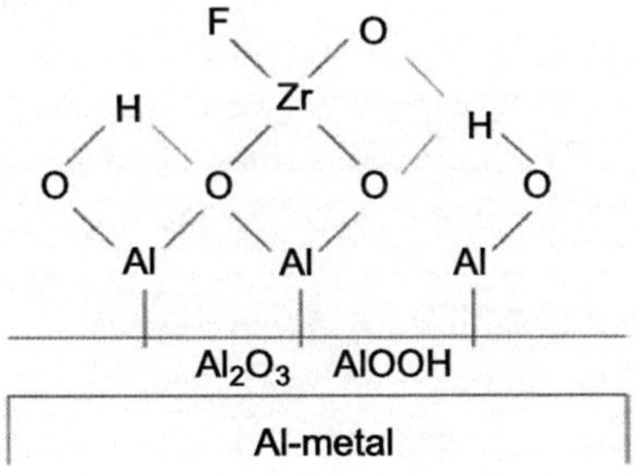

- layers are colorless
- process is sensitive to impurities or carryovers

5.9.5.2 Trication Zinc Phosphating

Trication zinc phosphating is the state of the art in the pretreatment of car bodies [46]. It is applied in combination with a cathodic electrodeposition coating, onto which the remaining paint layers are then applied. The treatment process, which can be carried out by dipping and spraying, comprises in principle the following process steps. Some of them can be designed to be multiple:

- Cleaning
- Rinse
- Activation
- Phosphatization
- Rinse
- Post-rinse (optional)
- Rinsing with deionised water

During activation, the surface is coated with titanium phosphate particles, on which the zinc phosphate crystals grow in the following stage. In addition to the cations of zinc, manganese and nickel, the acidic phosphating bath contains an accelerator (e.g. nitrite) and other additives that are required for optimum layer formation.

Zinc phosphating was originally designed for steel surfaces, later additionally for galvanized steel. When aluminium parts were introduced on the body-in-white, which were also to be coated with a zinc phosphate layer (nickel- and manganese-containing hopeite $Zn_3(PO_4)_2 - 4H_2O$), the process had to be modified, since free aluminium ions even in low concentrations interfere with phosphating on steel (phosphating poison). This problem can be solved by adding fluoride. This complexes the dissolved aluminium so that it no longer interferes with the formation of the phosphate layer and, ensures the removal of the dissolved aluminium from the phosphating solution by precipitating this complex as a sodium salt (ktyolite Na_3AlF_6). However, the amounts of cryolite formed can be quite considerable – 8 to 12 g/m^2 of aluminium surface area is expected, which makes the process difficult to handle as the aluminium content increases. Fig. 5.23 shows a schematic of zinc phosphating on aluminium; Fig. 5.24 shows a phosphated aluminium surface taken with scanning electron microscope.

The post-rinse solution in the phosphating process was changed from Cr-containing to a zirconium fluoride-containing solution.

5.9.5.3 The SAM Procedure

The abbreviation *SAM* stands for self assembling molecules [46, 47], i.e. molecules that self-assemble into a certain order on a surface. The principle of attaching organic molecules with functional groups to a surface so that, due to the spatial proximity, the

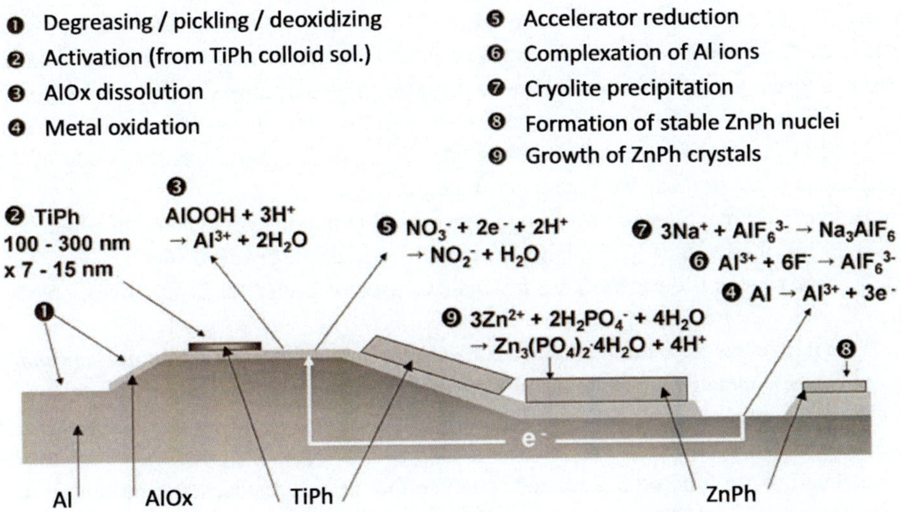

Fig. 5.23 Process of zinc phosphating on aluminium [46]

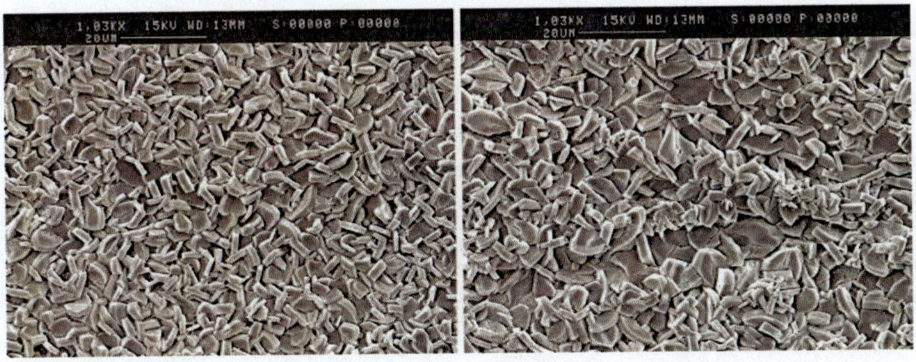

Fig. 5.24 SEM images of a zinc phosphate coating on aluminium; unground surface on the *left* and ground surface on *the right* [46]

interaction between the carbon chains is sufficient to align the molecules has been known for a long time. In contrast to the classical conversion layers, no pickling reaction takes place in these adsorption layers, but functional groups of an organic molecule, e.g. phosphonic acid groups, react with a hydrated metal oxide surface in an acid-base reaction according to Eq. (5.1) (schematic):

$$\text{Layering}: Me(OH)_3 + (HO)_3 P - R \rightarrow Me(O)_3 P - R + 3H_2O \tag{5.1}$$

In the strict sense, it is a chemisorption. Eq. (5.1) shows an essential advantage of such processes. Only water is formed as a reaction product; metal is not introduced by pickling and therefore does not pose a problem for disposal. It also becomes clear why these processes are mainly suitable for the treatment of aluminium surfaces: The hydrated metal oxide surface, to which the organic molecules are bound, in turn adheres firmly to the metal. This is not the case with other substrates, e.g. zinc.

An idealized model of the resulting layer, which is monomolecular on the aluminum matrix, while intermetallic precipitates are covered in multiple layers, is shown in Fig. 5.25. At the same time, it is clear from this figure that corrosion protection is effective by blocking the cathodic and anodic partial reactions.

If the molecules are equipped with a functional group at both ends, binding is possible not only to the metallic substrate but also to a subsequently applied lacquer.

The SAM process comprises alkaline or acidic cleaning of the aluminium surface, if necessary followed by acidic dip after a rinsing process. It is important to rinse several times, finally with fully demineralized water, so that no salt residues are introduced into the aqueous, acidic SAM solution. The adsorption of the molecules on the surface usually takes place so quickly that this technology is also suitable for coil coating with their usually short treatment times.

5.9.5.4 Silane Technology

A similar principle of bonding a pretreatment layer to a metal surface as in SAM coating is followed in silane technology [46, 48, 49]. Known for many years as an adhesion promoter in organic-based systems, this class of compounds has more recently been adapted for the surface treatment of metals and can in principle be used on all common surfaces (steel, zinc, aluminium or corresponding alloys).

In this context, silanes are molecules having the following formula: $R^2 - Si\ (OR1)_3$ R^1 is usually a methyl or ethyl group, R^2 represents an alkyl chain which can carry a functional group, e.g. an amino group, or a further alkoxy silyl group. Typical representatives of such silanes as used in surface treatment are:

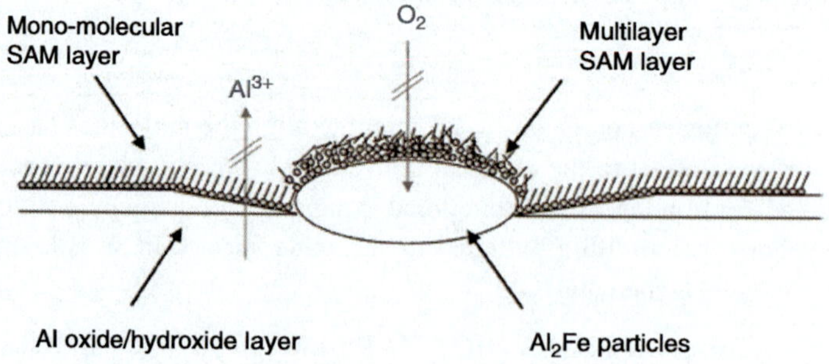

Fig. 5.25 SAM-treated AlMg1 surface [46]

γ-APS	$H_2N–CH_2CH_2CH_2–Si(OC_2H_3)_3$
BTSE:	$(C_2H_5O)_3Si–CH_2CH_2–Si(OC_2H_5)_3$

Single or multiple silanes are usually first hydrolyzed in a water-based solution to form Si(OH) groups that can provide good adhesion to a surface. Since, in addition to hydrolysis, condensation reactions can also occur with the formation of oligomers or polymers, careful selection of the raw materials and precise adherence to the reaction conditions are required. The treatment of the surface requires a water-wettable hydrated oxide layer, which is usually achieved by alkaline cleaning followed by rinsing. For the application of the silanes, a two-step process was first developed: Here, a non-functional silane (e.g. BTSE) is applied to the surface; strong Me-O-Si bonds are obtained. Subsequently, a functional silane (e.g. γ-APS) is brought into contact with the surface, which couples to the BTSE layer via Si-O-Si bonds and whose functional group (e.g. the amino group) is available for the adhesion of applied coatings, adhesives, etc. The functional group (e.g. the amine group) is then used for the adhesion of the BTSE layer.

In the meantime, however, suitable mixtures of silanes can be made available which are applied in a single step and still exhibit a firm bond to the metal surface. Furthermore, during drying of the silane layer, residual Si-OH groups are condensed, which leads to a more or less strong cross-linking of the layer (Fig. 5.26).

Usually, coating thicknesses in the order of 50–100 nm are targeted for corrosion protection under paint. Thicker layers can be set for the corrosion protection of the bare, not further coated sheet metal.

The concentration of the silanes in the treatment solution is decisive for the layer thickness, while treatment time, temperature and pH value only play a subordinate role.

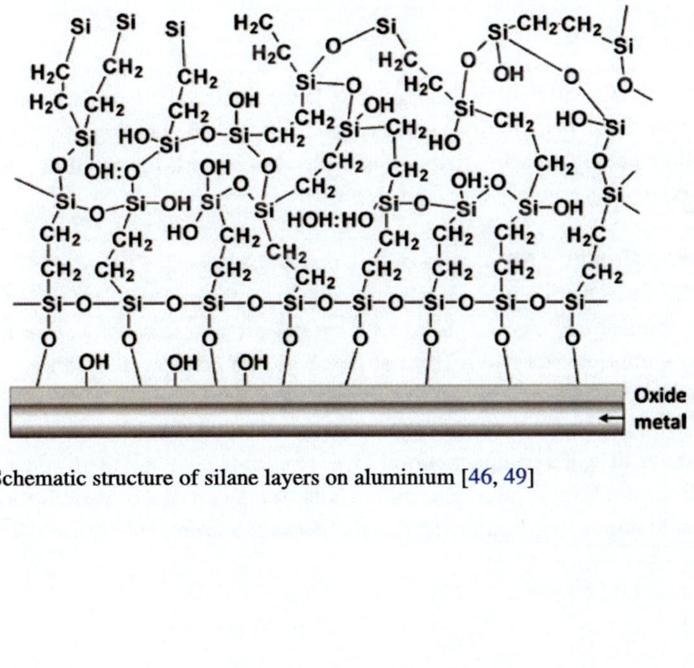

Fig. 5.26 Schematic structure of silane layers on aluminium [46, 49]

Experience shows that different silanes should be used depending on the substrate, organic coating and application. The most advanced are pretreatments for galvanized surfaces, especially in strip pretreatment plants and for aluminum castings in the wheel industry. In principle, however, these systems can be used for any type of pretreatment.Properties of the silane layers or the silane process:

- good adhesion properties on the metal surface as well as good adhesion primer for lacquers
- good formability
- layer thickness is in the range 20–100 nm
- to be used for light metal alloys (Al, Mg), zinc or galvanized surfaces, steel
- selection of special functional groups for good paint adhesion

5.9.5.5 Pre-Anodization

In organic and inorganic acids as well as mixed acids, a firmly adhering oxide layer can be produced on aluminium with the aid of electric current. These anodised layers have been known in technology for a long time and are used in a compacted state in layer thicknesses of 20–25 µm for the protection and decoration of aluminium in the construction industry.

For aluminum, the so-called pre-anodization is used as a conversion layer to produce an adhesion promoter layer.

Sulphuric acid concentrations of 180–200 g/l are selected, whereby the following parameters must be observed:

Max. aluminium content	15 g/l
Chloride content	<0.1 g/l
Temperature	20–30 °C
Current density	1–3 A/dm²
Layer thickness	3–8 µm

The parts are not sealed. A not sealed oxide layer reacts very sensitively to contamination – moisture, hand perspiration or grease and oils are absorbed and can no longer be removed with cleaning.. Clean gloves should always be worn for handling, e.g. transferring the workpieces.

5.9.5.6 Formation and Structure of the Oxide Layer

Aluminium surfaces form a firmly adhering oxide layer with a thickness of approximately 10 nm with atmospheric oxygen. It forms a natural protective and covering layer which protects the aluminium very well from atmospheric influences. According to Altenpohl [50], the natural oxide layer leads to a potential refinement from −1.69 V to about −0.47 to −0.55 V, which causes aluminium to behave more noble than zinc. The oxide layer formed in air consists of a compact barrier layer of alumina 3–5 nm thick and an overlying porous, hydrous top layer of alumina and bayerite. It may consist of various modifications of alumina or aluminum hydroxide depending on the prevailing environmental conditions.

Also, the environmental contaminants are incorporated into the layer. The oxide layer is practically insoluble in the pH range from 4.5 to 8.5.

Anodic Oxide Layers

Anodic oxide layers [51] can only be produced on a metal connected as an anode. During electrolysis atomic oxygen is produced at the anode and reacts with the aluminium, forming aluminium oxide. This is firmly anchored on the metal surface. This compact layer resists the passage of current and is penetrated as a result of the bath voltage. The electrolyte moves into this breakdown pore and a new, compact layer is formed at its base, i.e. where it meets the metallic aluminium. Since there are countless such pores next to each other, the layer continues to grow evenly into the metal. Only the layer that is freshly and directly formed on the metal has the compact, dielectric character. The part that has already been formed and penetrated continues to grow during anodisation, is loosened by the electrolyte advancing into the pores and is transformed into the porous and absorbent main layer. The oxide layer grows into the metal, i.e. the original metal surface is transformed into the oxide layer. One third of the layer grows into the metal surface and two thirds grow outwards. The outward layer growth is due to the larger volume of the oxide layer.

The layers consist of an impermeable barrier layer and a porous top layer, whose achievable thickness and pore diameter depend on the current density, time duration, temperature and acid concentration. The pore density is about one billion pores per mm^2 of external surface.

Figure 5.27 shows the oxide layer in the Keller, Hunter and Robinson [52] model, and Fig. 5.28 is a surface image of the oxide layer with SEM.

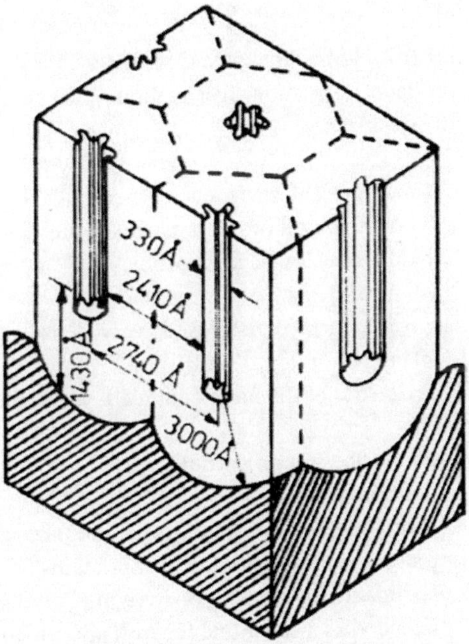

Fig. 5.27 Model representation of an oxide layer of aluminium produced by the DC process. Pore model according to Keller, Hunter and Robinson

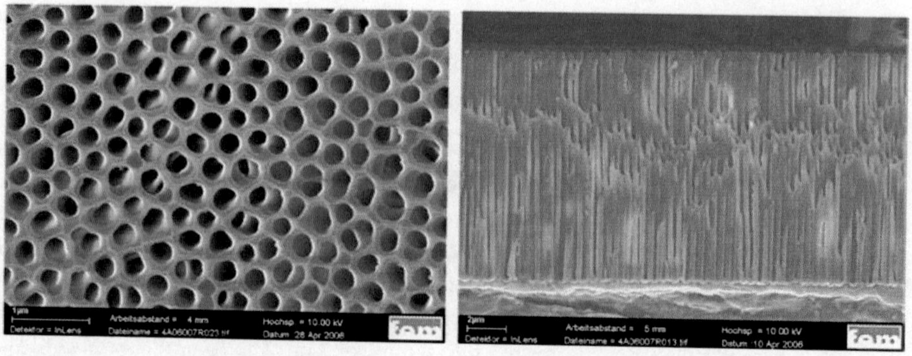

Fig. 5.28 Surface image of an aluminium oxide layer, base material AlMg1, production of the layer in the DC process; *left:* Oxide pores in top view, *right:* Pores in cross section

Properties of the layer or process:

- very good corrosion resistance
- variably adjustable layer thicknesses
- electrolyte contains only sulfuric acid
- is only possible in dip process
- possible with aluminium and magnesium (alloys)
- special plant technology necessary
- insufficient rinsing can lead to paint adhesion problems, especially after exposure to humidity.

5.9.5.7 Nano-Ceramic Coatings

The layer formation over ceramic-like metal oxides in nanoparticle size takes place in two steps:

1. Coating of the surface with metal oxide particles
2. Ti or Zr oxyfluoroglass particles are deposited or attached in the inter spaces of the oxide particles + Nanoparticles

The resulting layer is closed, very dense but thin. The thickness is in the range of 20 to 50 nm.

Properties of the nano-ceramic layer or process:

- good adhesion properties
- multi-metal
- short process time, layer formation takes place within 20–30 s.
- process is phosphate-free and contains no organic additives
- application by dipping or spraying possible
- intended as a substitute for iron phosphating

The process was newly developed or further developed as a replacement for zinc phosphating. With zinc phosphating, 11 zones including rinsing are the rule. With the nanoceramic process, activation and passivation can be eliminated and thus the process sequence is shortened to 7 zones.

The layers are amorphous and composed of Zr(OH)x, ZrOx and particles, not further described by the manufacturer.

5.10 Magnesium Materials

Magnesium and its alloys are the lightest, but also the least noble construction materials (normal potential of Mg: −2.34 V). Magnesium forms thin passive films in air and in water, but these are not resistant in aqueous solutions with a pH < 10. In aqueous, alkaline solutions with a pH value >10.5, the passive layer is very stable, since cracks formed in the passive layer are filled with $Mg(OH)_2$ due to compressive stresses; this is also referred to as healing of the layer.

In order to protect magnesium alloys for technical applications against corrosion, the surfaces are treated with passivation electrolytes. The redox process that takes place (without an external current source) forms a conversion layer consisting of oxides of the magnesium material and oxidic reaction products of the passivation electrolyte.

5.10.1 Pre-Treatment of Magnesium Alloys

Pretreatment [53–56]

For degreasing, the usual degreasing agents used in electroplating should be used, but not more strongly acidic solutions. $Fe(NO_3)_3/CrO_3/NaF$ mixtures are also recommended for bare metallic surfaces [55]. However, strongly alkaline media are to be preferred due to their lower corrosive effect.

For the pickling process, mild acidic pickles can be used, followed by sufficient rinsing and neutralisation. All pre-treatment steps must be adapted to the specific treatment process.

The problem with a chemical pretreatment of Mg surfaces is the strong attack of the metal surface, even in low-dose process baths. A pickle bast forms very quickly, which must be completely removed from the surface for the proper formation of a conversion layer.

The pretreatment of a magnesium surface before coating can be as follows:

- Cleaning strong alkaline – not pickling
- Rinse
- Pickling (optional, depending on surface finish). For AZ 91 alloy, pickling in a mixture of sulphuric and nitric acid is recommended. For AM 50/60 alloys, pickling in oxalic acid is recommended.

- Rinse
- Conversion treatment
- Rinsing in deionised water
- Drying

5.10.2 Conversion or Passivation Process for Magnesium

Chromating of magnesium surfaces can still be carried out today. The corresponding processes are described in MIL specifications M3171 Type 1 to Type 3 [57]. Chromic acid or its salts are used for passivation. Chemical passivation with electrolytes containing chromium(VI) is relatively easy to carry out, but has the disadvantage that the electrolyte and also the layers formed contain Cr^{6+} and special regulations therefore apply to its use.

In addition, the recyclability of chromated objects made of magnesium or its alloys is a problem, as they can only be recycled into so-called high-purity materials with great effort due to the heavy metals contained in the conversion layer.

Not least for reasons of health and environmental protection, the effort here was also to find and develop alternatives to chromate coatings. The EU Directive 2000/53 [58], which requires a limit of 2 mg/Cr^{6+} per vehicle from 2002, has also contributed to this. Today, the automotive industry demands Cr^{6+}-free conversion coatings on all surfaces.

The following systems are known as chromium(VI)-free aqueous passivation solutions for magnesium:

Phosphating, silane-based systems, titanium-zircon process.

5.11 Quality Assurance Measures

For the quality assurance of a pre-treatment before painting, the following are suitable,:

- Automatic bath analysis and chemical replenishment and preventive plant maintenance,
- Control of active baths, rinsing and system parameters
- Control of the layer formation
- Application of special QM methods, e.g. checklist, control chart, statistical process control.

5.12 Automatic Bath Analysis and Chemical Replenishment

The measurement of electrical conductivity and automatic titration are suitable for automatic bath analysis. The electrical conductivity serves, for example, as a control variable for the concentration in alkaline cleaners and in passivating rinses. It can also be used to determine the degree of contamination of rinsing baths with regard to the absorption of acids, salts and alkalis.

5.13 Preventive System Maintenance

The aim of preventive plant maintenance is to keep the plant in working order at all times. This requires the following measures:

- Provision of the necessary chemicals
- Testing of the equipment for bath analysis and chemical replenishment
- Cleaning and function control of nozzles, pipes, pumps, heating and desludging equipment
- Cleaning and function control of the transport systems, suspensions etc.
- Function control of peripheral equipment, e.g. production of demineralised water, extraction of exhaust air, treatment of waste water.

5.14 Selection of the Appropriate Pre-Treatment

Many factors influence the choice of the right pretreatment process for the application:

- Metal type and quality
- Surface condition
- Requirements for surface treatment and scope of application of the final product
- Economic and environmental considerations (mostly a question of the better alternative among more or less equivalent pre-treatment systems)

One criterion that can be used to classify and categorise suitable pre-treatment is the allocation of the product to a corrosion class.

The classification into corrosion classes or stress levels is described and explained in DIN EN 2081 for galvanized surfaces [59] and in DIN ISO 12944-2 [60]. The classification of the corrosion classes here was made on the basis of these standards.

Corrosion Class 0
(C1) Indoor climate with a relative humidity of less than 60%. Very low risk of corrosion, low aggressiveness, decorative application without stress.

Corrosion Class 1
(C2) Indoor exposure in warm, dry atmosphere, low risk of corrosion.

Corrosion Class 2

(C3) Indoor climate with temperature fluctuations and humidity variations.

Inland outdoor climate, away from industrial areas. Medium risk of corrosion.

Corrosion Class 3

(C4) Outdoor exposure under medium conditions. In agglomerations and near industrial areas. In water and near the coast. High risk of corrosion.

Corrosion Class 4

(C5-I/C5-M) In areas with permanent high humidity. In industrial areas with high pollution and in rooms where chemicals are processed or used, for example laboratories, swimming pools or thermal baths. Coastal and offshore areas with high salt load.

Depending on the base metal and the knowledge of the correct assignment of the product to the corrosion class, recommended pre-treatment methods can be found using Table 5.9, 5.10, 5.11, 5.12 and 5.13. In addition to the pre-treatment, the corrosion protection is strongly influenced by the coating system, paint structure and layer thickness.

The Cr-free processes given in the tables were based on silane-based or Zr-Ti processes.

Table 5.9 Recommended/not recommended pre-treatments of galvanized steel surfaces

Pretreatment of galvanized steel					
Type of treatment	Corrosion class				
	0	1	2	3	4
Cleaning/degreasing					
Blasting					
Iron phosphating					
Zinc phosphating					
Cr-free processes					

Table 5.10 Recommended/not recommended pretreatments of iron and steel

Pretreatment of iron and steel					
Treatment type	Corrosion class				
	0	1	2	3	4
Cleaning/degreasing					
Blasting					
Iron phosphating					
Zinc phosphating					

Table 5.11 Recommended/not recommended pretreatments of aluminium

Pretreatment of aluminum					
Type of treatment	Corrosion class				
	0	1	2	3	4
Cleaning/degreasing	Recommended pretreatment	Not recommended/insufficient	Not recommended/insufficient	Not recommended/insufficient	Not recommended/insufficient
Blasting	Recommended pretreatment	Recommended pretreatment	Not recommended/insufficient	Not recommended/insufficient	Not recommended/insufficient
Iron phosphating	Recommended pretreatment	Can be used	Can be used	Not recommended/insufficient	Not recommended/insufficient
Zinc phosphating	Recommended pretreatment	Can be used	Can be used	Not recommended/insufficient	Not recommended/insufficient
Chromating	Can be used, usually not required	Can be used, usually not required	Recommended pretreatment	Recommended pretreatment	Recommended pretreatment
Cr-free processes	Can be used, usually not required	Can be used, usually not required	Recommended pretreatment	Recommended pretreatment	Can be used
Pre-anodization	Can be used, usually not required	Can be used, usually not required	Recommended pretreatment	Recommended pretreatment	Recommended pretreatment

Table 5.12 Recommended/not recommended pre-treatments of magnesium

Magnesium pretreatment					
Type of treatment	Corrosion class				
	0	1	2	3	4
Cleaning/degreasing	Can be used	Can be used	Not recommended/insufficient	Not recommended/insufficient	Not recommended/insufficient
Blasting	Not recommended/insufficient	Not recommended/insufficient	Not recommended/insufficient	Not recommended/insufficient	Not recommended/insufficient
Iron phosphating	Not recommended/insufficient	Not recommended/insufficient	Not recommended/insufficient	Not recommended/insufficient	Not recommended/insufficient
Zinc phosphating	Not recommended/insufficient	Not recommended/insufficient	Not recommended/insufficient	Not recommended/insufficient	Not recommended/insufficient
Chromating	Can be used, usually not required	Recommended pretreatment	Recommended pretreatment	Recommended pretreatment	Can be used
Anodisation	Can be used, usually not required	Can be used, usually not required	Recommended pretreatment	Recommended pretreatment	Recommended pretreatment

[a]The natural oxide or cast skin of the magnesium surface must not be attacked by the cleaning or degreasing process.

Table 5.13 Recommended, not recommended pre-treatments of aluminium/zinc layers

Pretreatment of aluminum/zinc					
Type of treatment	Corrosion class				
	0	1	2	3	4
Cleaning/degreasing	Recommended pretreatment	Can be used	Not recommended/insufficient	Not recommended/insufficient	Not recommended/insufficient
Blasting	Not recommended/insufficient	Not recommended/insufficient	Not recommended/insufficient	Not recommended/insufficient	Not recommended/insufficient
Iron phosphating	Can be used, usually not required	Recommended pretreatment	Recommended pretreatment	Can be used	Can be used
Zinc phosphating	Can be used, usually not required	Recommended pretreatment	Recommended pretreatment	Can be used	Can be used
Cr-free processes	Can be used, usually not required	Can be used, usually not required	Recommended pretreatment	Recommended pretreatment	Can be used

- ☐ Recommended pretreatment
- ☐ Can be used
- ☐ Not recommended/insufficient
- ☐ Can be used, usually not required

References

1. Molz, Th.: Wässrige Reinigung metallischer Oberflächen, Referat, Technische Akademie Wuppertal, Vorbehandlung von Metalloberflächen (2001)
2. Stiles, M., Haasner, Th.: Überprüfung der Reinigungsqualität und Restschmutzbestimmung, Teil 1 und 2. In JOT 1998/6 und 7
3. Linford, J.B., Saubestre, E.B.: ASTM Bulletin 47, Mai (1953)
4. ISO 8502-7: Preparation of steel substrates before application of paint and related products – Tests for the assessment of surface cleanliness – Part 7: Field method for determination of oil and grease
5. ISO 8502-11: Preparation of steel substrates before application of paint and related products – Tests for the assessment of surface cleanliness – Part 11: Field method for titridimetric determination of water-soluble sulphate
6. ISO 8502-12: Preparation of steel substrates before application of paint and related products – Tests for the assessment of surface cleanliness – Part 12: Field method for titrimetric determination of ferrous ions (2000/04)
7. ISO 8502-13: Preparation of steel substrates before application of paint and related products – Tests for the assessment of surface cleanliness – Part 13: Analysis of soluble impurities on surfaces to be painted – Field method for soluble salts by conductometric measurement used surface salts meter
8. ISO 8502-5: Preparation of steel substrates before application of paint and related products – Tests for the assessment of surface cleanliness – Part 5: Measurement of chloride on steel surface prepared for painting (ion detection tube method) (1998/02)
9. ISO 8502-8: Preparation of steel substrates before application of paint and related products – Tests for the assessment of surface cleanliness – Part 8: Field method for refractrometric determination of moisture (1999/05)
10. ISO 8502-10: Preparation of steel substrates before application of paint and related products – Tests for the assessment of surface cleanliness – Part 10: Field method for the titrimetric determination of chloride (1999/12)
11. ISO 8503-5: Preparation of steel substrates before application of paint and related products – Surface roughness characteristics of blast cleaned steel substrates – Part 5: Method for the determination surface profile – Replica tape method (1999/05)
12. ISO 8504-4.: Preparation of steel substrates before application of paint and related products – Surface preparation methods – Part 4: Solvent cleaning
13. Baumgärtner, M.E., Gabe, D.R.: Elektrochemische Prüfverfahren zur Bestimmung der Reinheit von Metalloberflächen. In: Galvanotechnik, pp. 283–307. Leuze Verlag, Saulgau (1999)
14. DIN 53364: Ausgabe 1975, 06: Prüfung von Kunststofffolien, Benetzbarkeit von PE- und PP-Folien
15. ASTM D 2578-94.: Standard test method for wetting tension of polyethylene and polypropylene films
16. Simon, H., Thoma, H.: Angewandte Oberflächentechnik für metallische Werkstoffe. Hanser, München (1985)
17. Praktische Galvanotechnik. E. G. Leuze-Verlag, Saulgau, 133 (1984)
18. Burkart, W., Schmotz, K.: Handbuch für das Schleifen und Polieren. E. G. Leuze-Verlag, Saulgau (1984)
19. DIN-ISO 11124: Teile 1 bis 4, Vorbereitung von Stahloberflächen vor dem Auftragen von Beschichtungsstoffen, Anforderungen an metallische Strahlmittel (1997)
20. DIN-ISO 11125: Teile 1 bis 7, Vorbereitung von Stahloberflächen vor dem Auftragen von Beschichtungsstoffen, Prüfverfahren für metallische Strahlmittel (1997)

21. DIN-ISO 11126: Teile 1, 3, 4, 5, 6, 8, Vorbereitung von Stahloberflächen vor dem Auftragen von Beschichtungsstoffen, Anforderungen an nichtmetallische Strahlmittel (1997)
22. DIN-ISO 11127: Teile 1 bis 7, Vorbereitung von Stahloberflächen vor dem Auftragen von Beschichtungsstoffen, Prüfverfahren für nichtmetallische Strahlmittel (1997)
23. Kollek, H.: Reinigen und Vorbehandeln. Vincentz Verlag, Hannover (1996)
24. Grün, R.: Industrielle Teilereinigung, Galvanotechnik, Mai/Juni S. 875–888/1162–1175, Saulgau (2016)
25. Kresse, J.: Wässrige Reinigung von Aluminiumoberflächen vor der Konversionsbeschichtung. Referat, GSB Lehrgang (2001)
26. Specht, J.: Reinigen und Phosphatieren von Werkstücken aus Stahl, verzinktem Stahl und Aluminium als Vorbehandlung vor der Beschichtung mit organischen Stoffen. Referat; Technische Akademie Wuppertal, Vorbehandlung von Metalloberflächen (2001)
27. Rausch, W., et al.: Die Phosphatierung von Metallen. Eugen G. Leuze Verlag, Saulgau (1988)
28. Heitz, M., Fabian, W.: Renaissance der Eisen-Dickschichtphosphatierung. JOT. **11**, 94–96 (2000)
29. www.reach-service.de
30. Autorenkollektiv: Aluminium-Taschenbuch, Aluminium-Verlag Düsseldorf; 14. Aufl. 1988; 15. Aufl., Bd. 2. (1996)
31. Gebhard, M., Neuhaus, A., Blüthgen, W.: Chromatierschichten auf Aluminium. Fachberichte der Oberflächentechnik. **7–8**, 115–119 (1968)
32. Chemische Oxidation: Chromatieren, Phosphatieren von Aluminium, Merkblatt 02, 8th edn. Aluminium-Zentrale e. V, Düsseldorf (1987)
33. Arrowsmith, D.J.: Chromat-Phosphat-Überzüge auf Aluminium – Aufgabe der Ionen in der Lösung und Einfluss der Metallionen. Galvanotechnik. **78**(7), 1874–1882 (1987)
34. Sander, J.: Moderne Chromatierverfahren für Aluminium – Wirtschaftlichkeit und Umweltschutz. Tagungsberichtsband:„Entwicklungen in der Oberflächenbehandlung von Aluminium", 5. 19 ff., Aluminium-Verlag, Düsseldorf (1989)
35. Wernick, S., Pinner, R., Zurbrück, E., Weiner, R.: Die Oberflächenbehandlung von Aluminium, 2nd edn. Eugen G. Leuze Verlag, Saulgau (1977)
36. Roland, W.: Konversionsüberzüge auf Aluminium. Henkel-Referate. **24**, 20–25 (1988)
37. Wilkowski, W.: Chromatieren. Galvanotechnik. **70**(8), 712–719 (1978)
38. Brunner, H.: Chromatüberzüge auf Aluminium. Prod. Finish. (USA). **55**(2), 37–41 (1990)
39. de Riese-Meyer, L., Kintrup, L., Speckmann, H.-D.: Bildung und Aufbau Chrom-VI-haltiger Konversionsschichten auf Aluminium. Aluminium. **67**(12), 1215–1220 (1991)
40. Favilla, J.R.: No-Rinse-Behandlung für Aluminium. Prod. Finish. (USA). **55**(2), 45–55 (1990)
41. Puderbach, H., et al.: Analyse von dünnen Chromatierschichten auf Aluminium – Aufbau und Zusammensetzung von No-Rinse-Schichten. Henkel-Referate. **22**, 108–113 (1986)
42. Wernick, S., Pinner, R.: The Surface Treatment and Finishing of Aluminium and its Alloys, 5th edn. ASM International, Ohio (1987)
43. Dölling, H.: Übersicht über die Vorbehandlungsverfahren für das Beschichten von Aluminium-Oberflächen. Aluminium. **54**(11), 708–709 (1978)
44. Paatsch, W.: The danger of hydrogen embrittlement during electroplating. Oberfläche-Surface. **28**, 15 (1987)
45. Beschichten von Aluminiumguss – Lackieren, Arbeitsblatt der DFO. Deutsche Gesellschaft für Oberflächenbehandlung e. V., Düsseldorf (1991)
46. Schubach, P.: Chromfreie Vorbehandlungsverfahren für Aluminiumwerkstoffe. Galvanotechnik. **7**, 92 (2001)
47. Hughes, A.E., Nelson, J.H.: Interfinish 96. World Congress, The Institut of Metal Finishing (1996)
48. Child, T.F., van Ooij, W.J.: Application of silane technology to prevent corrosion of metals and improve paint adhesion. Trans. Inst. Of Metal Finish. **77**, 64–70 (1999)

49. Mulder, J.: Eine chromfreie Alternative. Metalloberfläche. **55**, 1 (2001)
50. Altenpohl, D.: Aluminium und Aluminiumlegierungen, vol. Bd. 19. Springer, Berlin (1965)
51. Hübner, W., Speiser, C.T.: Die Praxis der anodischen Oxidation des Aluminiums. Aluminium Verlag, Düsseldorf (1988)
52. Keller, F., Hunter, M.S., Robinson, D.L.: Structural features of oxide coatings on aluminum. J. Elektrochem. Soc. **100**, 411 (1953)
53. Gmelin: Handbuch der Anorgan. Chemie, Syst.-Nr. 27 Magnesium, Teil, 8. Aufl., S. 760–818 Verlag Chemie GmbH, Weinheim (und dort zitierte Literatur) (1952)
54. Dettner, H.W., Elze, J.: Handbuch der Galvanotechnik, vol. III, pp. 179–190. Hanser, München (1969)
55. Metals Handbook, 9. Aufl., Bd. 5, S. 628–649. ASM, USA (1982)
56. Emley, E.F.: Principles of Magnesium Technology, S. 670–735, 700–701. Pergamon Press, Oxford (1966)
57. SAE AMS-M-3171: Magnesium alloy: Processes for pretreatment and prevention of corrosion (1. Apr. 1998)
58. EU Altautoverordnung, Amtsblatt der Europäischen Gemeinschaften, Richtlinie 2000/53/EG vom 18. September (2000)
59. DIN EN ISO 2081:2018–07 Metallische und andere anorganische Überzüge- Galvanische Zinküberzüge auf Eisenwerkstoffen mit zusätzlicher Behandlung (ISO 2081:2018), Deutsche Fassung EN ISO 2081:2018
60. DIN ISO 12944-2 (2018-04): Korrosionsschutz von Stahlbauten durch Beschichtungssysteme, Teil 2: Einteilung der Umgebungsbedingungen

Further Literature on Pretreatment and Cleaning

Haase, B.: Wie sauber muss eine Oberfläche sein? JOT. **4**, 52–57 (1997)

Lohmeyer, S.: Silikonbedingte Benetzungsfehler beim Vorbehandeln, Lackieren und Löten. In: Galvanotechnik (1990), Nr. 3 Gerstenberg, K. A.: Qualitätsprüfung beim Entfetten. JOT (1995), **3**, 32–34

Lutter, E.: Die Entfettung – Grundlagen, Theorie und Praxis, 2nd edn. Eugen G. Leuze Verlag, Saulgau (1990)

Wood, W.: Metals Handbook. 9th Edition. VolumeSurface cleaning, finishing and coating. American society of metals, Handbook commitee, library of congress cataloging in publication data, A. S. of Metals (1982)

Kresse, J.: Säuberung technischer Oberflächen. Expert Verlag, Deutschland (1988)

Schreiner, A.: Bewertung der Rückstandsfreiheit von Metalloberflächen. JOT. **9**, 72–73 (1997)

Rahmel, A., Schwenk, W.: Korrosion und Korrosionsschutz von Stählen. Verlag Chemie, Weinheim (1977)

Kresse, J., Cornely, R., Pietsch, H., Vaeth, E.: Bestimmung von Restbefettungen auf Stahlblechen in niedrigen Oberflächenkonzentrationen. Meta. **37**(12), 500–503 (1983)

Böhm, H.: Glimmentladungsspektrometer zur Untersuchung von dekorativen Schichten. JOT. **4**, 124–128 (1994)

Hussain, A.: Wie sauber ist die Oberfläche? JOT. **4**, 120–122 (1994)

Bilger, G.: Messverfahren für die Oberflächenanalytik. Oberfläche + JOT. **8**, 32–37 (1989)

Kunze, E. (ed.): Korrosion und Korrosionsschutz, vol. Band 3. Korrosionsschutzverfahren. Wiley, Weinheim (2001)

Pietschmann, J., Pfeifer, H.: Filiformkorrosion auf organisch beschichtetem Aluminium, Teil 1 bis 3. Aluminium, **69**, **70**, 1019–1023, 1081–1084, 82–86 (1993/1994)
Dyllus, T.: Neue Möglichkeiten in der Vorbehandlung von Aluminium. Stahl und verzinkten Oberflächen durch die No-Rinse-Technologie. EPS-Pulvertagung, München (1999)
Pietschmann, J.: Chromfreie Vorbehandlungen im Vergleich – Welche Verfahren stehen Lohnbetrieben und Inhouse-Beschichtern zur Verfügung. EPS-Pulvertagung, München (2008)
Rauscher, G.: Neue chromfreie Vorbehandlungsverfahren für Aluminium. JOT. **2**, 54–56 (2005)
Rentsch, W.: Alternative Vorbehandlung in der Automobilindustrie. JOT. **51**(9), 34–39 (2011)
www.fit-online.org; Fachverband industrielle Teilereinigung e.V
www.technische-sauberkeit.de; Thema Bauteilsauberkeit
www.cleantool.org; Datenbank zum Thema industrielle Teilereinigung
www.cleaner-production.de; Projekt- und Literaturdatenbank zu Umweltthemen (Klima, Energie, Wasser, Luftreinhaltung u. a)

Images used in Chap. 5

FEM: 5.14 bis 5.17, 5.23, 5.24
Chemetall GmbH, Frankfurt: 5.6 bis 5.11; 5.18 bis 5.22

Modern Industrial Applications 6

6.1 Powder Coating of Wood and Wood-Based Materials

The industry's interest in the application of powder coatings for wood and wood-based materials was already aroused in the 1980s. At the beginning of the 1990s, the first formulations were available. However, experience with these systems was only gained in the test laboratories of some paint manufacturers from a purely technical point of view.

The first industrial application of powder coating of wood-based materials was in 1994 for MDF parts with structural powder coating. The process became known as the Wood Dual process [1, 2]. What was required here was a surface design adapted to the process and an electrically conductive water-based primer with careful intermediate sanding of the lacquer. However, the high demands on surface quality and the necessary reproducibility were not met. Even the slightest fluctuations within the MDF board production led to failures. A cost factor not to be neglected was the necessary surface quality of the MDF boards and the costly priming before powder coating [3]. In the meantime, only a few applications have been added. Well known are phono furniture made of MDF and beech wood toys as well as woven seating furniture made of bamboo or peddigree cane (rattan furniture) [4]. The coating material used here is a low temperature powder. These systems now have baking temperatures of 100 to 130 °C.

6.1.1 Material Requirements

The electrostatic coating process [5] requires a specific surface resistance Rn ≤ 1010 Ω, or in the case of a conductive backing, a specific electrical contact resistance ρ ≤ 108 Ωm. The resistances of dry wood-based materials (for example, after storage in a room climate of 23 °C and 50% relative humidity) are significantly above these limit values.

J. Pietschmann, *Industrial powder coating*,
https://doi.org/10.1007/978-3-658-37592-8_6

The relatively high temperatures for the melting and curing process of commercially available powder coating systems place great demands on the substrate material. It must be temperature-resistant (at least 110 to 140 °C for low-temperature epoxy resins (NT-EP), for higher requirements 160 °C for epoxy/polyester (EP/PE) or 180 °C for PE or PUR powder coatings) and must not outgas during film formation.

To prevent cracking in later use, the hydrophilic properties inherent in the wood should be suppressed, i.e. in the event of changes in ambient humidity, the wood-based material should absorb or release only a small amount of water and thus swell or shrink only slightly.

6.1.1.1 Solid Wood

Wood differs significantly in its structure from the substrates previously used for powder coating. It is a naturally grown anisotropic raw material. The main component is the fibers, which essentially run in an axial direction. They enclose vessels that are visible to the naked eye as pores on the cut surfaces from 150 μm in diameter. On average, their diameter is only 10 to 80 μm, but in hardwoods they can be up to 350 μm wide. In most native woods, this cavity system occupies 50 to 60% of the total volume and is filled with water, water vapor and air.

In order to avoid stresses in the solid wood, the heating rate would have to remain below the gradient necessary for good flow of thermoreactive powder coating films, with the consequence of poorer flow and the formation of an orange peel.

The resinous softwoods are not suitable for powder coating. The resin is pressed to the surface at temperatures as low as 70 to 80 °C. The wood is not suitable for powder coating. Hardwoods behave more favourably. The maximum permissible temperatures for baking the coating are 110 to 130 °C, otherwise drying cracks, warping or cell collapse may occur. Problems are caused by cross-sectional areas (end grain) as well as anomalies in growth or secondary wood changes.

6.1.1.2 Wood-Based Materials

Due to their more homogeneous structural composition, which can be influenced in the manufacturing process, wood-based materials, in particular medium- and high-density fibreboard (MDF and HDF), but also particle materials such as fine particleboard, offer more favourable conditions for powder coating.

In the case of fibrous materials, the production process involves mechanical/thermal treatment to break down the starting material into wood fibres or fibre bundles, with certain components such as the resins being washed out. After mixing with the binder (usually 8 to 12% urea or phenol-formaldehyde resin) and additives to improve water repellency (0.5 to 1.5% paraffin) and, if necessary, additives to protect against pest infestation and to increase fire retardancy, the pressing process follows. The temperatures on the upper and lower side of the boards reach about 260 °C, the pressure can reach values of up to 10 MPa. The pressing time is 10 to 15 s per millimetre of board thickness.

The greatest difficulty for industrial use lies in the formation of cracks in the powder coating film on the cut surfaces or in the profiles during subsequent use. Changes in the ambient humidity also cause water absorption or release of the hydrophilic substrate due to the limited water diffusion resistance of the coating or via direct moisture ingress at uncoated areas or damage.

Moisture-dependent swelling and shrinkage behavior perpendicular to the plane of the board is particularly pronounced in particleboard and fiberboard; the differential swelling value for MDF bonded with UF resin is 0.8 to 1% elongation per 1% increase in moisture content. For example, a change in relative humidity from 30% to 80% can shift the moisture content by 5% and cause a 1 mm change in thickness for a 20 mm thick board. A hermetic envelope is difficult to achieve in practice. But even then, temperature gradients in the board (for example, exposure to sunlight on one side) can cause shifts in the moisture content and thus partial changes in thickness that must be absorbed by the coating film.

6.1.2 Application Techniques

Three different technologies can be considered for the powder coating of wood-based materials [5]:

- the one-coat powder coating application directly on the MDF surface, with low bake or radiation curing systems,
- application of a conductive primer based on liquid paint and subsequent powder top coat,
- a direct powder coating primer with filler effect and subsequent liquid coating.

Single-Layer Powder Coating

Here, a substrate that is as smooth as possible is required. Mechanical processing methods such as sawing and milling only meet the requirements for a single-layer coating to a limited extent. Cut edges must be smoothed and compacted. Another problem is the correlation between moisture content and electrical conductivity. On the one hand, a uniform moisture content (5–8%) is required for the conductivity necessary for electrostatic application; on the other hand, this moisture can considerably disturb the process of paint film formation during the heating phase due to escaping water vapor. One approach to solving this problem is the use of fine structure powders in the low temperature range.

Conductive Liquid Paint Primer and Powder Topcoat

With the liquid paint primer, a smoother powder coating surface can also be achieved than with direct coating due to the smooth substrate after sanding. However, the actual advantages of powder coating are reduced, as the waste problem remains and processing steps such as intermediate sanding cannot be eliminated.

Powder Coating Primer and Liquid Coating Finish
Powder coatings are in principle also suitable as a primer. The micro-bubbles formed can be concealed by sanding and overpainting. The overpainting after an intermediate sanding can be done with liquid lacquers.

6.1.3 Powder Coatings

Thermoreactive Powder Coatings
Epoxy resins or epoxy/polyesters are mainly used as low-temperature powder coatings. The process window for the melting and crosslinking process is between 110 and 140 °C for epoxy resins and between 130 and 150 °C for epoxy/polyesters.

In addition, a pure polyester system is available today, which has a stoving temperature of 130 °C and 3 min holding time. The wood material is preheated to 60–80 °C, painted and baked at 130 °C for 3 min.

A significant reduction of the processing temperature is not to be expected due to the necessary storage stability of the powder coatings. Furthermore, with the lowering of the melting temperature, a deterioration of the flow and the adhesion is to be expected. A decisive step towards temperatures below 100 °C requires the development of crystalline binders, but this is not currently foreseeable (Fig. 6.1).

6.2 In-mould Coating

With in-mould coating (IMC), the plastic moulded part is already coated in the mould [6]. For this purpose, a suitable coating system is first introduced into the mold and then the carrier material, the plastic, is added. After the reaction time, the part can be removed from the mold already coated. The surface finish can range from high-gloss to matt or textured,

Fig. 6.1 Powder coating of MDF boards

depending on the shape of the mould. In addition to greater design freedom in shaping, the IMC process enables cost savings by eliminating or shortening individual process steps.

In-mould powder coating is not new. The first generation of powder coatings was used in the sanitary sector, e.g. for the direct coating of shower trays and sinks. In the automotive industry, however, higher demands are made in terms of durability, technical properties and process speed. One example for automotive applications are powder-coated storage compartments for truck cabs made of SMC (sheet-moulding compound).

In-mould powder coatings have now established themselves as effective systems for coating numerous moulded fibre-reinforced plastics. While initially only SMC, a composite material based on glass fiber reinforced polyester, was used on a technical scale, other applications have now also been successfully introduced. Examples are BMC (bulk moulding compound) or prepreg composites.

A prerequisite for the application of fiber-reinforced plastics in the automotive industry is good recoatability with high demands on durability and optical appearance. In wet painting, pinholes or porosity caused by the micropores in the cured fiber composite are often the cause of high rework or rejects. The PIMC process (powder-in-mould-coating) offers a solution.

Powder Coating

The powder coating is applied directly into the hot mold using electrostatic guns. The powder melts and undergoes pre-crosslinking already here. The composite material is placed in the mold and the powder coating and plastic are cured together. Depending on the size and complexity of the mould, the powder application takes between 10 and 40 s. New developments in the field of powder coatings have made curing temperatures of 105 °C and a holding time of 3 min possible.

During joint pressing and curing, the plastic reacts with the unsaturated polyester in the coating material. This results in a very good bond and very good adhesion of the paint. Figure 6.2 shows the cross-section of the bond between the powder coating layer and the SCM.

Depending on the complexity of the mould, special additives are added to the powder coating to allow easy and clean release of the workpiece from the mould.

The surface of the workpiece is completely painted. During the curing process of the powder, the surface texture of the mold is mapped. The roughness of the mold, smooth or textured, causes the appearance of the coating. Glossy, matte or textured surfaces can be produced.

Figure 6.3 shows the process flow of the PIMC method [7, 8].

The powder coating film forms an extremely thick layer that covers the air pockets in the substrate material. When overcoating an in-mould powder coating with a wet paint, the pores in the material thus have no negative effect on the paint finish. To achieve a perfect high gloss, the matrix must be polished and chromed.

Most pre-mould powder coatings are based on unsaturated polyester resins. The high reactivity allows the plastic compound to be applied immediately after powder coating.

Fig. 6.2 Section of SMC with in-mould powder coating [6]

High cross-linking densities of the coatings result in finished parts with high scratch and chemical resistance.

The coating is available as a primer coat or as a top coat and offers, in addition to the advantages already mentioned, further advantages such as:

- high degree of hardness (4H-5H)
- excellent adhesion
- very good edge coverage
- barrier effect
- good chemical resistance
- very good recoatability when used as a primer
- very good scratch resistance

With the coating the plastic gets a higher resistance. Without a protective coating, aging processes in the polymer lead to fiber separation and minute cracks in the system. The relatively thick and uniform powder coating (100 to 200 μm) reliably protects the porous surface of the plastic composite material [8].

The application of in-mould powder coating for motor vehicles includes the priming of exterior parts for the automotive sector, mouldings and air deflectors for trucks or body parts directly in topcoat for agricultural machinery. One advantage of the process is that the coating has an exceptionally good barrier effect; as a result, the volatile substances – which evaporate from SMC after a certain time – are stopped. The use of SCM moulded parts is therefore also possible for interior applications.

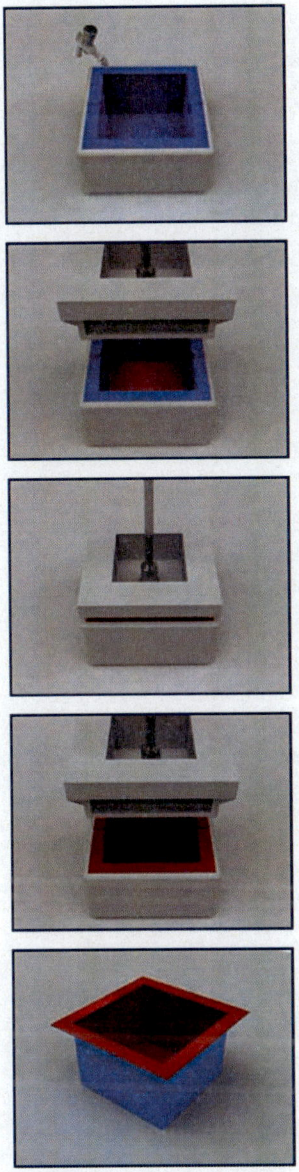

Fig. 6.3 Process steps in the PIMC process [6]

6.3 Powder-in-Powder Technology

A new powder coating process is the powder-in-powder technology [9, 10] where a primer and the colored top coat are applied in one step without baking the first powder layer. The process offers advantages due to

- Saving of one baking and one cooling process and thus savings in energy consumption and production time
- Higher corrosion protection due to two layers and better edge coverage.
- Lower investment and maintenance costs due to reduction of conveyor technology, oven and cooling zone and thus less production area (Fig. 6.4)

The process is mainly used in the field of massive construction and agricultural machinery. The following parameters must be observed during application:

- Good grounding of the components
- Application without excess charge. Excessive charging of the primer or topcoat leads to repulsion effects which can cause mixing of the two layers.
- low conveying air, soft cloud to avoid blowing off the first layer.

6.3.1 Minimising Mixing Effects

There should not be a too great difference in grain size distribution between the primer and the topcoat in order to minimise the risk of mixing effects when overcoating.

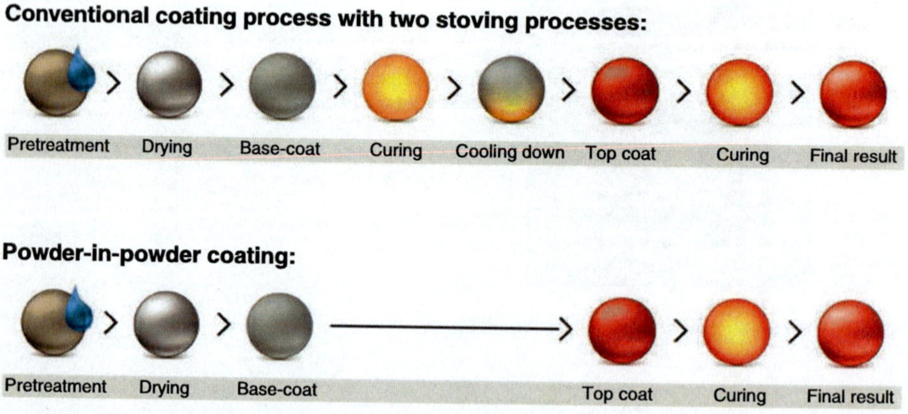

Fig. 6.4 Comparison of the coating process with conventional coating and in comparison with the "powder in powder" technology

Co-crosslinking of the powder coating layers prevents the migration of the primer to the surface, which impairs the light resistance. Furthermore, the intercoat adhesion is improved in this way.

The primer and the topcoat must have a matching charging behaviour. The primer should have a high charging potential in order to achieve a good deposition efficiency without trickle effects on the substrate. At the same time, a high degree of deposition efficiency is also required with the topcoat, without trickling and mixing effects during overcoating.

The charging behavior is adjusted via the powder coating formulation. The selection of binders, pigments, fillers and additives has a considerable influence.

6.3.2 Back Spray Effect Due to High Charge

The application technique should charge the powder coating without excess charge. Care should be taken to ensure very good grounding of the components to be coated. Excessive charging of the primer and/or the top coat can lead to repulsion effects, which can cause both layers to mix with each other. This leads to an inhomogeneous surface, poorer weathering stability and also to lower corrosion resistance.

The air volume to the gun should be set as low as possible ("soft cloud"), otherwise blow-off effects of the primer and thus also mixing effects may occur during overcoating. Particularly in the case of tribo application, ensure uniform coating conditions.

In practice, charging by means of tribo-technology has meanwhile become established. With unequal charging (plus/minus or minus/plus), the deposition of the top coat can take place with lower repulsion effects, but there are also discharge effects, since these opposite charges cancel each other out.

A reduction of the primary adhesion of the powder coating on the substrate, as well as an increased trickling behaviour, and thus material loss, are the consequences.

6.3.3 Edge Cover for Better Corrosion Protection

In order to achieve the highest possible corrosion protection, the primer should reliably cover the edges of the component in particular. This can be achieved by adapting the flow properties of the primer to the application. If the flow properties of the primer are too good, the topcoat will flow away from the edge together with the primer during the melting phase.

When using a primer with very good edge coverage, however, the layer thickness of the primer must not be too high (max. 80 μm), otherwise the flow will be significantly worse.

This means that the coating window becomes much narrower when using a primer with high edge coverage, which is difficult to maintain in practice for complex components. Here it is necessary to find a good compromise between flow and edge coverage.

In order to achieve optimum edge corrosion, the edges of the components to be coated should also be sufficiently mechanically processed or blasted.

6.3.4 Application

Different application options are available for this procedure, see Table 6.1.

Depending on the application or the components to be coated, the tribo lance should be mentioned [12]. Compared to a conventional tribo gun, the tribo lance offers the following advantages:

- Significantly higher charging potential even with low air volume
- Higher area output (at least comparable to a Corona gun) due to larger output of charged powder
- The lance is telescopic for manual coating of large machines
- Compact design and angle adjustment of the nozzle head

Figures 6.5 and 6.6 show the components of the lance and an example in use.

6.3.5 Different Coating Technologies on the Market

Depending on the powder coating manufacturer and the user, different solutions are used for the coating structure. Table 6.2 describes the various possible combinations of powder coating systems and their advantages and disadvantages.

In order to be able to carry out a visual layer thickness check and for the purpose of a meaningful assessment of the edge coverage, the primer and top coat should be applied in different colour tones. This means that the topcoat cannot be applied in a closed circuit, as overcoating the primer would result in primer entering the topcoat circuit, which would lead to color specks in the topcoat.

If the primer and topcoat have the same colour, the topcoat must still be run at loss if an epoxy or hybrid primer is used, as its lightfastness is impaired by the primer powder addition.

The primer and the topcoat can only be circulated if they have the same colour and the same binder chemistry. Polyester systems should be used for weather-resistant systems.

Table 6.1 Application procedure for the powder-in-powder technique [11]

	Primer		Top coat	
	Application	Charging	Application	Charging
Procedure 1	Tribo	Positive	Tribo	Positive
Procedure 2	Corona	Negative	Corona	Negative
Procedure 3	Tribo	Positive	Corona	Negative
Procedure 4	Corona	Negative	Tribo	Positive

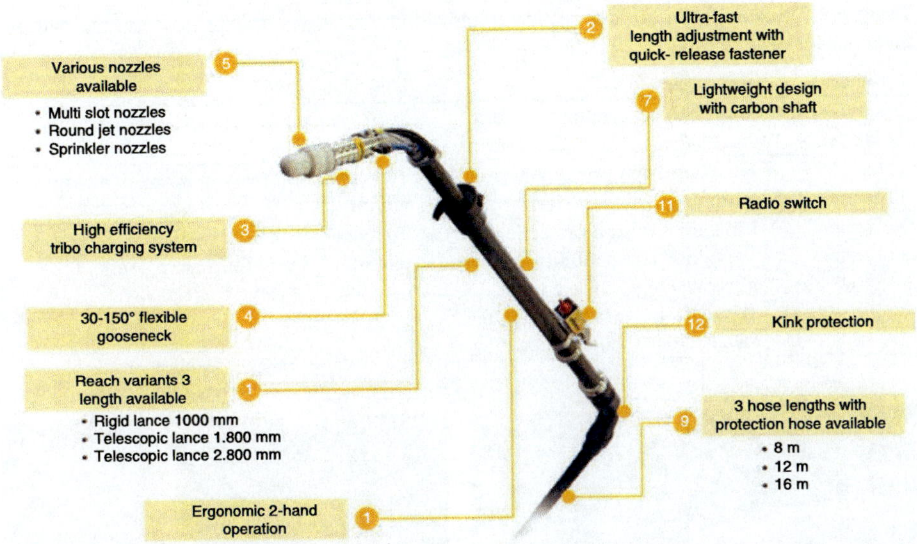

Fig. 6.5 Structure and components of a tribological lance

Fig. 6.6 Application of large workpieces with the tribo lance

Table 6.2 Possibilities of powder coating combinations for primer and top coat and their advantages and disadvantages [11]

Structure	Advantages	Disadvantages
Epoxy/ Polyester	Very high corrosion protection	Top coat cannot be processed in a close circuit
Hybrid/ Polyester	High corrosion protection	Top coat cannot be processed in a close circuit
Polyester/ Polyester	Limited corrosion protection Very good compatibility, for high gloss systems Topcoat can be circulated in the same colour as the primer	Topcoat cannot be processed in a close circuit in case of different colours

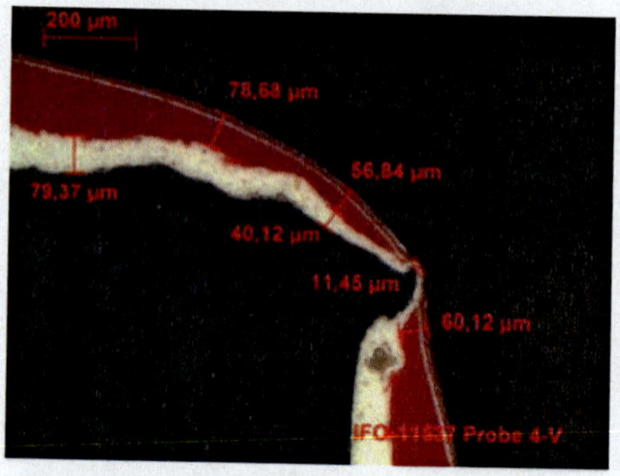

Fig. 6.7 Cross section over the edge on the front side of the workpiece

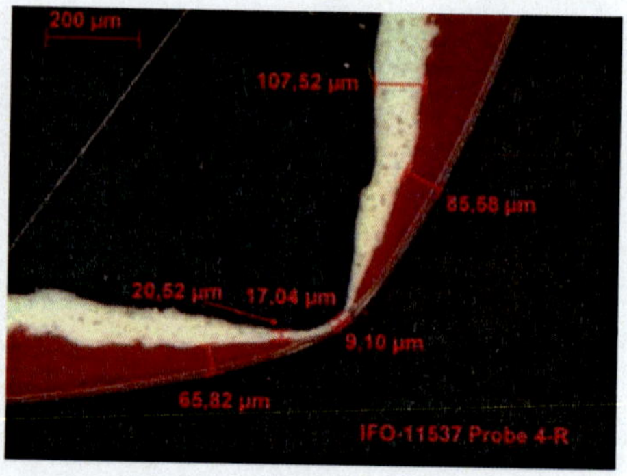

Fig. 6.8 Cross-sectioning over the edge on the back of the workpiece

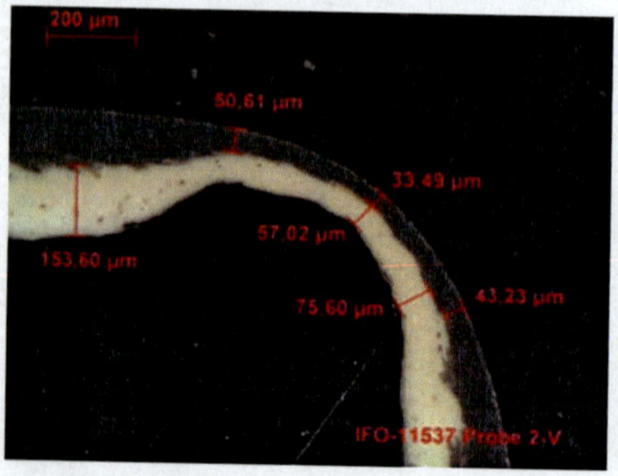

Fig. 6.9 Cross section over the edge on the front side of the workpiece

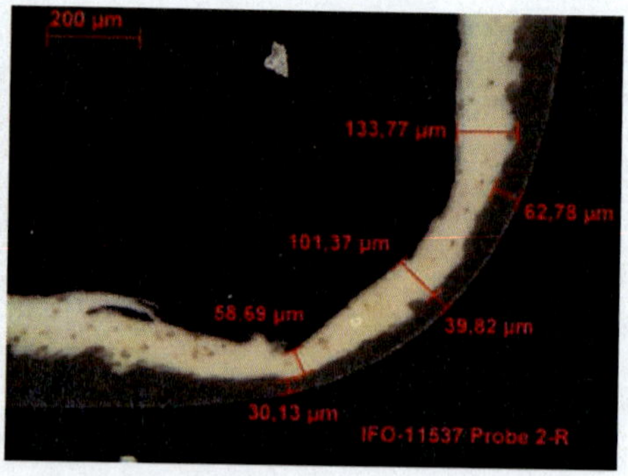

Fig. 6.10 Cross section over the edge on the back of the workpiece

6.3.6 Problems with the Layer Structure

Repulsion effects, mixing and blow-off effects lead to problems with the coating. Poor edge build-up and edge coverage is the result, see Figs. 6.7 and 6.8.

Optimum powder-in-powder layer build-up on the front and back of the workpiece can be seen in Figs. 6.9 and 6.10.

6.4 Contactless Coating Thickness Measurement Before Baking [13]

An important instrument for controlling the coating quality is the coating thickness. This is influenced by various factors, such as the geometry of the workpiece, the coating system – powder coating and its charging behavior, the ambient temperature, the humidity and the application itself. Despite optimal settings, fluctuations occur during the day, so that even with a carefully adjusted coating system, the coating thicknesses exceed the tolerance limits.

With conventional contact coating thickness gauges, measurement and control can only take place after the coating has been baked and cooled, which means a time lag between coating and the detection of undercoating or overcoating. Thus, contact coating thickness gauges are not suitable for quasi-online process monitoring of coating systems.

For continuous process monitoring of coating systems, the coating thickness measurement must be carried out promptly after the coating has been applied. A non-contact measurement is a basic requirement here in order not to destroy the coating. A robust and reliable coating thickness measurement must be tolerant to tilting and swaying of components. A distance tolerance of at least 10 cm and a tilt tolerance of at least ±70° are useful guidelines. The measuring range should cover a range of at least 20 µm and 300 µm and be suitable for all colors. A movement of the measuring object often leads to incorrect measured values. This can be eliminated if the measurement time is as short as possible, i.e. ideally less than 200 ms.

In thermal coating testing (TSP), Fig. 6.11, the surface of a coating is heated over a large area with a short light pulse of 1 ms in length. The temperature difference generated in this way produces a heat wave which travels through the coating structure. The travel time is determined by means of non-contact temperature measurement and based on heat conduction equations. The travel time of the heat wave is converted into a coating thickness value. Due to similar thermal properties of powder coatings, calibration is not required for thermal coating testing [13].

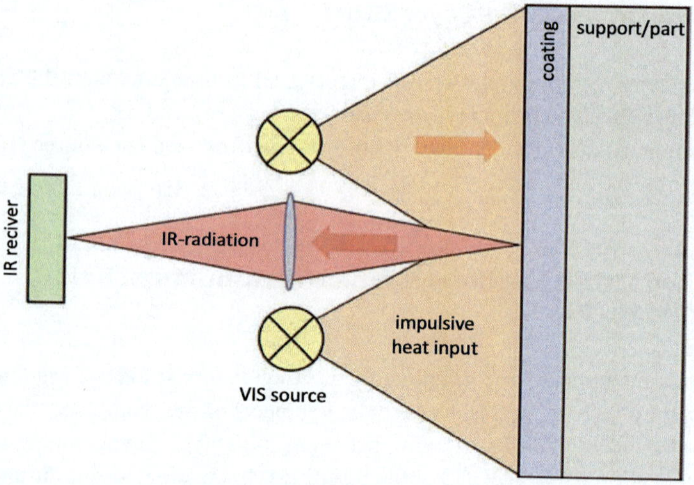

Fig. 6.11 Function of the thermal coating test (TSP)

6.4.1 Possibilities of Coating Thickness Measurement

Depending on the application, either a hand-guided, Fig. 6.12, or a fixed-integrated coating thickness measurement Fig. 6.13 immediately after the coating application is possible. A fixed integrated solution offers the advantage of an automated and thus continuous readjustment and logging of the coating process. A hand-held measurement can be used flexibly and does not require any integration effort.

Imaging layer thickness measurement (Fig. 6.14) is an alternative to point-based measurement. In this method, the coating thickness distribution on complex geometries and extensive surfaces is determined in a single measurement process. With the imaging coating thickness distribution, turning points of coating guns can be detected as well as locally too thin or too thick coated areas.

For continuous process monitoring of coating systems, the coating thickness measurement must be carried out directly after the coating has been applied.

Fig. 6.12 Hand-held device for punctual coating thickness measurement during operation

Fig. 6.13 Measuring device for automated coating thickness measurement during operation

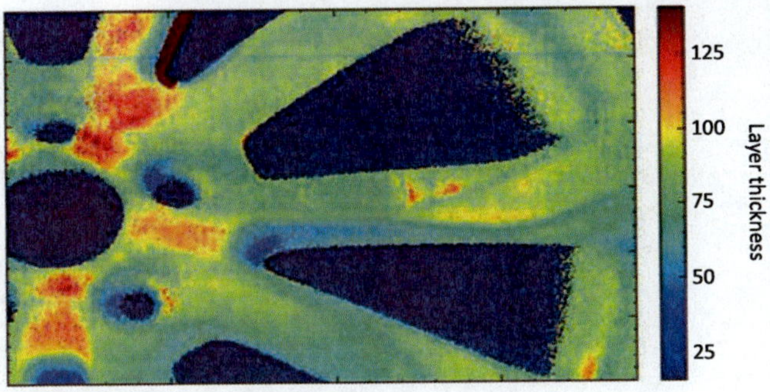

Fig. 6.14 Imaging layer thickness measurement using the example of a rim

6.5 Tailor-Made System Layout in Wheel Painting

The powder coating of wheels for the automotive industry places high demands on the application and quality of the coating. Increasing individualization, different wheel sizes, highest quality of surfaces and more and more frequent color changes are only some of the challenges. The flexible and expandable C-Line coating unit from Wagner [14] is an example of highly automated coating in a small space.

The C-Line powder coating system was developed specifically for wheel coating in the automotive industry. The focus was on the smallest possible footprint for easy integration into existing production lines. The compact design with a booth length of 2 m requires only about 1/3 of the space compared to today's systems, Fig. 6.15.

To achieve the high demands on the coating quality, the booth and application are equipped with the latest control technology, whereby this has been specially designed for the application. The parameters are entered purely visually using simple graphics. This means that the positions of the guns can be assigned to the wheels shown to scale using a

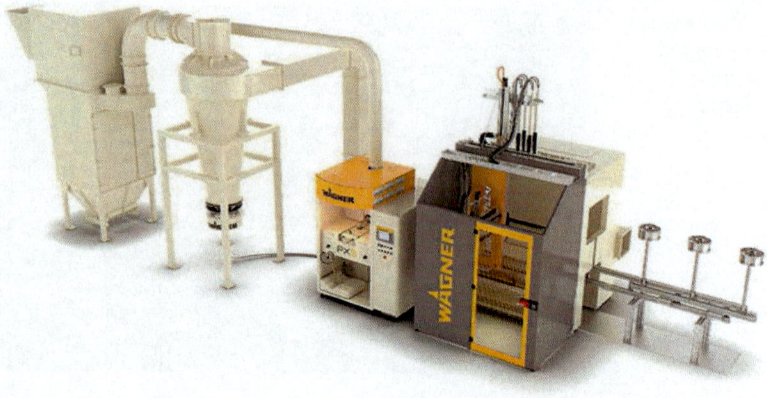

Fig. 6.15 Compact coating plant for powder coating of wheels

Fig. 6.16 Arrangement of the powder guns from below with immersion in the wheel; from above and the side

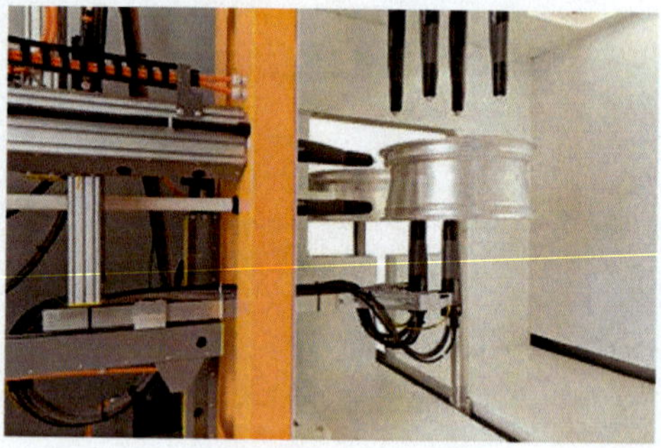

Fig. 6.17 Moving powder cloud while driving through the cabin

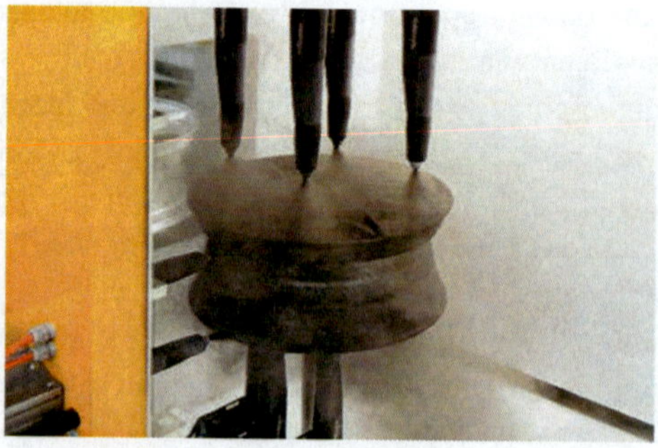

visual "teach-in" procedure. Complex parameterization is not necessary, programming knowledge is not required.

During the coating process, the 3D-axis system moves all powder guns simultaneously with the wheel through the booth. The guns are arranged opposite each other (Fig. 6.16) in order to make the best possible use of the "moving powder cloud" effect (Fig. 6.17). In this way, a wide and stable powder cloud can be generated even at high conveying speeds of up to 8 m/min. This leads to a uniform coating quality for wheels between 13″ and 24″. At the same time, each individual powder gun can be program-controlled and fully automatically positioned according to the wheel type.

The booth itself is about 2/3 smaller than conventional cabins in this application area. Cleaning takes place from the outside via a large opening. With the contactless spindle sealing at the conveyor slot, the booth design with stepped floor as well as one-sided suction, color changes can be carried out in less than 10 min. The guns are automatically blown off by moving them from the outside along the blowing nozzles.

The coating booth has a filter, cyclone and pipe systems in a compact design. Space and up to 30% energy can be saved compared to conventional coating lines, according to the manufacturer.The system is available in modular form as a single or multiple booth system and allows the combination of several booths connected in series.

References

1. Hödelmoser, S. P.: Pulverbeschichtung von MDF-Platten in der Praxis. „Der Pulvertreff '95", München (Februar 1995)
2. Radoczy, I.: Das Wood-Dual-Verfahren zur MDF-Veredlung. I-Lack **63**(3), 96–98 (1995)
3. Rijkse, K.: Neue Möglichkeiten mit Niedrigtemperatur-Pulver. JOT **38**(5), 36–39 (1998)
4. Foitzik, B.: Pulver auf Holzwerkstoffen – Wer wagt den nächsten Schritt? JOT. **41**(2), 14–15 (2001)
5. Bauch, H.: Pulverbeschichten von Holzwerkstoffen. JOT **10**, 40–51 (1998)
6. Kok Enno, Powder In-Mould Coatings (In-mould-Pulverbeschichtungen); Eine effiziente Methode für die Beschichtung von SMC bei Fahrzeuganwendungen; AkzoNobel Powder Coatings, Sassenheim (NL); enno.kok@akzonobel.com, tel: +31 653 677 268; www.Interpon.com
7. Impress – In-Mould Powder Coatings, Impress Verfahrens Broschüre der AkzoNobel Powder Coatings
8. Groch, A.: In mould Pulverbeschichtung von Kunststoffteilen, JOT Fachtagung EPS 2010, Tagungsband Seite 109 bis 114
9. Schelling, A.: Die Pulver-in-Pulver Technik, interne Dokumentation 2018, Emil Frei GmbH & Co. KG
10. Capizzi, M.: interne Informationen, Karl Bubenhofer AG, CH-Gossau.
11. Schelling, A.: Pulver-in-Pulver – Herausforderungen und Perspektiven. JOT. **56**(4), 70–74 (2016)
12. Topp, M.: Tribolanze PEM TL-1 mit Teleskop Schaft, Beschichtung in neuen Dimensionen, interne Produktinformation. Wagner Group, Markdorf (2016)

13. Reinke, N.: interne Informationen, Winterthur Instruments AG, CH-8404 Winterthur
14. Haller, R.: WAGNER C-Line: spezifische Lösung für die Räderbeschichtung, interne Information. Wagner Group, Markdorf (2018)

Further Literature

Bauch, H.: Fortschritte bei der Pulverlackierung von Holzwerkstoffen. Pulverlack-Kongress 1996, Hamburg, Tagungsband S. 95–110, 24. September 1996

Sandig, C., et al.: Reibglätten – ein rationelles Verfahren zur Feinbearbeitung von Profilen in MDF. Holz-Zentralblatt, **6**, 81–82 (Januar 1996)

Sandig, C.: Rollglätten – Alternative zum Schleifen? HK Holz- und Kunststoffbearbeitung. **33**(5), 64–67 (1998)

Bauch, H.: Verfahren zum elektrostatischen Beschichten von Holz und Holzwerkstoffen. DE 195 33 858 A1

RAVEL-Dokumentation „Energiesparmöglichkeiten in der industriellen Lackiertechnik" Nr. 724.397.21.07 D von Rolf Gloor

Bayerisches Landesamt für Umwelt (2006) Energieeinsparung in Lackierbetrieben – Langfassung, Augsburg

Bayerisches Landesamt für Umwelt (2009) Leitfaden für effiziente Energienutzung in Industrie und Gewerbe, Augsburg

BUBW (2012) Lackiertechnik – Grundsätzlicher Verfahrensablauf, Betrieblicher Umweltschutz, Baden Württemberg (BUBW)

Images used in Chap. 6

Akzo Nobel Powder Coatings, Sassenheim 6.2 und 6.3
Coatmaster AG, Winterthur, CH: 6.11, 6.12, 6.13, 6.14
Freilacke GmbH&CoKG, Bräunlingen, D: 6.4
Gema, St. Gallen, CH: 6.1
Karl Bubenhofer AG, Gossau, CH: 6.7, 6.8, 6.9, 6.10
J. Wagner GmbH, Markdorf, D: 6.5, 6,6, 6.15, 6.16, 6.17

Measurement and Test Engineering

7

7.1 Tasks of the Test Engineering

Any paint or coating should contribute to this,

- protect the painted material from the attacks of the environment.
- to maintain the visual impression throughout the entire service life.
- fulfil the functional task for which a surface or component was designed.

To achieve these goals, all materials and working methods must be tested and coordinated with each other. The ready-to-use coating is of course in the foreground. Finally, the actual product, the coated object, must be tested. For the control of the coating result and the production, most tests are performed after the final curing of the coating and a few can be used already during the coating process. The requirements for the coating result from its intended use and are usually specified by the customer or in quality guidelines. Standards do not usually contain specifications on minimum requirements to be met. They do, however, describe in detail the procedure for testing and documentation of the test results. In the testing technology of paint coatings, there is a multitude of procedures and procedure implementations. This chapter presents some of the most important and most frequently used test methods for coatings and coated surfaces.

7.1.1 Testing of Coating Materials

The inspection of the delivered coating materials is in principle an incoming goods inspection. In the accompanying safety data sheet or specification data sheet, the coating supplier provides information on the composition, the chemical-physical, application-related and

J. Pietschmann, *Industrial powder coating*,
https://doi.org/10.1007/978-3-658-37592-8_7

safety-related properties and characteristic values. He is responsible for the classification and labelling and for the information on handling, storage and disposal of the products.

The user checks the conformity of the stated properties with the actual properties or characteristic values. The rapid incoming inspection of coating materials is complicated by the fact that most of the technical application properties cannot be tested on liquid or solid (in the case of powder coatings) coating materials, but that only the finished coating shows whether the mechanical, optical or resistance properties meet the requirements. This can mean a time-consuming incoming inspection. Another possibility is an agreement between supplier and user, in which the supplier guarantees the required quality, or only those materials are used that bear an independent quality seal, which is awarded by independent organizations and is constantly monitored. An example in the field of unit coating of components are the quality marks Al 631, for the coating of aluminium components or ST 663 in the field of galvanized steel of GSB-International [1].

Further quality associations are the QIB [2] in the field of industrial coatings and Qualicoat [3] in the field of architecture.

In principle, not as much as possible is tested, but as little as necessary. The tests used always depend on the requirements and the intended use of the coated object.

7.1.2 Testing the Substrate

The same applies to the inspection of the supplied materials to be painted. The removal of inherent and foreign surface impurities, release agents or other substances that impair the adhesion of the coating material to the substrate is a prerequisite for an optimum and durable coating.

Checking the cleanliness of a workpiece surface before coating can be carried out using various methods as already described in Chap. 5.

A quick test for the pre-treatment of the workpiece surface and adhesion of the paint layer is the boiling test. It is anchored in the quality guidelines of GSB-International [1] as a test for checking the quality of the pretreatment, see Sect. 7.3.16.

7.1.3 Checking the Application and Drying (Stoving Process)

When checking the application and drying, the focus is on the film thickness, visual assessment for uniformity of application and checking for contamination in the paint film and checking the curing conditions.

In order to achieve uniform melting and sufficient cross-linking of the powder coating film in the curing oven, the workpieces must reach and maintain a certain object temperature over a certain time. Only with an optimum curing of the coating layer can the properties be achieved, such as adhesive strength, sufficient elasticity, resistance to weather and environmental stresses, gloss and colour consistency, hardness. The exact temperature distribution in the curing oven must be known in order to avoid rejects and to guarantee consistent quality.

In coating operation, the proper function of the baking oven should be checked at regular intervals by means of oven temperature measuring devices.

7.1.4 Testing of the Coating

Properties of coating materials that can be expressed by chemical and physical key figures, e.g. the density or the particle size distribution spectrum, can be checked on powder coatings relatively quickly and with little effort. The situation is different for a number of application-related properties. The lightfastness of an automotive paint or the corrosion protection effect of a coating can only be definitively determined once the object has fulfilled its intended purpose over a certain period of time.

With the help of defined periods and locations in the outdoor weathering and short-term weathering, one tries to make statements about the long-term behaviour of the coating. However, the repeatedly asked question of a correlation between the results of the short-term test and those of the long-term weathering or even the practice – for example according to the formula: x minutes short-term test, e.g. weatherometer, = y years, outdoor weathering = xy years practice – cannot be answered.

Accelerated test methods are suitable for establishing a relationship. If a proven coating material is to be replaced by another, both products can be tested side by side. From the results obtained in this way, it can be deduced whether the resistances are the same, different or very different. Such a statement can be obtained very quickly by means of ever better developed and specified test equipment, but it does not make a definitive statement about the behaviour of the system under practical conditions.

Test specifications exist in the form of

- Standards,
- Guidelines and rules, especially industry-specific,
- Factory standards and
- Agreements between suppliers and customers.

Within the company, all test specifications must be implemented in binding procedural instructions, because tests are an essential part of a company's quality assurance. They must be described in a comprehensible manner. Their results must be carefully documented, because they can be helpful in the event of complaints.

7.2 Appearance

Perception of the Appearance

A uniform appearance is considered an important quality feature for many products. Gloss effects are based on the interaction of light and physical surface properties. However, they also depend on physiological evaluation standards.

The human eye is still the best instrument for assessing differences in gloss and color. Nevertheless, visual surface inspection is insufficient, since

- usually no defined sampling conditions are available and
- different examiners see and evaluate differently.

In addition, the subjective perception of gloss and color is also influenced by personal experience. The following factors are involved in the visual evaluation of appearance:

Surface Texture

Material (e.g. lacquer, plastic or metal) and structure (e.g. smooth, rough, wavy)

Type of Lighting

A prerequisite for the assessment is directional lighting. With diffuse lighting, the reflection is also diffuse and the gloss impression is reduced.

Observer

In visual perception, the vision and mood of the observer play a decisive role. However, what we focus our eye on is also decisive. If the gaze is focused on a reflected light source, the imaging quality of the surface, i.e. the ability to reflect objects recognizably, is evaluated. The surface may appear brilliant or dull (gloss). When an edge is reflected, the dark area may be somewhat lightened (gloss haze) and the edge boundary may be sharp or blurred (DOI – distinctness of image, brilliance) [4].

If the view is focused on the surface itself, an additional impression of the size and shape of structures is obtained. This is seen as a wavy pattern of light and dark fields. This waviness is often referred to as orange peel or gradient interference.

Figure 7.1 shows both forms of observation again, which contribute to the overall impression of the appearance.

7.2.1 Gloss Measurement

Gloss is a visual perception that occurs when looking at surfaces. The perception of gloss is more pronounced the more directionally the light is reflected.

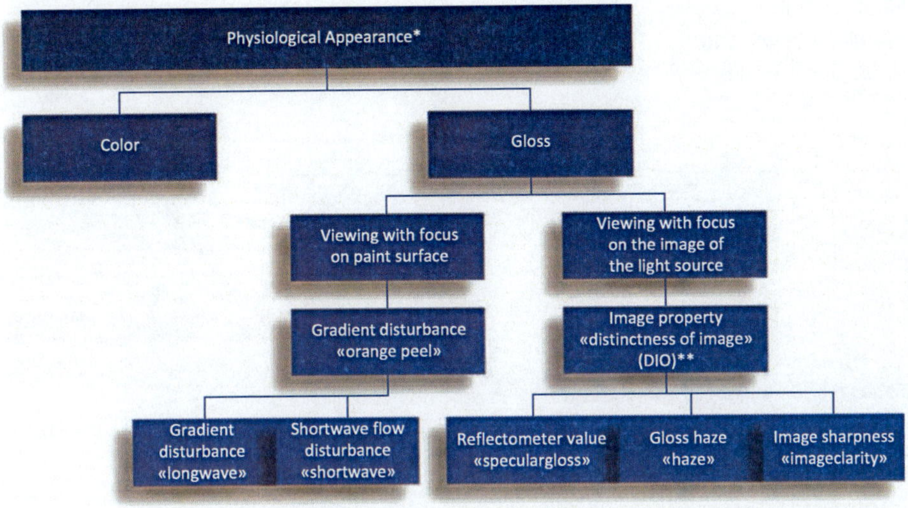

Fig. 7.1 Schematic of the surface description. *) Since there are no uniform terms in the German language, the internationally common terms according to Hunter [5] were listed. **) The term DIO is used in English as a synonym for image sharpness

Fig. 7.2 Principle of gloss measurement on high-gloss surfaces: Angle of incidence is equal to the angle of reflection

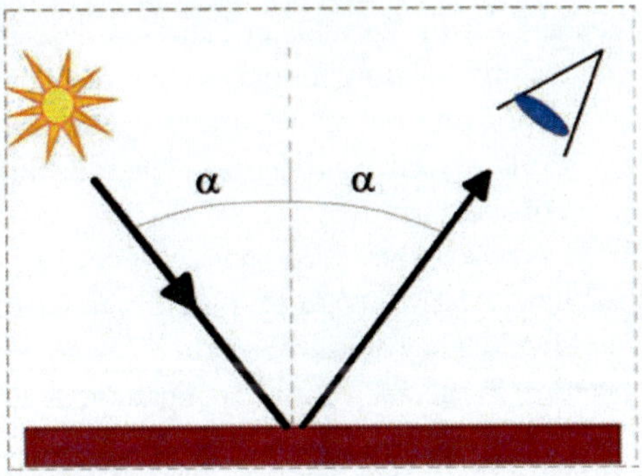

High Gloss
Mirror images can be clearly recognised on flat, completely smooth, polished surfaces. The incident light is reflected directionally at the surface, i.e. only in the main reflection direction. The angle of reflection is equal to the angle of incidence, see Fig. 7.2.

Medium to Matt Gloss
From rough surfaces, the light is not only reflected in the main reflection direction, but also diffusely in other directions. This reduces the imaging quality of the surface. A reflecting object is no longer imaged brilliantly here, but blurred, see Fig. 7.3.

Fig. 7.3 Principle of gloss measurement on rough surfaces: diffuse reflection of incident light

Table 7.1 Standardized geometries for gloss detection

Gloss level	60° value	Is measured with
Medium gloss	10 to 70	60° geometry
High gloss	> 70	20° geometry
Matte gloss	<10	85° geometry

Measuring Principle

The so-called reflectometer value, as described in detail in DIN 67530 [6], is used as a comparative measure for the measurement of the gloss of a surface. Its value is determined by the proportion that the surface contributes to the development of the gloss impression due to its reflective properties. The reflectometer value is the optical parameter for the gloss of a surface.

In practice, the reflectometers described in the standards DIN 67530, DIN EN ISO 2813 and ASTM D 523 [6–8] with the measuring geometries 20°, 60° and 85° are used worldwide. They are sufficient to measure the gloss types of various surfaces that occur in practice. In special areas of application, devices with 45° and 75° geometry are also used.

The principle of the reflectometer is based on the measurement of the directional reflection. For this purpose, the intensity of the reflecting light is measured in a narrow range of the angle of reflection. The intensity of the reflecting light depends on the material and the angle of incidence. In the case of painted and plastic surfaces, more light is reflected as the angle of incidence increases. The rest of the irradiated light penetrates the material and is partially absorbed or diffusely scattered there, depending on the colour tone.

The measurement results of the reflectometer are referred to a black, polished glass standard with a defined refractive index and not to the irradiated light quantity. For this standard, the measured value is set = 100 gloss units.

Since the angle of incidence used has a strong influence on the reflectance value, three geometries (measuring ranges) were standardised to differentiate between highly specular and matt surfaces. Table 7.1 shows the three measuring ranges.

7.2.2 Haze

A clear, brilliant appearance is expected from high-quality surfaces. Microstructures, e.g. due to poor dispersion, cause a milky appearance. This effect is called gloss haze or haze.

If microscopically fine disturbances are present on a highly specular surface, low-intensity scattered light occurs in addition to the main reflection. The majority is reflected in the direction of the mirror, so that the surface appears highly specular and image-forming, but is overlaid by a milky cloudy veil.

Measuring Principle

Goniophotometers are used to measure the gloss haze. The basic amount, which remains after the steep drop at the measured reflection indicatrix, represents a measure of the gloss haze [9–11]. However, the handling of goniophotometers is quite complicated and time consuming. Therefore, a combination of reflectometer and additional sensors was developed to detect the gloss haze.

Since gloss haze only occurs with high-gloss surfaces, the 20° geometry was used for these measuring instruments in accordance with the reflectometer. The angular range of the reflectometer at 20° is 1.8°. Two additional sensors next to the gloss detector measure the intensity of the scattered light for haze evaluation. In order to make the obtained values more consistent with human perception, the haze value is displayed logarithmically. The smaller the haze value, the better the surface quality.

7.2.3 Waviness: Orange Peel

Wavy structures in the order of approximately 0.1 to 30 mm are referred to as flow defects or orange peel. Such effects are usually assessed visually and described with the terms greasy or gritty. Gradient defects are seen on high gloss surfaces as a wavy pattern of light and dark fields. Light is reflected in different directions according to the angle of inclination of the structural elements. Only the surface elements that reflect light toward the eye are seen as bright. The appearance and detectability of the structures depend on:

Structure size: Fig. 7.4 shows the visual differences of surfaces with different structure sizes.

Viewing distance: Long-drawn waves of e.g. 30 mm wavelength are better visible from a great distance. Short wavelength structures in the range of 0.1 to 1 mm are only visible at close observation.

Resolution power of the eye: Structures below the human resolving power are no longer visible as light/dark patterns, but lead to a reduction in brightness.

Imaging quality: Structures are perceived more clearly the higher the contrast and the sharper the edges are reflected in the surface.

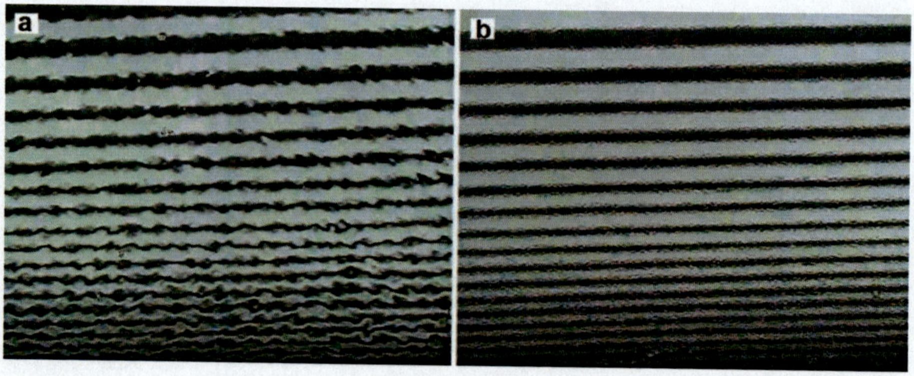

Fig. 7.4 Influence of structure size on visual appearance; (**a**) larger structures, (**b**) small structures

Fig. 7.5 Schematic representation of ripple scanning on the surface

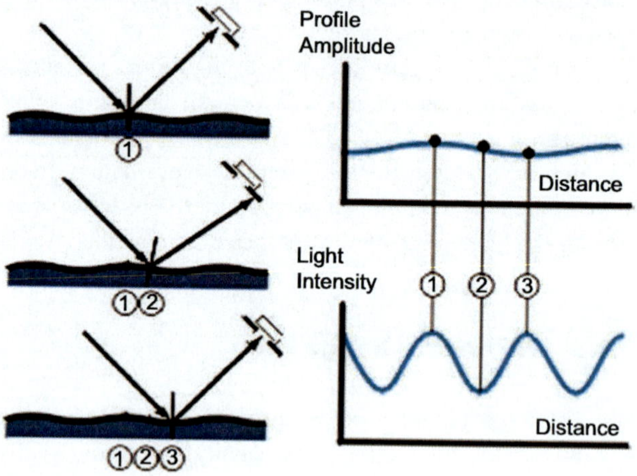

Measuring Principle

The measuring device is used to simulate the visual observation of the human eye.

The wavy brightness pattern on the surface is optically scanned. A laser point light source illuminates the sample at an angle of 60° and a detector on the opposite side measures the reflected light. Figure 7.5 shows the scanning of the surface and Fig. 7.6 shows the result of a measurement.

The measuring device is moved over a defined distance on the sample surface and the optical brightness profile is measured every 0.08 mm.

The measuring device analyzes the ripple according to the structure size. To take into account the resolving power of the eye at different distances, the measurement signal is divided into several parts:

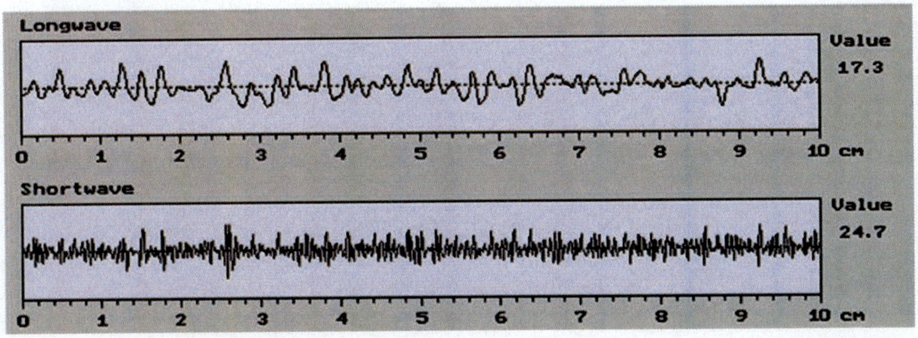

Fig. 7.6 Result of an Orange Peel measurement

Wa 0.1 to 0.3 mm wavelength
Wb 0.3 to 1 mm wavelength
Wc 1 to 3 mm wavelength
Wd 3 to 10 mm wavelength
We 10 to 30 mm wavelength

The measured values form a structure spectrum which allows a differentiated evaluation of the paint flow and the influence of material and processing conditions.

7.2.4 Image Sharpness: Distinctness of Image (DIO)

The clearer and more distinct an object is reflected in the surface, the more brilliant, sharper the gloss effect is perceived. The mirror image is disturbed by structures, edges are then perceived as blurred.

The image sharpness always comes to the fore in the visual evaluation if the paint surface has hardly recognizable surface structures. This may be the case, for example, with very good flow or semi-glossy surfaces.

Measuring Principle
In addition to the structure spectrum, the brilliance is evaluated here with the aid of a CCD camera. The camera records the scattered light caused by fine structures smaller than 0.1 mm wavelength. This measurand is called dullness. However, DOI or brilliance are not only influenced by these fine structures, which the human eye can no longer detect. Therefore, in addition to the dullness, a part of the ultra-short waves Wa and Wb is also included.

7.2.5 Color

A uniform colour tone plays a decisive role in many products today. Achieving this is often not easy, especially when, for example, the individual parts of a product are manufactured in different departments. It becomes much more complicated when several suppliers are involved. In the end, however, the colour must match.

The visual assessment of color is strongly influenced by the individual color perception of the observer and external impressions. In addition, the possibilities of describing a color impression in words and documenting color differences are limited. This is also not surprising with ten million color tones that the human eye can distinguish.

The remedy here is the use of color measuring devices with internationally standardized color systems. This ensures the objective description of colored objects.

Color perception is influenced by the light source, the observer and the sample.

7.2.5.1 Light Source

Colours change with the illumination. For this reason, the light sources used have been defined. A light source suitable for color measurement must continuously emit energy over the entire visible spectrum (400 to 700 nm).

The CIE (Commission Internationale de l'Eclairage) defines a light source by the amount of energy emitted at each wavelength (= relative spectral energy distribution).

The most important types of standard light are:

- Daylight D 65, C
- Green light A
- Fluorescent tubes F2, F11

7.2.5.2 Observers

The light reflected from a colored object falls through the lens of the human eye onto the retina. Three different types of light-sensitive receptors are located there: one type reacts to red light, the second to green light and the third to blue light. In addition, there are brightness-sensitive sensory cells. All together, they create the impression of color in the brain.

To determine the sensitivity of the receptors, the CIE carried out systematic tests in 1931 and 1964. Based on the results, the 2° and 10° normal observers were defined. The 2° normal observer stands for small field of view, the 10° normal observer for a large field of view.

When looking at an object, the eye integrates over a large area. This is most consistent with the 10° normal observer.

7.2.5.3 Object

Light source and observer are standardized by the CIE and the spectral functions are stored in the colorimeters. The only variable to be determined is the optical properties of the object.

Modern colorimeters measure the amount of light reflected from a colored object at each wavelength. This data is called spectral data. For example, no light is reflected from a black object across the entire spectrum (0% reflectance), whereas an ideal white surface reflects nearly all light (100% reflectance).

All other colors reflect light only in certain parts of the spectrum. Thus, each color has its specific spectral curve.

7.2.5.4 Colour Systems

Color systems combine the information of the three elements: light source, observer and object. They form the basis for being able to talk about and document colour and colour differences.

In the CIE system (DIN 5033 [12]), each colour is uniquely identified by the three standard colour values X, Y, Z. Two samples with the same standard colour values X, Y, Z therefore look the same. However, the difference in the standard colour values of two samples cannot be used to directly infer the magnitude of the visually perceived colour difference between the two samples, since the CIE system is not "perceptually equidistant".

For example, a difference in color value Y of $\Delta Y = 0.5$ can be perceived visually very differently depending on the color of the samples. In contrast, a system of colorimetric values is "perceptually equidistant" if equal differences in the colorimetric values always correspond to perceived color differences of equal magnitude, regardless of the position of the colors of the samples in the color space and regardless of the direction of the color difference. Therefore, a measure of the visually perceived color difference must first be calculated from the measured standard color values X, Y, Z of two samples. This is done with the help of a color difference formula, which calculates a color difference called color difference ΔE.

In the past, a variety of color distance formulas were developed and used to agree on color tolerances. Since the formulas are based on mathematically very different methods, color distances calculated according to different formulas cannot simply be converted into each other. In order to standardize the use of color distance formulas, the CIELab formula was defined. For the agreement of color tolerances, this formula should preferably be used.

The system recommended by the CIE, the CIELab system (Fig. 7.7), has now become established worldwide. It consists of the two axes a* and b*, which are at right angles to each other and define the hue. The third axis designates the brightness L*. This is perpendicular to the a*b* surface.

In this system, each color can be determined by the coordinates L*, a*, b*. Alternatively, L*, C*, h° are used. C* (= chroma) represents the saturation or chromaticity of the color, the angle h° is another name for the actual hue (= hue angle).

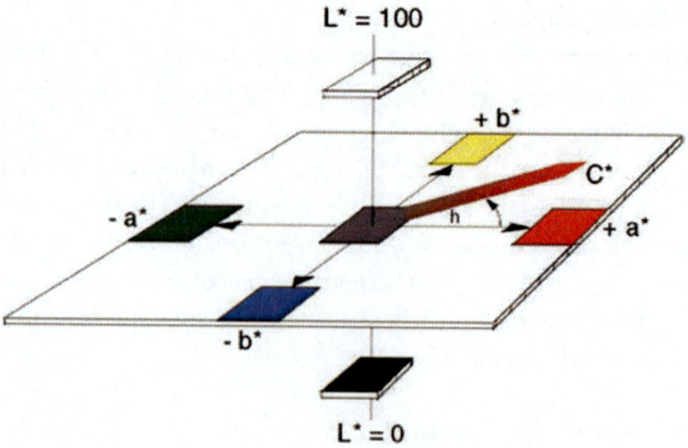

Fig. 7.7 CIELab color space; L* = 0 corresponds to black, L* = 100 corresponds to white, + a* = redder, – a* greener, + b* yellower, – b* bluer

The total color distance ΔE* is calculated from the L*, a*, b* values of two colors according to DIN 6174 [13] and ISO 7724-3 [14].

$$\Delta E_{ab}{}^{*} = \sqrt{\left(\Delta L^{*}\right)^{2} + \left(\Delta a^{*}\right)^{2} + \left(\Delta b^{*}\right)^{2}} \tag{7.1}$$

where

$$\Delta L^{*} = L_{P^{*}}, - L_{B^{*}}; \Delta a^{*} = a_{P^{*}}, - a_{B^{*}}; \Delta b = b_{P^{*}}, - b_{B^{*}} \tag{7.2}$$

are the differences of the L*, a*, b* values of a sample (index P) and a reference or standard (index B).

In order to always deliver a color shade consistently, a standard must be defined and the current production must be compared with this standard. Generally, the permissible colour differences are defined.

The color distance ΔEab* can be formed in different ways from contributions corresponding to different color directions in the color space. These contributions can be used to agree on different tolerances for different color directions.

For achromatic colors and for colors with low chromaticity (e.g., color-cast gray), the split into the contributions ΔL*, Δa*, and Δb* conveys statements about the visual significance of the direction of the color difference.

ΔL*	is positive if the sample is lighter than the reference and negative if it is darker than the reference
Δa*	is positive if the sample is redder than the reference and negative if it is greener than the reference
Δb*	is positive if the sample is yellower than the reference and negative if it is bluer than the reference

For chromatic colors, DIN 6174 [13] defines a splitting in the direction of lightness, chromaticity and hue with the contributions ΔL*, ΔCab* and ΔHab*.

$$\Delta C_{ab}^{*} = C_{ab,p} - C_{ab,B} = \sqrt{\left(a_{p}^{*2} + b_{p}^{*2}\right)} - \sqrt{\left(a_{B}^{*2} + b_{B}^{*2}\right)} \tag{7.3}$$

$$\Delta H_{ab}^{*} = \sqrt{\left[\left(\Delta E_{ab}^{*}\right)^{2} + \left(\Delta L^{*}\right)^{2} + \left(\Delta C_{ab}^{*}\right)^{2}\right]} \tag{7.4}$$

Equation (7.1) writes itself with the contributions given by Eqs. (7.3) and (7.4):

$$\Delta E_{ab}^{*} = \sqrt{\left[\left(\Delta L^{*}\right)^{2} + \left(\Delta C_{ab}^{*}\right)^{2} + \left(\Delta H_{ab}^{*}\right)^{2}\right]}$$

ΔH_{ab}* describes as a chromaticity contribution approximately the size of a perceived color difference in the direction of the chromaticity. This color difference cannot be calculated directly from the chromaticity angles $h_{ab,\,B}$ and $h_{ab,\,P}$. This is because the perceived color difference between two colors with constant angle difference increases with the chromaticity of the colors. For the evaluation of the components of the color difference in the direction of chromaticity and hue is true:

ΔC_{ab}* is positive if the sample is more colorful (brilliant, clearer) than the reference and negative if it is less colorful (duller, cloudy) than the reference.

ΔH_{ab}* has a positive sign if the chromaticity (colour) tone line in the a*-b* diagram must be rotated to the left from the reference to the sample (counterclockwise), i.e. from e.g. red towards yellow, from yellow towards green or from green towards blue.

Two pairs of samples can have the same ΔE* value but visually look different, as shown in the following example:

Sample pair 1 Sample pair 2

Table 7.2 contains the results of the colour measurement on sample pair 1 and 2.

Therefore, the individual components ΔL*, Δa*, Δb* or ΔL*, ΔC*, ΔH* shall be used to determine the actual colour deviation.

Table 7.2 Measured values of the colour measurement of sample pair 1 and 2

	Sample pair 1	Sample pair 2
ΔL*	0.57	0.0
Δa*	0.57	0.0
Δb*	0.57	1.0
ΔE*	1.0	1.0

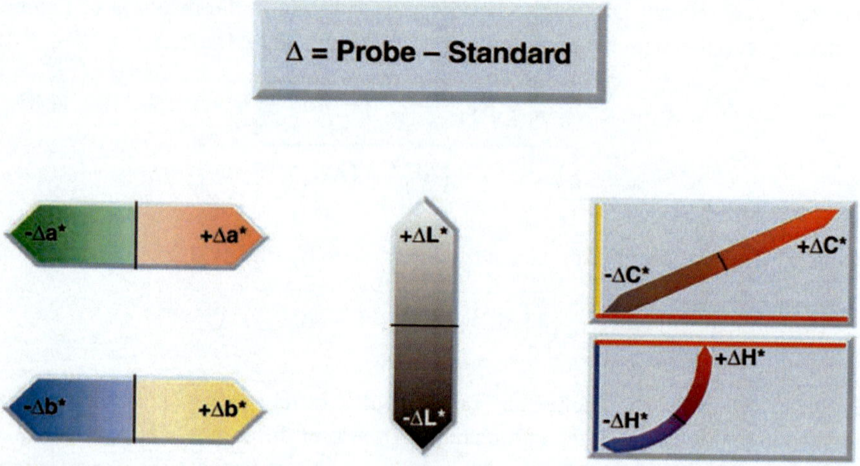

Fig. 7.8 Illustration of the significance of the contributions ΔL*, Δa*, Δb*, ΔCab* and ΔHab*

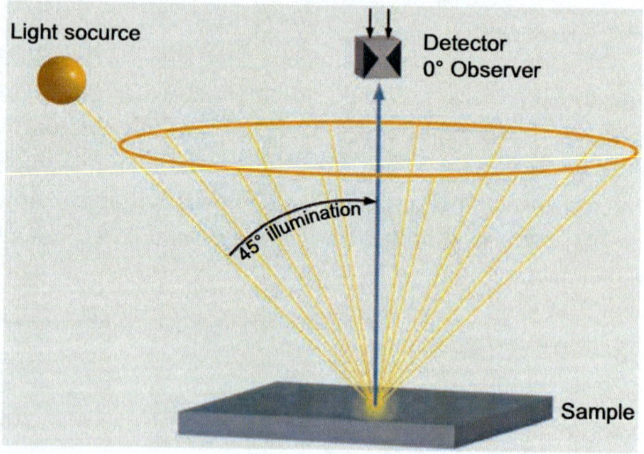

Fig. 7.9 Illustration of the 45/0 geometry

Figure 7.8 illustrates the relationships once again.

Measuring Principle

Two measuring arrangements have become established in industry: the 45/0 geometry and the spherical geometry. In the 45/0 geometry, circular illumination is used at an angle of 45° and measurements are taken perpendicular to the surface at 0°, Fig. 7.9.

With this geometry, color is evaluated as under normal sampling conditions. For example, when comparing a high-gloss object with a sample that has the same pigmentation but is matte or textured, the eye perceives the high-gloss object as darker. This phenomenon is measured with the 45/0 geometry. Here, differences in gloss/structure also lead to color differences.

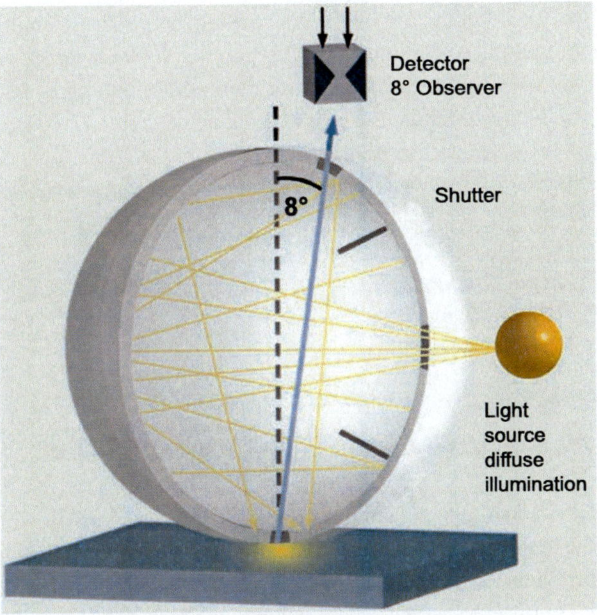

Fig. 7.10 Representation of the sphere geometry

Circular illumination is used to ensure better repeatability on textured and directional surfaces.

When comparing different batches in production and checking the colour consistency of products that are assembled from several different parts, consistency with the visual impression is important.

In the sphere geometry, Fig. 7.10, the sample is diffusely illuminated by means of a white-coated sphere. Shaders inside the sphere prevent the light from falling directly onto the sample surface. The measurement is made at an angle of 8°. A sphere device can be used under two different measuring conditions:

Gloss included (spin)
Gloss excluded (spex)

In "spin" mode, all reflective light is measured: diffuse reflection (color) + directional reflection (gloss). This allows color to be measured independently of surface gloss or texture. The "spin" mode is suitable for the following applications:

- Colour strength as a function of dispersion time
- Influence of weathering and temperature
- Colour formulation

In "spex" mode, a specular trap swallows the directionally reflected light (specular). This configuration simulates the 45/0 geometry. However, deviations between 45/0 and spherical "spex" geometry occur with medium to matt gloss surfaces, as the gloss trap does not completely exclude the gloss component.

Only measurements carried out under the same conditions can be compared with each other. Therefore, a colorimetric report must contain the following information:

- Colorimeter (geometry)
- Light type/observer
- Colour system
- Sample preparation

7.2.6 Color Measurement of Metallic Paints

Effect coatings play an important role in coating today. They change their appearance with the angle of observation and the lighting conditions. With interference paints, not only the brightness changes with the viewing angle, but also the hue.

In order to obtain an objective color assessment of metallic coatings, ASTM, DIN and ISO standards for multi-angle color measurement are specified. Investigations have shown that at least three angles, optimally five angles are necessary for a reliable evaluation. In multi-angle color measurement, the measurement geometry refers to the difference from the specular angle. The viewing angle is measured away from the specular reflection in the illumination plane. The angle is positive when measured from the specular reflection toward the normal. Measurements are taken at 15°, 25°, 45°, 75° and 110°.

The colorimetric values L*, a*, b* (or L*, C*, h*) and ΔE* are used for quality control. The tolerances for the angles near the gloss (15°, 25°) and the flop angles (75°, 110°) are usually larger than the tolerance at the 45° angle. Weighted factors must be used to give unique tolerances regardless of hue. Automotive companies have established specifications for ΔECMC or ΔE′ according to DIN 6175-2 [15] for 3- or 5-angle color measurement. Another helpful metric is the flop index. This evaluates the change in brightness of metallic paints over the entire range of observation angles.

In order that the entire hue change of interference pigments can be fully recorded, it is necessary to add observation and illumination angles. For the portable measuring instruments, an observation angle "behind the gloss" at −15° was added.

7.2.7 Assessment of Color Differences

The question of the tolerance agreement plays a decisive role, especially in the case of paint. The same question comes up again and again between supplier and buyer: Is a batch or a delivery accepted or rejected with regard to its color? In the case of rejection, the additional question may arise: Is the rejection justified or unjustified within the framework of the agreements made?

A simple color distance agreement as a tolerance specification is usually not sufficient. Rather, agreements must also be made on the measurement and evaluation conditions. There is no generally valid, tolerable color distance. This must always be worked out between supplier/customer specifically for colour and surface.

DIN Technical Report 49 [16] describes procedures for agreeing colour tolerances. DIN EN ISO 3668 describes the performance of color matching in a visual comparison of the color of coating materials [17]. In DIN 6175 [18], the automotive industry has specified color differences for plain coatings. However, this agreement cannot be applied to powder coated surfaces. In Code of Practice St02 [19] of the Window and Facade Association, information is given on the assessment of painted surfaces. This lists criteria for assessing colour differences and damage (e.g. faulty coatings, disturbances in the painted surface caused by spots, bubbles or craters). Permissible colour tolerances for solid-coloured powder coatings in architectural applications, published by the German Paint Industry Association, are described in the guideline VdL – RL 10. This guideline describes reference samples, test methods and permissible colour tolerances for differences between the sample and the delivery as well as between two deliveries. However, the guideline does not apply to the assessment of end-coated objects and components. Here, reference is to be made exclusively to sample sheets produced under standardized conditions (laboratory conditions) [20].

7.3 Adhesion

In the practice of testing paints, varnishes and other coatings, serviceability tests were introduced a long time ago for assessing adhesion. Although they are not based on physically clear principles, they are relatively simple to perform and provide largely reproducible measurement results for the corresponding areas of application. Because of their simple applicability, the cross-cut test, the scratch test and the application of an adhesive tape to a scored or cut film with subsequent peeling off are used for the rapid assessment of adhesive strength. The latter methods are essentially yes/no decision tests.

When assessing the adhesion strength, it must be taken into account that the test result is not only determined by the mechanical properties of the system to be tested, but can also be influenced by the type and pre-treatment of the substrate, by the application procedure, by the curing or drying conditions and by the test conditions.

Each method has its specific application. Improper application can lead to incorrect results. All test methods for assessing adhesion require a certain amount of practice on the part of the tester. For most methods, the test results depend on the thickness of the coating under test. In addition, for some methods, differences between tests on a sample sheet and in practice may occur due to differences in roughness of the substrate.

In most cases, the bond strength cannot be adequately assessed using only one method. An overview of the most common methods is given in Table 7.3.

Table 7.3 Procedure for assessing the adhesive strength of coatings [21]

Procedure	Inspection according to	Application	Assessment	Comments
Pull-off test	DIN EN ISO 4624 [22]	Universal for laboratory tests and field trials	Measured value and visual	Stamps must not be glued on tilted; at least 5 measured values
Mandrel bending test with conical mandrel	DIN EN ISO 6860 [23]	Industrial coatings and special applications such as plastic and thick-film coatings	Visual	The type of crack formation allows conclusions to be drawn about the adhesive strength. Application also combined with adhesive tape tear
Mandrel bending test with cylindrical mandrel	DIN EN ISO 1519 [24]	Industrial coatings	Visual	The type of cracking allows conclusions to be drawn about the adhesive strength. Also combined with adhesive tape tear
Crosscut	DIN EN ISO 2409 [25]	Universally applicable as laboratory test and field test	Visual	Result dependent on the viscoelasticity of the coating
Cross cut with tape test	Coatings on steel, aluminium, plastic and wood	Visual		
Ball impact test	DIN EN ISO 6272 [26]	Coatings for special impact and shock loads	Visual	Cracking and delamination of the coating
Shot peening test	DIN 53154 [27]	Furniture varnishing	Visual	
Stone impact test, single impact test	DIN EN ISO 20567-3 [31]	Automotive paints in the case of stone chipping	Visual, if necessary also measured value, image analysis	
Stone impact test, multi-impact test	DIN EN ISO 20567-1 [30]	Automotive paints in the case of stone chipping	Visual, image analysis	
Depression star pattern		Coatings in the aerospace industry	Visual	
Cupping test	DIN EN ISO 1520 [28]	Industrial coatings	Visual	Also combined with grid cut or tape tear-off

(*continued*)

Table 7.3 (continued)

Procedure	Inspection according to	Application	Assessment	Comments
Steam jet test	Visual	Series control for automotive coatings		Certain chemicals can affect the bond strength
Three-point bending test	NF T 30-010 [29]	Coatings on metal and plastic, not on mineral substrates and concrete	Measured value	
Scratch test		For single and multi-layer industrial coatings	Visual	Result dependent on the hardness of the substrate and the individual application of force by the tester
Etching sample		Especially for coatings on wood and plastics	Visual	
Peeling test 1		Especially for testing coated thermoplastic elastomers	Measured value	
Peeling test 2		Plastic coatings	Visual	Test with yes/no decision

Increasing importance is being attached to assessing the adhesion strength when exposed to moisture (so-called wet adhesion), usually in conjunction with pull-off test.[32].

Adhesion strength is defined in DIN EN 971, Part 1 [33] as the totality of the bonding forces between a coating and its substrate.

7.3.1 Pull-Off Test

The method can be used for single and multi-layer coatings. Different penetrations are specified for the test depending on whether the substrate is deformable, e.g. thin metal, plastic or wood, or rigid, e.g. thick concrete and metal panels.

The test result is the minimum tensile stress that must be applied to break the weakest interface (adhesion failure) or the weakest layer/component (cohesion failure) of the test assembly. Adhesion and cohesion fractures can occur simultaneously.

For performance and evaluation, see DIN EN ISO 4624 [22, 34]. The fracture pattern is evaluated according to the following scheme:

A	Cohesive fracture in the subsoil
A/B	Adhesion failure between substrate and first coating
B	Cohesion failure of the first coating
B/C	Adhesion failure between the first and the second coating
–/Y	Adhesion failure between last coating and adhesive
Y	Cohesive fracture in the adhesive
Y/Z	Adhesion failure between adhesive and test stamp

7.3.2 Mandrel Bending Test with Conical Mandrel

The conical mandrel bend test is a method of evaluating the resistance of coatings to cracking and/or peeling from a metallic substrate or plastic when bent around a conical mandrel under standard conditions.

In a multi-layer system, each layer can be tested on its own or the entire system can be tested.

The evaluation is carried out according to DIN EN ISO 6860 [23].

7.3.3 Mandrel Bending Test with Cylindrical Mandrel

The cylindrical mandrel bend test is a method of assessing the resistance of coatings to cracking and/or peeling from a metallic substrate or plastics when bent under standard conditions around a cylindrical mandrel (or plate with a suitably rounded edge).

Depending on the agreement, the mandrel bending test can be carried out as

(a) a yes/no test, where the test is performed with a single mandrel of specified diameter – to assess compliance with specification (specified mandrel diameter).
(b) a repeated test by applying successively smaller diameter mandrels – to determine the diameter of the first mandrel above which the coating cracks or becomes detached from the substrate.

In a multi-layer system, each layer can be tested on its own or the entire system can be tested.

The mandrel bending test is suitable for comparative measurements of brittle coatings. Elastic coatings usually do not crack even with the smallest mandrel diameters.

Execution and evaluation, see DIN EN ISO 1519 [24]. Figure 7.11 shows the result when bending around a 5 mm mandrel.

Fig. 7.11 Cracks in the area of the bending radius down to the base material (white sample, top left). Second sample no cracks in the paint film (blue sample, bottom right)

7.3.4 Cross-Cut Test

The cross cut is a method of estimating the resistance of a coating to spalling from the substrate when a grid is cut into the coating down to the substrate. The property measured by this empirical method depends, on the adhesion of the coating to the preceding layer or to the substrate.

However, this method should not be considered only as a method of evaluating bond strength. It is used for comparative measurements and for testing serviceability.

Although the test is primarily intended for use in the laboratory, it is also suitable as a field test. However, temperature changes and changes in relative humidity then influence the test result.

The described method can be applied either as a yes/no test or as a classification test with 6 steps. Applied to a multilayer system, the resistance of the individual layers of the coating to separation from each other can be estimated. The test is possible on finished articles and/or on separately manufactured sample plates.

The process is suitable for coatings on hard (e.g. steel) and soft (e.g. wood and plaster, stucco) substrates. However, these different substrates require appropriate implementation. With soft materials, such as wood, reproducible results are often not achieved.

The method is only suitable for smooth coatings with a total layer thickness of up to 250 µm. Coatings with a total film thickness above 250 µm can be tested by a single cross cut. If the method is used on coatings that have a rough surface (e.g. textured coatings), the results may vary considerably.

Performance and evaluation, see DIN EN ISO 2409 [25]. Figures 7.12 and 7.13 show two results of a cross-cut test: with good adhesion and with extremely poor adhesion.

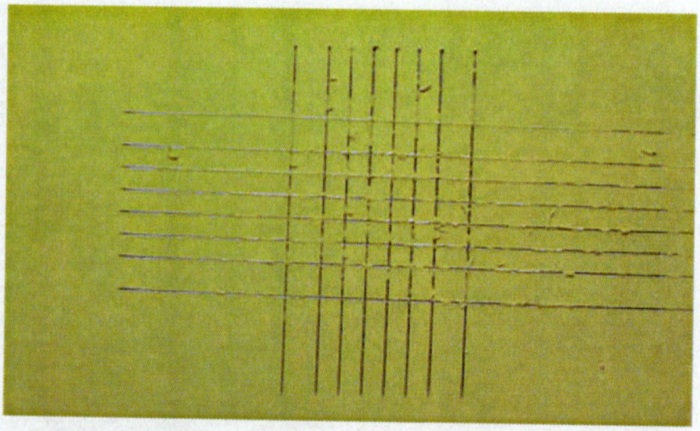

Fig. 7.12 Result of the cross-cut test on a powder coating film with good adhesion, cross-cut characteristic Gt0

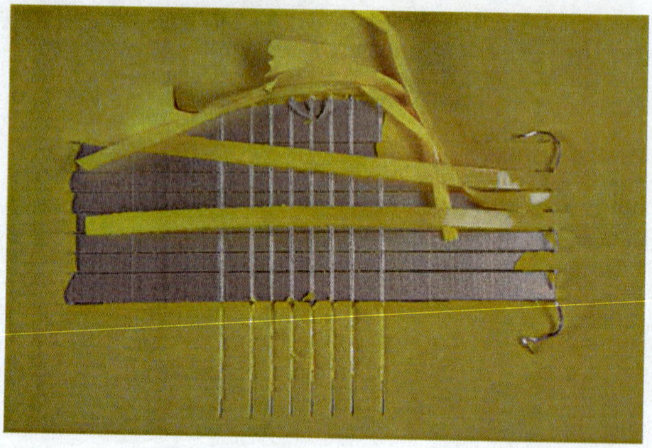

Fig. 7.13 Result of the cross-cut test on a powder coating film with extremely poor adhesion, cross-cut characteristic Gt5

7.3.5 Cross Section with Adhesive Tape Pull Off

The cross cut with tape pull off [35] is a method for determining the adhesive strength of coatings on metals, e.g. steel, aluminium, galvanised steel and plastic. It is more of an orienting test, but nevertheless gives results that are readily usable in practice.

Devices
Scriber or pocket knife; ruler; adhesive tape, 25 mm wide, adhesive force on steel at least 11 N/25 mm.

Implementation
After conditioning, the coating is scored in a cross shape down to the substrate. An adhesive tape of about 150 mm length is stuck onto the cross section and pressed on well. After 5 to 10 s, the adhesive tape is pulled off at a uniform speed by taking the free end and pulling it off uniformly within 0.5 s to 1.0 s at an angle of about 60° to the direction of pull.

Evaluation:

Assessment according to the following levels:

0	Very good adhesive strength. The coating is not peeled off
1	Good adhesive strength. The coating is peeled off max. 0.5 mm
2	Poor adhesive strength. The coating is peeled off 0.5 mm to 3 mm
3	Very poor adhesion. The entire coating is peeled off

7.3.6 Ball-Impact Test/Falling Ball Test

The ball impact test (falling ball test) is a method of assessing the resistance of coatings to cracking and/or peeling from a metallic substrate when the substrate is in the form of a thin sheet and is subjected to a spherical impact by a falling weight under standard conditions.

In addition, the cross-linking of the coating film, i.e. the curing conditions, can be checked with the method of the ball impact test for powder coating films.

The ball impact test can be carried out with

(a) indent, where the dropping body falls onto the coated side of the sample or
(b) bulge where the dropping body falls onto the back of the sample.

The ball impact test can be performed as

(a) a yes/no test in which the drop body is dropped onto the sheet from an agreed falling height and an assessment is made as to whether or not the coating is cracking and/or peeling off;
(b) a repeated test in which the stress is increased until the coating cracks and/or peels.

The ball impact test can also be carried out on a multi-layer system. For performance and evaluation, see DIN EN ISO 6272 [26].

The following Figs. 7.14 and 7.15 show the result of a ball impact test on a powder coating film with insufficient adhesion and a powder coating film which has not been sufficiently cured.

7.3.7 Shot Peening Test

The shot peening test is a method for assessing the behaviour of coatings against stresses such as those that occur in practice as a result of small, frequently repeated blows or impacts. In the case of a multi-layer system, the base coating can be tested on its own or in conjunction with subsequent coatings. For performance and evaluation, see DIN 53154 [27].

Fig. 7.14 Result of the ball impact test with 20 inch-p on a specimen with insufficient paint adhesion

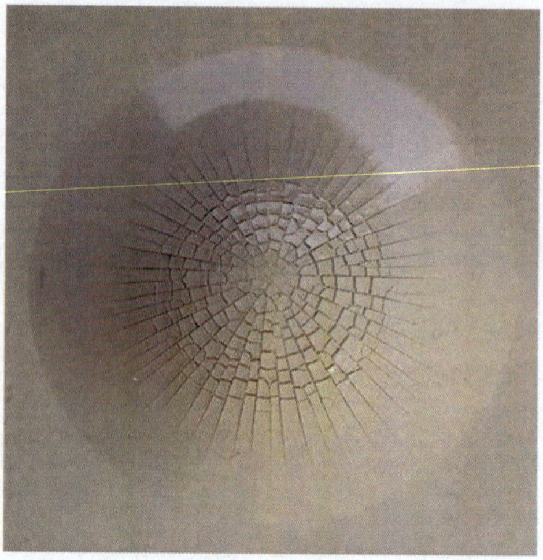

Fig. 7.15 Result of the ball-impact test with 20 inch-p on a specimen with an not suffiecent crosslinked paint film

7.3.8 Stone Impact Test, Single Impact Test

The purpose of the test is to determine the separation level of multi-layer coatings when subjected to stone chipping [31, 36–39].

The single impact test is a method for assessing the resistance of multi-layer coatings to stone impact, such as those used for motor vehicles. Not only the size of the damage but also the level of delamination can be determined. The method provides results that tend to be similar to those obtained after multi-impact exposure, but are more quantifiable.

Implementation:

A steel ball under defined conditions (velocity, angle of impact) or an impactor with a wedge-shaped cutting edge, possibly damped by an intermediate spring, can be fired at the coated sheet. The devices are partly transportable, so that in these cases tests in chambers with different climates are also possible, or they can be tempered.

Evaluation:

After the bombardment, the damaged area is evaluated quantitatively. For multi-layer coatings, the delamination level is also determined.

7.3.9 Stone Impact Test, Multi-Impact Test

The multi-impact test, e.g. according to VDA (German Association of the Automotive Industry) [40] or ISO 20567-1 [30], is a method for assessing the resistance of automotive paints and similar coatings to impact with defined impact material (ISO 11124-2 description of a defined chilled cast material as impact material) as simulated stone impact.

The evaluation is performed according to predefined images or with the help of image analysis [41]. The scheme has characteristic values from 0 to 5.

In ISO 20567-1, regulations are made on the influencing variables of the test in order to obtain a better comparability of the tests and results.

7.3.10 Cupping Test

The cupping test is used to test the resistance of coatings to cracking and/or peeling on a metallic substrate by appling increasing deformation by cupping (Fig. 7.16) under standardised conditions. In the case of a multi-layer system, each layer can be tested individually or the entire system can be tested.

The specified procedure may be carried out

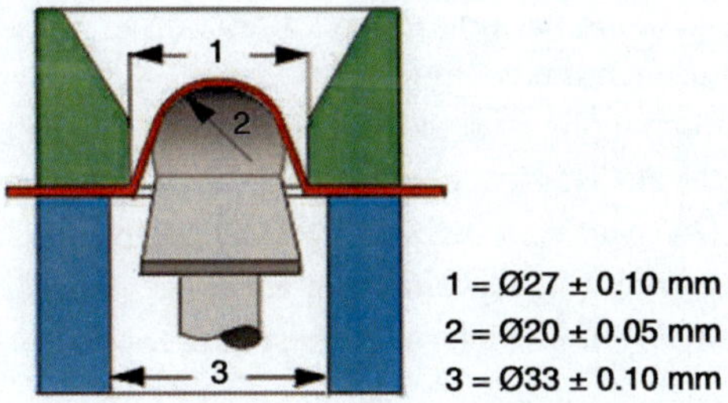

Fig. 7.16 Schematic figure of the cupping test

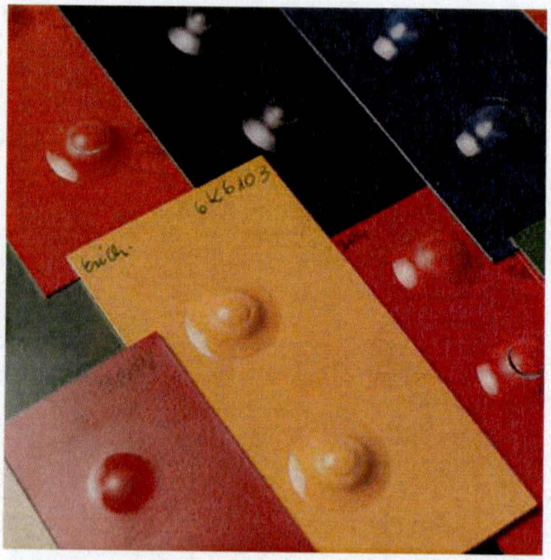

Fig. 7.17 Result of the cupping test on different powder coatings

(a) as a yes/no check by checking for the fulfillment of a certain requirement at a specified depth value;
(b) by determining, with increasing deformation, the smallest cupping value at which the coating cracks and/or detaches from the substrate.

The core of the cupping machine is a die with hardened surface and plane-ground pressure surface. This is accompanied by a blank holder with eccentric clamping device. The penetration tool is a polished, hardened ball cap with a diameter of 20 mm. The testing process is observed through a stereo microscope or magnifying glass. Careful attention must be paid during the test to determine when the coating begins to crack. Execution and evaluation, see DIN EN ISO 1520 [28], result see Fig. 7.17.

7.3.11 Star Cut with Impact According to Randel

Star section with impact [42] is a method for determining the adhesive strength of coatings that is used particularly in the aerospace industry (DIN 65046-4, method 4-4-3) [43].

Using the very sharp knife point of a pocket knife, four cuts are made in a star shape through the coating down to the substrate. The coating is then clamped in the Erichsen cupping jig with the star cut point in the centre of the die. Through the cuts in the coating, the stresses that occur during the deepening process are released in a clearly visible form. One can see, which is important for the evaluation of the adhesive strength, whether in the coating the adhesion (the adhesive strength to the substrate) is greater or less than the cohesion (the internal cohesion of the coating).

7.3.12 Scratch Test

The scratch test [44] represents a utility value test for assessing the adhesive strength of a coating. From the edges of the resulting scratch mark, conclusions can be drawn about the adhesion of the coating:

Well-adhering coatings give a uniform, trough-shaped trace with smooth edges; poorly adhering coatings show strongly jagged or irregular, broken edges. Very hard coatings are difficult and irregular to scrape off. Very brittle coatings sometimes show weakly jagged edges without necessarily showing poor adhesion.

Using the sharp side of a pocket knife blade (held at right angles to the sample sheet), the coating is scraped off under pressure by pulling (moving sideways, not cutting) the knife down to the substrate or, in the case of multi-layer systems, through the layer to be assessed. The coating must be scraped off in a width of several millimetres. In the case of hard coatings, it is sometimes necessary to press on the knife with both hands. The knife must not cut (by being held at an angle).

Assessment is done according to the levels:

K 0	The coating adheres very well and shows perfectly smooth edges on the scraped areas
K 1	The coating adheres very well, but shows jagged edges
K 2	The coating adheres well, can be scraped off with the entire contact surface of the knife; smooth edges
K 3	The coating adheres well, can be scraped off with the entire contact surface of the knife; serrated edges
K 4	Worse than K 3, flaking
K 5	Insufficient adhesion, flaking of the coating

7.3.13 Eraser Test

The eraser test is a utility value test for assessing the adhesion strength of a coating to a substrate and/or the intercoat adhesion of individual layers to each other. The eraser test is preferably carried out on soft substrates such as wood and plastics.

The coating is erased under pressure with the sharp side of a pocket knife blade. This checks whether the bond between the individual layers is perfect or whether the coating adheres well.

Assessment is done according to the levels:

R 1	No jagged edges become visible during erasing. The transition to the individual layers is flawless. The coating shows smooth edges
R 2	When erasing slightly jagged edges become visible. However, the adhesive strength is good
R 3	Individual layers can be seen when erasing. Jagged edges, slight flaking
R 4	The individual layers or the coating clearly flakes off from the substrate during erasing

7.3.14 Scratch Hardness Test

This test is used to evaluate the adhesion of organic coating systems on flat smooth specimens. The scratch hardness test has proven to be useful for the comparative testing and evaluation of test sheets whose coatings exhibit significant differences in their adhesive strength.

The sample sheets are produced with uniform coating thickness. After cross-linking of the paint layers, the adhesive strength is determined by moving the sheets against a rounded pin or chisel loaded with weights until the paint layer detaches from the substrate.

In addition to the adhesive strength, the scratch resistance can also be tested.

7.3.15 Steam Jet Test

Steam jet testing is a method of assessing the resistance of coatings to the effects of a pressurised water jet. The test method simulates the effects of a pressurized water jet on a coating. The test method is mainly used for testing vehicle coatings, but is also used in other areas. The test of the resistance of coatings to pressurized water jets is also referred to in German as the "Dampfstrahltest", DIN EN ISO 16925 [45]. The test provides information for example, coating damage occurs during the de- waxing of motor vehicles; on the other hand, serious application defects of a incorrect coating are quickly detected (series control).

The evaluation is carried out by assessing the adhesive strength on the St. Andrew's cross in accordance with DIN EN ISO 9227 [46] or DIN EN ISO 4628-8 [47].

7.3.16 Cooking Test

The boiling water test [1] is used as a quick test of pretreatment or paint adhesion under the influence of water diffusion in the paint film.

Residues on the surface of the workpiece, such as grease, dust, salts, lead to bubbles in the paint film under the influence of moisture.

Samples or parts from production are stored in boiling deionised water for 2 h. Then the samples are taken and after cooling a visual assessment for bubbles and/or infiltrations is carried out. In addition, after conditioning the samples for 1 h at room temperature, a cross-cut test can be performed.

7.4 Elasticity/Fexibility

In practice, three different empirical test methods are used to assess the resistance of paints and similar coating materials to cracking and/or peeling from a metallic substrate due to deformation.

- Mandrel bending test with cylindrical mandrel acc. to DIN EN ISO 1519 [24] and conical mandrel acc. to DIN EN ISO 6860, ASTM D 522, [23, 48], see Sects. 7.3.2 and 7.3.3.
- Ball impact, DIN EN ISO 6272 and ASTM D 2794 [26, 49], see Sect. 7.3.6
- Cupping test, DIN EN ISO 1520 [28], see Sect. 7.3.10.

7.5 Hardness

The term hardness has often led to misunderstandings in the coatings industry. Since almost all coatings exhibit viscoelastic behavior, the term hardness is defined in DIN 55945 [50] as follows: "Hardness is the resistance of a coating to a mechanical action, such as pressure, rubbing or scoring".

There are essentially three different hardness testing methods: the scratch, the vibration, and the indentation test methods [73].

As hardness is a complex parameter, the term hardness should always be used in conjunction with the test method used.

7.5.1 Pendulum Hardness

This test method for hardness assessment uses the damping of a swinging pendulum as a measure. The pendulum with two stainless steel balls is placed on a lacquer film. There is a physical relationship between the duration of the pendulum oscillation, amplitude and the geometric dimensions of the pendulum. The viscoelastic behaviour of the coating is decisive for the hardness determination.

When the pendulum is set in oscillation, the balls roll on the surface and thus exert pressure on it. The more or less strong springback depends on the elasticity. The absence of elastic forces causes a strong damping of the pendulum movements. High elastic forces, on the other hand, cause only slight damping.

Two pendulum types have been standardised for this test method. Table 7.4 lists the differences between the pendulums.

The pendulum damping device according to *König* is specified in DIN EN ISO 1522 [51] (vibration number 179 ± 3). Important for the reproducibility of the results is the humidity, temperature and cleanliness of the balls.

Table 7.4 Differences of the pendulums according to König and Persoz

	King	Persoz
Weight	200 g ± 0.2	500 g ± 0.1
Ball diameter	5 mm	8 mm
Deflection start	6°	12°
Deflection end	3°	4°
Period of oscillation	1.4 s	1 s
Attenuation time on glass	250 ± 10 s	430 ± 10 s

7.5.2 *Buchholz* Intendation Test

A widely used and proven method in practice is the *Bucholz* indentation test according to DIN EN ISO 2815 [52]. Prerequisites for the applicability of the method are:

- the varnish layer must have a sufficient thickness
- the substrate must not yield under the test load
- the paint layer must be plastically deformed, i.e. it must not behave elastically under the test load
- under load, a paint film of at least 10 μm must remain under the indenter
- the varnish layer must be smooth and clean.

In the penetration test method, a suitably shaped body penetrates the coating to a certain depth during a defined time of 30 s and leaves an indentation after relief. The indentation resistance according to Bucholz, which is defined as 100 times the reciprocal value of the indentation length, serves as the measured value.

$$\text{Indentation resistance to Buchholz} = \frac{100\,\text{mm}}{\text{Indentation length}}$$

The method is best suited for hard lacquers, whereas with very soft lacquers there is a risk that the skid of the device will rest on the substrate. With very hard but elastic coatings, a higher hardness is simulated. Measurements with indentation lengths below about 0.6 mm must be evaluated critically.

7.5.3 Universal Hardness Measurement Using the Force-Penetration Method

The measurement procedure is the same for all known indentation depth methods. An indenter is pressed into the coating by applying a constant or variable force over time and then lifted off again. The distance covered by the indenter in the coating, the indentation depth, is recorded as a function of time or the acting force. A hardness value is defined from the measured variables force and penetration depth according to the following equation:

$$\text{Hardness H} = \frac{\text{Power F}}{\text{Impression surface A}}$$

With the measured indentation depth and the known geometry of the indenter, the area produced by the indenter can be determined. Hard metal balls or diamond pyramids according to Vickers, Berkovich or Knoop are used as indenters.

In contrast to the conventional hardness measurement for metals, the universal hardness measurement, the H100 [53, 54], determines the area generated by the indenter in the coating under an acting test force. This evaluation method is necessary and useful for materials with viscoelastic behavior, because with a determination method as with metals, only the permanent plastic deformation would be recorded.

In relation to paint layers, this hardness measurement can be used to obtain information about the degree of polymerization, hardness changes due to temperature influences, as a result of embrittlement or UV irradiation, or about changes in viscoelastic properties caused by weathering.

If the Vickers pyramid is used as the indenter, the hardness determined is to be designated as the universal hardness HU or, since June 2000, as the Martens hardness HM.

In addition to the hardness of the paint film, this test method can also be used to determine the plastic hardness, the modulus of elasticity and the creep tendency at minimum and maximum force. Figures 7.18 and 7.19 show the H 100® measuring device and a typical curve of the hardness measurement of a paint layer.

Fig. 7.18 Universal hardness tester H 100

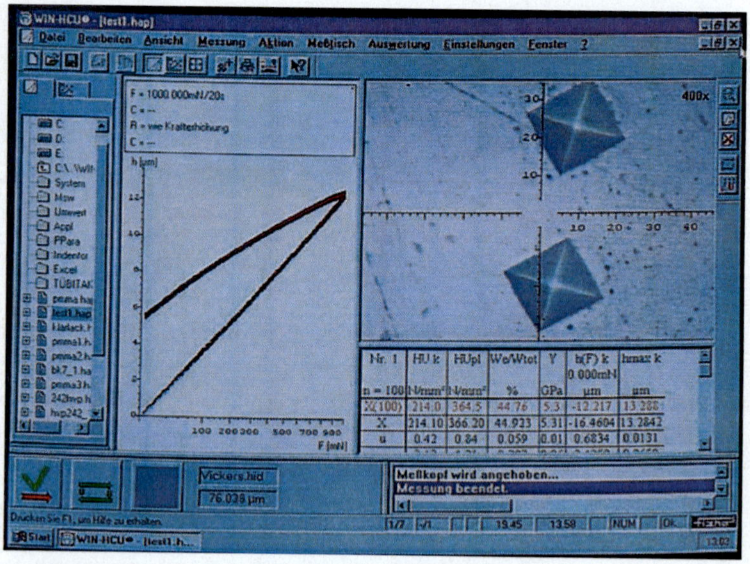

Fig. 7.19 Hardness curve during loading and unloading and view of the hardness impression with the video image

7.6 Layer Thickness

Precise measurement of the coating thickness helps to maximize quality and minimize material costs. The determination can be carried out non-destructively or, e.g. for multi-layer systems, destructively.

Non-destructive measurement of the cross-linked paint layer involves the use of electronic measuring devices that measure and digitally display the thickness of insulating coatings on non-magnetic metallic substrates (NFe) and of non-magnetic coatings on steel or iron (Fe). Two different measuring principles are used:

- Magnetic induction measurement on Fe substrates
- Eddy current measurement on NFe substrates.

7.6.1 Magnetic Induction Method

This method is based on the use of two magnetic coils whose magnetic field is changed by approaching a ferromagnetic substrate. The magnetic field depends on the distance of the probe to the substrate – and thus on the layer thickness. The second coil picks up the magnetic current. This magnetic coupling between the two magnetic poles is a measure of the layer thickness. Electromagnetic induction uses alternating magnetic fields produced by a ferromagnetic coil. Modern Fe probes now use Hall-effect semiconductors with high measurement accuracy. Coating thickness measurement is described in DIN EN ISO 2178 [55].

7.6.2 Eddy Current Method

This method is used for the measurement of non-conductive layers (NFe) on non-ferromagnetic base metal (NFe), such as aluminium. The eddy current method is based on the principles of electromagnetic induction. A high-frequency alternating current sets up a magnetic field in a coil of fine wire, which changes its direction according to the applied alternating current. When the probe is brought close to a conductive substrate, eddy currents are generated in the substrate which act back on the magnetic field of the coil. The magnitude of the feedback depends on the nature of the carrier and the distance – and thus the layer thickness – between the probe and the substrate. The measurement of the layer thickness is described in DIN EN ISO 2360 [56].

7.6.3 Measurement of Powder Layer Thickness Before Baking/Crosslinking

With the non-contact and non-destructive measurement of powder coatings before baking, it is possible to quickly optimize the coating process if necessary.

Three different methods are available today, some of which differ fundamentally in their application and use. The spectrum ranges from portable hand-held devices to complete systems that automatically control the layer thickness via PLC.

Measuring method for non-contact coating thickness measurement before baking [54]:

- Photothermal process, pulsed
- Photothermal process, thermal wave
- Ultrasonic measuring method

In photothermal measurement, the characteristic course of the temperature rise of a powder layer is evaluated when it is slightly heated with a certain excitation radiation.

When measuring with ultrasound, focused ultrasonic signals are directed at the object, which reflect at the layer boundaries (air/powder layer and powder layer/base material) and are received again in the measuring head.

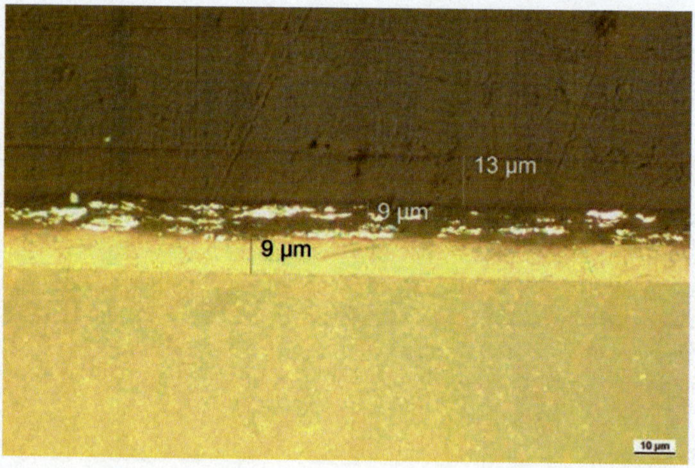

Fig. 7.20 Determination of film thickness in cross-section; paint build-up: primer, metallic basecoat, clearcoat

7.6.4 Destructive Coating Thickness Measurement: Cross-Section Method

The arbitration method for coating thickness measurement is the microscopic measurement of coating thickness – cross-section method according to DIN EN ISO 1463 [57].

For the preparation of the cross section, a sample is taken from the part to be examined, embedded, ground and polished. The thickness of the coating can then be measured in the microscope at a certain magnification.

An additional advantage of the method is that the course of the coating, e.g. over edges, can be precisely observed. In addition, inclusions in the coating layer, defects in the coating and spalling can also be assessed. The measurement of several lacquer layers is also possible here, provided that these can be visually distinguished.

The measurement is independent of the composition of the layer and the base material. Figure 7.20 shows the cross-section with layer thickness data for the individual layers of the multilayer structure.

7.7 Corrosion Tests

7.7.1 Corrosion Test Methods

In Germany, the following climates with different sulphur dioxide (SO_2) and sodium chloride (NaCl) contents are generally used for long-term corrosion tests, which often extend over years (outdoor weathering) [58]:

- Coastal climate (Cuxhaven): 1 µg/m³ SO_2, 200 mg/m³ NaCl
- Industrial climate (Duisburg): >100 µg/m³ SO_2
- Urban climate (Nuernberg): 60 µg/m³ SO_2
- Land climate (Olpe): 30 µg/m³ SO_2.

Such tests are, because of the duration of the investigation, useful and important for basic research, but usually not useful for development and production.

For this reason, short-term test methods have been developed with which various environmental conditions can be simulated in a more stringent form, but for a shorter test duration. They are nationally or internationally standardised, but always only give an indication of the corrosion resistance of a protection system – never a 100% statement.

In the corrosion testing of coatings, the following procedures have proven themselves in practice and have become established as quality tests with defined conditions:

Resistance to humidity DIN EN ISO 6270 [59].

Resistance to humid atmospheres containing sulfur dioxide (SO_2) DIN EN ISO 3231 [60].

Corrosion tests in artificial atmospheres – Salt spray tests DIN EN ISO 9227.

In addition, however, combination tests are often used which correlate better with the actual environmental conditions and long-term results than tests in a monoclimate. For example, a combination of climate change and salt spray tests, as is common practice in the automotive sector and is described in VDA change test 621-415 [61].

In the automotive industry, motor vehicles or their components are tested with cyclic changing stresses (1 test cycle lasts 7 days).

1 day = 24 h salt spray test.

4 days = 4 h cycles Condensation alternating climate.

2 days = 48 h Room temperature 18 to 28 °C.

A similar combination test is used, for example, by Stiftung Warentest to determine the corrosion resistance of consumer products.

The standard DIN EN ISO 11997-1 [62] defines cyclic stresses, including additional UV irradiation.

Table 7.5 Test conditions according to DIN EN ISO 6270

Name	Cycle duration	Temperature (h)	Relative humidity	
Constant humidity	Total	24	40 ± 3 °C	About 100%
	Total	24	–	About 100%
Alternating climate – change in humidity and temperature	Section 1	8	40 ± 3 °C	Below 100%
	Section 2	16	18 to 28 °C	
	Total	24	–	About 100%
Alternating climate – change in temperature	Section 1	8	40 ± 3 °C	About 100%
	Section 2	16	18 to 28 °C	

7.7.2 Condensation Test Climates DIN EN ISO 6270

This form of testing simulates the behavior of materials in humid environmental climates.

Heating the specimens in air saturated with water vapor causes condensation, which is typical for humid climates. The exposure can take place either in constant climates or in alternating climates (with intermediate cooling). For coatings, blistering and infiltration of the paint film are assessed. All tests are performed in the same type of test chamber, only the individual test sections change.

Constant humidity climate DIN EN ISO 6270-1-continuous condensation.

Aternating humidity climate DIN EN ISO 6270-2 with change of humidity and temperature and with change of temperature only.

A summary of the test conditions is shown in Table 7.5.

7.7.3 Stress in a Condensation Water Alternating Climate with an Atmosphere Containing Sulphur Dioxide DIN EN ISO 3231

The condensation water atmosphere is more aggressive when corrosion-promoting gases are supplied. This test therefore makes it possible to detect problems in corrosion protection systems more quickly. Since metallic base materials corrode to varying degrees, the time it takes for the base material to start corroding is a good measure of the corrosion protection provided by the coating. To perform the test, the bottom trough of the instrument is filled with distilled (deionized) water, and the sample chamber is heated to 40 ± 3 °C. The sample chamber is then filled with water.

It has long been known that atmospheric corrosion of metals is strongly dependent on the sulphur dioxide content of the atmosphere. In the case of organic coatings, the protective effect of the paint layer can thus be tested. This dependence is simulated in the *Kesternich* method. Two variants of the test are common: with 0.2 and with 2.0 L sulfur dioxide per 300 l test chamber volume. Each cycle lasts 24 h, Table 7.6.

Table 7.6 Summary of test conditions according to DIN EN ISO 3231

Humidity climate	Cycle duration	Temperature	Relative humidity
Test conditions	Total: 24 h	–	–
	1st test section: 8 h	40 ± 3 °C	about 100%
	2nd test phase: 16 h	18 to 28 °C	max. 75%
SO_2 addition per cycle with 300 l test chamber	0.2 l	1.0 l	2.0 l

Fig. 7.21 Test equipment for condensation water constant and condensation water cycling tests

1. Section: 8 h at 40 ± 3 °C and condensation
2. Section: 16 h ventilation at room temperature and humidity below 75%.

The sulphur dioxide gas introduced into the closed test chamber dissolves rapidly in the water of the bottom trough (2 L). The sulphur dioxide concentration in the gas chamber is then only 1/7 of the theoretical concentration of 0.067 vol.%. Due to interaction with the samples and oxidation, the sulfur dioxide concentration drops to zero at the end of the test section. After the required number of cycles, the samples are removed, rinsed and assessed after 1 h (Fig. 7.21).

7.7.4 Salt Spray Test with Various Sodium Chloride Solutions DIN EN ISO 9227

The spray tests are used to simulate marine and coastal climates and corrosion caused by de-icing salts. They are test methods with a very short time span, which contain a 5% sodium chloride solution as the main component, which is sprayed continuously. A distinction is made between three spray tests, which are usually sprayed onto the test specimens in a continuous test. The exposure is recorded in hours [74].

Prior to corrosion aging, the specimens are damaged in a defined manner using a scribing tool. Depending on the specification, a St. Andrew's cross or two parallel scribes and one at a 90° angle to it are applied. The type of scribing tool and the design have a decisive influence on the result. After the test, the infiltrations starting from the scribe marks and the blister formation are usually assessed (Fig. 7.22).

The specimen sheets shall be placed in the test chamber at an angle of 60 to 75° to the vertical. The scribed side shall face upwards and the specimens shall not touch each other. After the required test duration, the test specimens are removed, rinsed with deionised water and assessed 1 h later.

Neutral salt spray test	DIN EN ISO 9227-NSS
Acetic acid salt spray test	DIN EN ISO 9227-AASS
Copper accelerated acetic acid salt spray test	DIN EN ISO 9227-CASS

Fig. 7.22 View into a spray chamber, arrangement of the test specimens in the sample chamber

Table 7.7 Test conditions in the salt spray test

Short name	Solution to be sprayed		Temperature in the test chamber	Captured solution	
	NaCl g/l	Additions		NaCl g/l	pH value
NSS	50 ± 10	None	35 ± 2 °C	50 ± 10	6.7–7.2
AASS[a]	50 ± 5	[a]	35 ± 2 °C	50 ± 10	3.1–3.3
CASS[b]	50 ± 5	[b]	50 ± 2 °C	50 ± 10	3.1–3.3

[a]Acetic acid for adjusting the pH value to 3.1–3.3
[b]0.26 ± 0.02 g/l $CuCl_2 \times 2\ H_2O$ and acetic acid to adjust the pH to between 3.1 and 3.3

The addition of acetic acid and copper chloride intensifies the corrosive conditions compared to the neutral salt spray test (SS), so that the CASS test is particularly used in the testing of the most corrosion-resistant material systems. Table 7.7 shows a summary of the test conditions.

7.7.5 Test of Resistance to Filiform Corrosion

According to the definition in DIN EN 3665 [63], filiform corrosion is corrosion that occurs under coatings applied to metal substrates. It generally originates from unprotected edges or local damage that passes through the coating.

The test method is intended to simulate the environmental conditions under which this type of corrosion can primarily occur.

These include:

- Interference points in the coating → Scoring of the specimens
- Presence of starter salts, so-called initiators → Inoculation of the samples in HCl vapour or with 24 h salt spray test

- A certain relative humidity is necessary for the waxing of the corrosion threads → storage of the samples in a climatic cabinet (test chamber) at a temperature of 40 ± 2 °C and a relative humidity of 82 ± 3%.

The evaluation is carried out either on the basis of the comparison images in the standard or the average thread length is determined.

In the quality guidelines of GSB-International [1], in addition to the filament length, the filament frequency is also determined. Experience with the evaluation of the filament frequency shows that it is precisely with this that a clear differentiation and classification of the resistance of a coating system to filiform corrosion is possible.

The standards DIN EN ISO 4623, Parts 1 and 2 describe the filiform corrosion test for the substrates steel (Part 1) and aluminium (Part 2) [64, 65].

The automotive industry has not defined any uniform implementation regulations for the filiform corrosion test on coated components. Individual procedures are used here, depending on the manufacturer.

7.8 Testing Cross-Linking

There are basically two methods available for testing the cross-linking of the powder coating film:

1. Simple on-site testing at the coating plant using a type of solvent resistance test
2. Analytical methods such as thermomechanical analysis (TMA) or differential scanning calorimetry (DSC) [66].

For an in-operation test, a cotton swab is soaked in solvent (in the case of powder coatings, Xylene or an agent specified by the powder manufacturer). Subsequently, 30 double strokes (back and forth movements) are carried out on the surface to be tested within 30 s without any particular pressure. After 30 min the surface is evaluated [3]:

1. The layer is very matte and soft.
2. The coating is very matt and scratchable with the fingernail.
3. Slight decrease in gloss (less than 5 E).
4. No visible changes to the surface, lacquer layer cannot be scratched with the fingernail.

Since paint layers show different resistances to the various solvents, it is important to use a solvent for the test which does not immediately cause permanent damage to the paint layer. The selection of the solvent should be clarified with the powder manufacturer. Also the quantity and especially the execution of the back and forth movement can also strongly influence the result.

If possible, this test should always be carried out as a comparative test. This means that a sufficiently cross-linked paint film of the same powder is to be tested as a reference sample.

7.9 Weather Resistance: Outdoor Weathering and Short Term Tests

Free weathering and weathering in equipment will only be briefly described here, as a large number of publications and book contributions are available on this subject.

7.9.1 Outdoor Weathering

The weather resistance of coatings (outdoor weathering) primarily requires time. Outdoor weathering takes place at selected locations under their ambient conditions and provides information on the durability of the coating against atmospheric influences. Unless otherwise agreed, sample plates according to DIN EN ISO 1514 [67] with dimensions 300 × 100 mm are used. For long-term weathering, coated specimens are stored on large weathering stands at a defined angle to the solar radiation. The exposure can take place behind window glass (for coatings in the interior) or directly. After the required storage time, the colour tone, gloss and chalking are measured and compared with the values of the unweathered coating.

Solar radiation causes the paint films to degrade and the influence of air pollutants changes the gloss and colour shade. Cracks and loss of adhesion occur when the paint surface is heavily weathered. The evaluation of the resistance includes the recording of meteorological data such as temperature (air temperature and, if necessary, sample temperature), irradiance, amount of precipitation. Furthermore, the location has an influence on the result; e.g. in industrial areas the sulphur dioxide content or near the coast the salt load increases compared to a medium atmosphere. Outdoor weathering could be optimised in Florida because, in contrast to Central Europe, very constant weather conditions prevail here (Figs. 7.23 and 7.24).

The alternative to Florida weathering is accelerated outdoor weathering [68–70] using so-called sunlight concentrators, EMMAQUA® or Q-Trac® testing. At test stations in Phoenix/Arizona, weathering tests are carried out with weathering machines. Ten mirrors reflect sunlight onto the specimens. This exposes the specimens to approximately eight times the sunlight. The surfaces are cooled by a fan and wetted with deionized water in a specific cycle. The equipment is intended to allow accelerated outdoor weathering testing. The desired radiation dose in MJ/m^2 is specified here as a measure of the weathering period. 1 Florida standard year corresponds to approx. 280 MJ/m^2 TUVR (total UV-radiation) (Fig. 7.25).

Fig. 7.23 Outdoor weathering area of the company Q-Lab site in Homestead Florida

Fig. 7.24 Test area in Arizona. Test rack for carrying out accelerated weathering

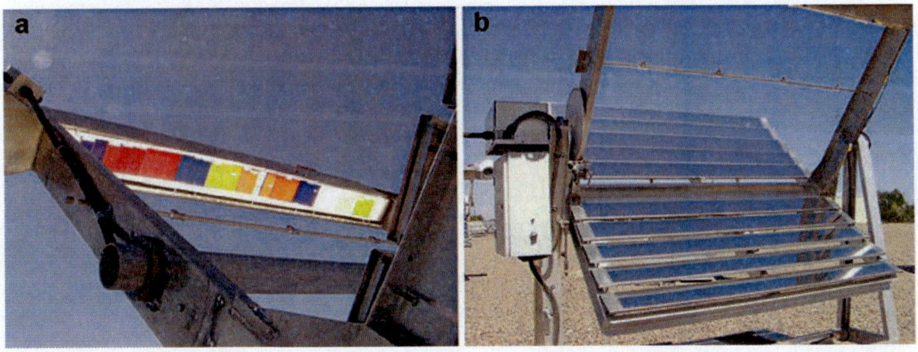

Fig. 7.25 (**a**) Arrangement of the samples opposite the mirrors, (**b**) Illustration of the mirrors for focusing the sunlight

7.9.2 Short-Term Weathering

In order to accelerate the outdoor weathering tests, devices have been developed in which test influences act uniformly and consistently on the samples. Short weathering means optimizing the effects of light and water in such a way that the chemical degradation reactions that occur proceed more quickly. In the case of artificial weathering in equipment, the ageing behaviour is shown by the change in the properties of the coating. Criteria for the change can be, for example, loss of gloss or colour changes.

Generally valid correlations between the ageing processes during artificial and natural weathering cannot be established due to the large number of influencing variables. Certain correlations can only be expected if important influencing variables, such as spectral distribution of irradiance in the photochemically effective range, sample temperature, type and cycle of humidification and relative humidity, are the same. The better it is possible to precisely define and maintain the test conditions, the better the reproducibility and correlation between the results of outdoor weathering and the exposure in devices.

In short weathering there are basically two types of devices. Those that use xenon arc lamps as the radiation source, or those that work with different fluorescent tubes. Another difference can be found in the type of humidification. Some devices work with flooding of the sample chamber, but most work with humidification via condensation or spraying.

Xenon emitters in combination with a filter system emit a light spectrum that is very similar to sunlight. Below the emitter is a quartz shell with a selective reflective layer. The radiation passes through the quartz shell except for the infrared part; the infrared radiation is reflected and leaves the test chamber through a corresponding exit point. The UV light components of the xenon emitter are reflected by mirrors above the emitter and fed to the sample surface. The radiation can be modified by inserting filters made of special UV glass (artificial weathering) or window glass (artificial irradiation). The tested specimens are moistened after repetitive cycles (DIN EN ISO 11341) [71] (Table 7.8; Figs. 7.26 and 7.27).

Table 7.8 Cycles in short weathering according to DIN EN ISO 11341

Procedure	Cycle A	Cycle B	Cycle C
Humidification time	18 min	3 min	Not applicable
Drying time	102 min	17 min	Permanent

Fig. 7.26 Table-top unit for testing with xenon arc lamps

Fig. 7.27 WOM for the tests according to DIN EN ISO 11507

Cycle A is preferably used for artificial weathering of coloured coatings, cycle B for white coatings and process C for irradiation behind window glass.

The QUV® device can also be used to simulate the damaging effects of outdoor weathering in the laboratory. Rain and dew can be imitated by means of a dew system. The effects of sunlight are simulated with UV fluorescent lamps. The weathering temperature is automatically controlled, as is the daily alternation between UV irradiation and dewing. The usual parameters for tests with fluorescent tube devices are specified in DIN EN ISO 11507 [72]. A test cycle may look as follows:

Dew time:	4 h
Dew temperature:	40 °C
Lighting time:	4 h
Illumination temperature:	50 °C
Lighting data:	313 nm – 0.75 W/(m^2 h)

Two types of lamps are available with the QUV® device: UVA and UVB tubes. All lamps produce primarily ultraviolet light in addition to visible or infrared light. This differs in the total amount of UV energy emitted and in the wavelength spectrum. Due to the

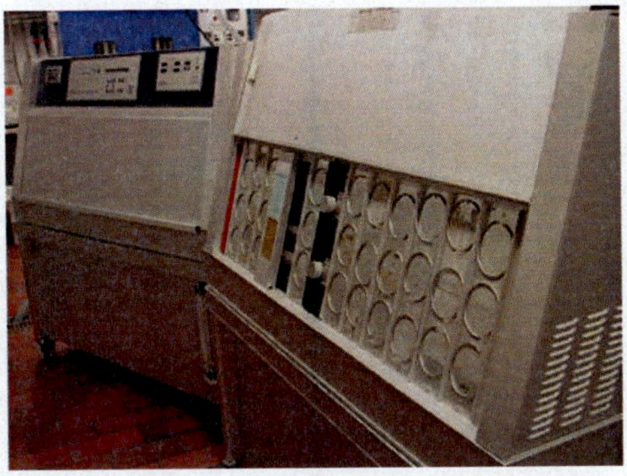

Fig. 7.28 QUV tester with view into the sample chamber

differences, the results may vary. If you want to check the degradation of the binder, i.e. loss of gloss and chalking, you use UVB lamps. When testing for colour changes, UVA tubes are used, as their spectrum is more similar to that of sunlight (Fig. 7.28).

References

1. Qualitätsrichtlinien der GSB-International, Al 631, GSB-International, Düsseldorf
2. Qualitätsgemeinschaft Industriebeschichtung QIB, Ifo Schwäbisch Gmünd
3. Qualicoat, Postfach 10, CH- 8027 Zürich
4. Byk Gardner, Instrumente 2002, physikalische Testgeräte, Geretsried, 2002
5. Hunter, R.S.: The Measurement of Appearance. Wiley, New York (1975)
6. DIN 67530: Ausgabe 1982-01, Reflektometer als Hilfsmittel zur Glanzbeurteilung an ebenen Anstrich- und Kunststoff-Oberflächen, zurückgezogen.
7. DIN EN ISO 2813, Ausgabe 2015-02, Beschichtungsstoffe – Bestimmung des Reflektometerwertes von Beschichtungen (außer Metallic-Beschichtungen) unter 20°, 60° und 85° (ISO 2813: 2014, einschließlich – Deutsche Fassung EN ISO 2813: 2014
8. ASTM D 523, Ausgabe 1989, Spiegelglanzmessung; ASTM D523-14 (2018)
9. Becker, H., Noven, H., Rechmann, H.: Messung des Glanzschleiers an hochglänzenden Lackoberflächen. Farbe und Lack. **74**(2), 145–160 (1968)
10. Ladstädter, E., Geßner, N.: Die quantitative Erfassung von Reflexionsvermögen, Verlaufsqualität und Glanzschleier mit dem Gonioreflektometer GR-COMP. Farbe und Lack. **85**(11), 920–924 (1979)
11. Lex, K.: Die erweiterte Glanzmessung und die Messung von Oberflächenstrukturen, Prüftechnik bei Lackherstellung und Lackverarbeitung. Vincentz Verlag, Hannover (1992)
12. DIN 5033–1, 7, 8 und 9, Farbmessung, 2017–12
13. DIN 6174, Ausgabe 2007-10, Farbmetrische Bestimmung von Farbmaßzahlen und Farbabständen (bei Körperfarben nach der CIELab-Formel) im angenähert gleichförmigen CIELAB-Farbraum ersetzt durch DIN EN ISO 11664-4:2011–07
14. ISO 7724-3, Ausgabe 1984-10; Lacke und Anstrichstoffe; Farbmessung; Teil 3: Berechnung von Farbabständen (zurückgezogen)
15. DIN 6175-2, Ausgabe 2001-03, Farbtoleranzen für Automobillackierungen – Teil 2: Effektlackierungen Teil 1: Unilackierung 2009-07

16. DIN Fachbericht 49, Verfahren zur Vereinbarung von Farbtoleranzen. Beuth Verlag, Berlin (1995)
17. DIN EN ISO 3668, Ausgabe 2001-12, Beschichtungsstoffe, Visueller Vergleich der Farbe von Beschichtungen, ISO 3668: 1998
18. DIN 6175-1, Ausgabe 2009-07, Farbtoleranzen für Automobillackierungen; Unilackierungen
19. Visuelle Beurteilung von organisch beschichteten (lackierten) Oberflächen auf Stahl. Merkblatt St02, April 1997. Verband der Fenster und Fassadenhersteller, Frankfurt
20. Richtlinie VdL-RL 10: Zulässige Farbtoleranzen für unifarbene Pulverlacke bei Architekturanwendungen, Ausgabe April 2003. Verband der deutschen Lackindustrie e. V., Karlstraße 21, 60329 Frankfurt a. M.
21. DIN Fachbericht 63, Lacke und ähnliche Beschichtungsstoffe. Beuth Verlag Berlin (1997)
22. DIN EN ISO 4624, Beschichtungsstoffe – Abreißversuch zur Beurteilung der Haftfestigkeit (ISO 4624: 2016); Deutsche Fassung EN ISO 4624: 2016-08
23. DIN EN ISO 6860, Beschichtungsstoffe – Dornbiegeversuch (mit konischem Dorn) (ISO 6860: 2011); Deutsche Fassung EN ISO 6860: 2011
24. DIN EN ISO 1519, Beschichtungsstoffe – Dornbiegeversuch (zylindrischer Dorn) (ISO 1519: 2011); Deutsche Fassung EN ISO 1519: 2011
25. DIN EN ISO 2409, Beschichtungsstoffe – Gitterschnittprüfung (ISO 2409: 2013); Deutsche Fassung EN ISO 2409: 2013-06
26. DIN EN ISO 6272-1, Ausgabe 2011-11, Beschichtungsstoffe – Prüfung der Widerstandsfähigkeit bei schlagartiger Verformung (Schlagprüfung) – Teil 1: Prüfung durch fallendes Gewichtsstück, große Prüffläche (ISO 6272-1: 2002); Deutsche Fassung EN ISO 6272-1: 2011; Teil 2: Prüfung durch fallendes Gewichtsstück, kleine Prüffläche (ISO 6272-2: 2011)
27. DIN 53154, Prüfung von Anstrichstoffen und ähnlichen Beschichtungsstoffen – Kugelstrahlversuch an Anstrichen und ähnlichen Beschichtungen (zurückgezogen)
28. DIN EN ISO 1520, Ausgabe 2007-11, Beschichtungsstoffe – Tiefungsprüfung (ISO 1520: 2006); Deutsche Fassung EN ISO 1520: 2006
29. NF T 30–010 F, Peintures et vernis – Evaluation de l'adherence ou de la cohesion – Methode par flexion trois points
30. Beschichtungsstoffe – Prüfung der Steinschlagfestigkeit von Beschichtungen – Teil 1: Multischlagprüfung (ISO 20567-1:2017); Deutsche Fassung EN ISO 20567-1:2017
31. Beschichtungsstoffe – Prüfung der Steinschlagfestigkeit von Beschichtungen – Teil 3: Einzelschlagprüfung mit frei fliegendem Schlagkörper (ISO 20567-3:2012); Deutsche Fassung EN ISO 20567-3:2012
32. Hemmelrath, M., Funke, W.: Bestimmung der Nasshaftung von organischen Beschichtungen, XIX. FATIPEC-Kongreß, Aachen 1988, Kongressband IV, S. 137 ff.
33. DIN EN 971-1BW 1, Ausgabe 1996-09, Lacke und Anstrichstoffe – Fachausdrücke und Definitionen für Beschichtungsstoffe – Teil 1: Allgemeine Begriffe – Erläuterungen
34. EN ISO 4624, Ausgabe 2016-08, Beschichtungsstoffe; Abreißversuch zur Beurteilung der Haftfestigkeit (ISO 4624: 2016)
35. VOLVO STD 1029, 5475 (Ausgabe 1980-02)
36. Harlfinger, R.: Einzelschlagprüfung für Mehrschichtlackierungen. Farbe und Lack **94**(3), 179 ff. (1988)
37. Zorll, U.: Die Spannungsverteilung im Grenzbereich Lackschicht/Substrat bei Stoßbelastung. Farbe und Lack **91**(12), 1123 ff. (1985)
38. Sickfeld, J.: Prüfung der Haftfestigkeit von Anstrichen und ähnlichen Beschichtungen. Industrie-Lackierbetrieb **58**(2), 53 ff. (1990)
39. Sickfeld, J., Raabe, H.-J.: Materialprüfung. **29**, 203–209 (1987)
40. VDA Prüfblatt 621-427, Steinschlag-Prüfung, Multischlag

41. DIN EN ISO 21227-2, Ausgabe 2006-11, Beschichtungsstoffe – Beurteilung von Beschichtungsschäden mittels digitaler Bildverarbeitung – Teil 2: Auswertung der Multisteinschlag-Prüfung, (ISO 21227-2: 2006)
42. Wilborn, F.: Physikalische und technologische Prüfverfahren für Lacke und ihre Rohstoffe, Bd. II, S. 426 f. Berliner Union, Stuttgart (1951)
43. DIN 65046-4, Ausgabe 2010-10, Luft- und Raumfahrt – Prüfverfahren für Oberflächenschutzschichten – Beschichtungsstoffe – Prüfungen der Beschichtungen
44. Kratzprobe DBL 7399
45. Beschichtungsstoffe – Prüfung der Beständigkeit von Beschichtungen gegen Druckwasserstrahl (ISO 16925:2014); Deutsche Fassung EN ISO 16925:2014
46. DIN EN ISO 9227, Ausgabe 2006-10, Korrosionsprüfungen in künstlichen Atmosphären Salzsprühnebelprüfungen (ISO 9227: 2006)
47. DIN EN ISO 4628-8, Ausgabe 2013-03, Beschichtungsstoffe – Beurteilung von Beschichtungsschäden – Bewertung der Menge und der Größe von Schäden und der Intensität von gleichmäßigen Veränderungen im Aussehen – Teil 8: Bewertung der von einem Ritz ausgehenden Enthaftung und Korrosion (ISO 4628-8:2012); Deutsche Fassung EN ISO 4628-8:2012
48. ASTM D 552, Mandrel Bend Test; ISO 1519:2011; ASTM D1055 und D1056
49. ASTM D 2794-93 (2010), Prüfung von organischen Beschichtungen auf Beständigkeit gegen schnelle Verformung, Schlagbeanspruchung
50. DIN 55945, Ausgabe 2016-08, Beschichtungsstoffe und Beschichtungen – ergänzende Begriffe zu DIN EN ISO 4618
51. DIN EN ISO 1522, Ausgabe 2007-04, Beschichtungsstoffe – Pendeldämpfungsprüfung (ISO 1522:2006), Deutsche Fassung EN ISO 1522:2006
52. DIN EN ISO 2815, Ausgabe 2003-10, Beschichtungsstoffe – Eindruckversuch nach Buchholz (ISO 2815:2003), Deutsche Fassung EN ISO 2815:2003
53. Neumaier, P.: Bestimmung physikalischer und mechanischer Eigenschaften von Lackschichten, Prüftechnik bei Lackherstellung und Lackverarbeitung. Vincentz Verlag, Hannover (1992)
54. Kecht, J.: Pulver-Schichtdicken-Messgeräte genauer betrachtet. JOT. **8** (2000)
55. DIN EN ISO 2178, Ausgabe 2016-11, Messen der Schichtdicke – Magnetverfahren
56. DIN EN ISO 2360, Ausgabe 2017-12, Nichtleitende Überzüge auf nichtmagnetischen metallischen Grundwerkstoffen – Messen der Schichtdicke – Wirbelstromverfahren (ISO 2360:2017)
57. DIN EN ISO 1463, Ausgabe 2004-08, Metall- und Oxidschichten – Schichtdickenmessung – Mikroskopische Verfahren (ISO 1463:2003); Deutsche Fassung EN ISO 1463:2004
58. Jehn, H. A., et al.: Galvanische Schichten, Bd. 406. Expert Verlag, Kontakt & Studium, Tübingen (1999)
59. DIN EN ISO 6270-2, Ausgabe 2018-04, Beschichtungsstoffe – Bestimmung der Beständigkeit gegen Feuchtigkeit – Teil 2: Verfahren zur Beanspruchung von Proben in Kondenswasserklimaten (ISO 6270-2:2017); Deutsche Fassung EN ISO 6270-2:2018
60. DIN EN ISO 3231, Ausgabe 1998-02, Beschichtungsstoffe – Bestimmung der Beständigkeit gegen feuchte, Schwefeldioxid enthaltende Atmosphären (ISO 3231:1993); Deutsche Fassung EN ISO 3231:1998
61. VDA-Wechseltest, Prüfblatt 621-415, 2.82
62. DIN EN ISO 11997–1, Ausgabe 2018-01, Beschichtungsstoffe – Bestimmung der Beständigkeit bei zyklischen Korrosionsbedingungen. Teil 1: Nass (Salzsprühnebel)/trocken/Feuchte Teil 2: Nass (Salzsprühnebel)/trocken/Feuchte/UV-Strahlung
63. DIN EN 3665, Ausgabe 1997-08, Luft- und Raumfahrt – Prüfverfahren für Anstrichstoffe – Prüfung der Beständigkeit gegen Filiformkorrosion von Aluminiumlegierungen; Deutsche Fassung EN 3665:1997

64. Beschichtungsstoffe – Bestimmung der Beständigkeit gegen Filiformkorrosion – Teil 1: Stahl als Substrat (ISO 4623-1:2000); Deutsche Fassung EN ISO 4623–1:2002
65. Beschichtungsstoffe – Bestimmung der Beständigkeit gegen Filiformkorrosion – Teil 2: Aluminium als Substrat (ISO 4623-2:2016); Deutsche Fassung EN ISO 4623-2:2016
66. Zorll, U. (Hrsg.): Lacke und Druckfarben, Römpp Lexikon. Thieme, Stuttgart (1998)
67. DIN EN ISO 1514, Ausgabe 2016-12, Beschichtungsstoffe – Norm-Probenplatten (ISO 1514:2016); Deutsche Fassung EN ISO 1514:2016
68. Scott, J.: Seasonal variations – nemesis of repeatable/reproducible accelerated outdoor durability tests. J. Coat. Technol. **49**(633), 27–36 (1977)
69. Wootton, A.B.: Natural and accelerated outdoor weathering. Polym. Paint Colour J. **177**(4199) (1987)
70. Pietschmann, J., Junkert, M., Blecher, A.: GSB-Systemprüfungen – Kurz- und Freibewitterung im Vergleich. JOT. (2000)
71. DIN EN ISO 11341, Ausgabe 2004-12, Beschichtungsstoffe – Künstliches Bewittern und künstliches Bestrahlen – Beanspruchung durch gefilterte Xenonbogenstrahlung (ISO 11341: 2004); Deutsche Fassung EN ISO 11341:2004
72. DIN EN ISO 11507, Ausgabe 2007-05, Beschichtungsstoffe – Beanspruchung von Beschichtungen durch künstliche Bewitterung – Beanspruchung durch fluoriszierende UV-Strahlung und Wasser (ISO 11507:2007); Deutsche Fassung EN ISO 11507:2007
73. Michalzik, G.: Instrumentierter Eindringversuch – Härte und weitere Werkstoffkenngrößen, Seminarunterlagen Härtemessung an Schichtsystemen, ZOG (2002)
74. DIN EN ISO 9227, Ausgabe 2017-07, Korrosionsprüfungen in künstlichen Atmosphären – Salzsprühnebelprüfungen (ISO 9227:2017); Deutsche Fassung EN ISO 9227:2017

Further Literature on Chap. 7

Lückert, O.: Prüftechnik bei Lackherstellung und Lackverarbeitung. Vincentz Verlag, Hannover (1992)

Berger-Schunn, A.: Praktische Farbmessung, 2., überarbeitete Aufl. Muster-Schmidt Verlag, Göttingen (1994)

Moeller, E., Ritter, R.: Industrielle Lackiertechnik – Stoffe, Verfahren. Anlagen. Weka Verlag, Augsburg (1997)

Liberto, N. (Hrsg.): Powder Coating – The Complete Finisher's Handbook, 2. Aufl. The Powder Coating Institute, Alexandria, USA (1999)

Kittel, H.: Lehrbuch der Lacke und Beschichtungen, Band VIII, Teil 1, Untersuchung und Prüfung, W. A. Colomb Verlag, Berlin (1980)

DIN 50014, Ausgabe 2018-08, Klimate und ihre technische Anwendung – Normklimate

Images used in Chap. 7

FEM 7.11 bis 7.15, 7.18, 7.21 bis 7.23, 7.25 bis 7.26
Q-Lab: 7.24
Fischer Messtechnik: 7.19, 7.20
Byk-Gartner: 7.2 bis 7.10, 7.16, 7.17
Atlas Material Testing Technology BV: 7.28

Coating Defects

8

The variety and properties of material surfaces decisively determine the utility value of the objects in our environment. The importance of painted surfaces is correspondingly great in almost all areas of the economy, but also for private consumers.

Improper and/or incorrect processing, incorrect selection or surface treatment of the material, not correct application of the paint systems or improper storage of the finished parts will lead to defects or damages to the painted surface and thus, depending on the utility value and service life of the object, to complaints.

Causes of Defects

- Errors in paint production
- Defects in the base material
- Error in coating operation
- Errors during transport and storage of coated workpieces

Defects often only become apparent after coating on the finished product, but often also after installation and a certain period of use.

Despite a functioning in-house monitoring system, it is not always possible to prevent disruptive, i.e. decorative and functional, defects on the finished painted component. The cause can often be narrowed down by the appearance of the defect, although individual damage patterns can be caused by different reasons.

J. Pietschmann, *Industrial powder coating*,
https://doi.org/10.1007/978-3-658-37592-8_8

8.1 Disturbances in the Coating Film

In this chapter, the different types of defects in the coating film will be described, the causes will be shown and possibilities for repair will be explained.

8.1.1 Crater in the Coating Film

Craters, also called fish eyes, are circular cavities with a diameter of 0.5 to 3 mm. Their appearance reaches from very flat hollows in the las paint layer (multi-coatings) up to serious wetting disturbances, which go through to the substrate.

Despite a uniform appearance, there are several reasons for craters.

To identify them by their appearance is mostly impossible. The cater formation is based on two different mechanisms:

- During the film formation, when volatile compounds are localized and spontaneously released, but the viscosity of the coating layer has increased so much, that the coating layer can not flow together at the bursting blisters.
- Before the actual film formation phase, when drops of a substance with a low surface tension hit the still liquid paint film and will not get adsorbed or delayedly absorbed.

(Figs. 8.1, 8.2, 8.3, 8.4 and 8.5).

Reasons for Crater Formation

- Oil, grease, wax and silicone residues
- The powder coating itself. Incompatibility with other powder coatings (the compatibility of different systems is influenced by the respective chemical properties and furthermore significantly by the surface tension, reactivity and viscosity differences).
- Insufficient cleaning of the substrate surface

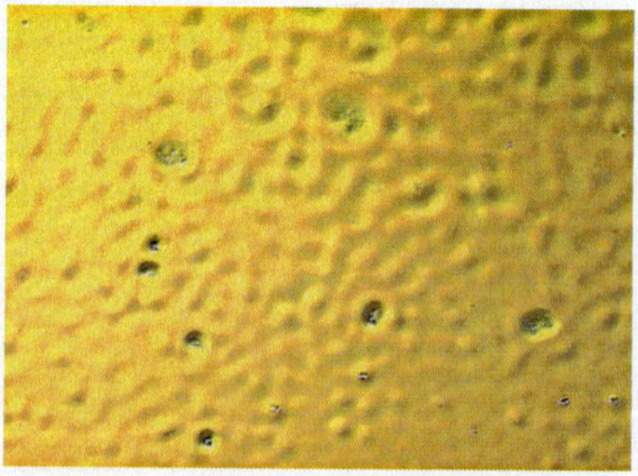

Fig. 8.1 Microscopic image of craters in a polyester powder coating film

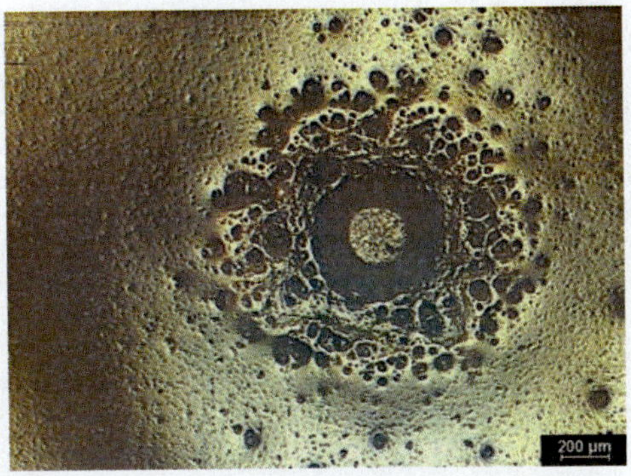

Fig. 8.2 Craters in a powder coating film caused by residues of oil or grease

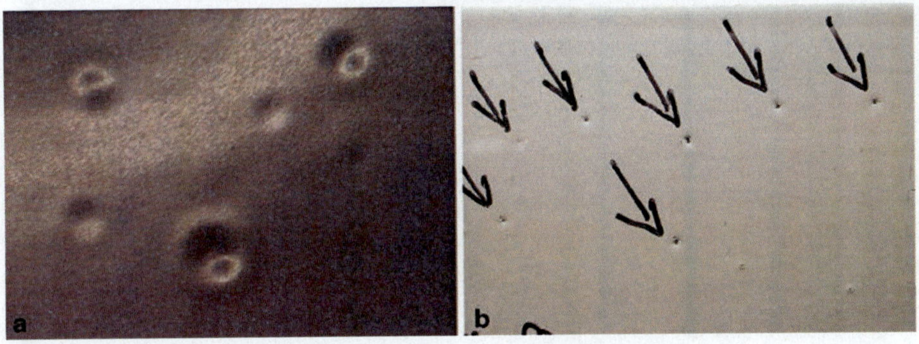

Fig. 8.3 Craters in the paint film due to silicone contamination

- Corrosion products (e.g. rust on the substrat surface)
- Residues from pre-treatment
- Insufficient final rinsing, drying up of rinsing water remainders and contaminations (carry-over)
- Contaminations from the surrounding, which deposit on the substrate or on the still not cross-linked paint film and result to disturbances
- wet surfaces (insufficient drying)
- Outgassing through the substrate (holes and pores in the substrate in which residues from the pre-treatment can collect and escape explosively during the curing process and lead to disturbances)
- Application of wet paint and powder coatings in one booth – and drying in one oven.

Reasons for the Entry of Oils, Greases or Silicone Residues

- Worker: work clothes, rubber gloves, skin and hair care products

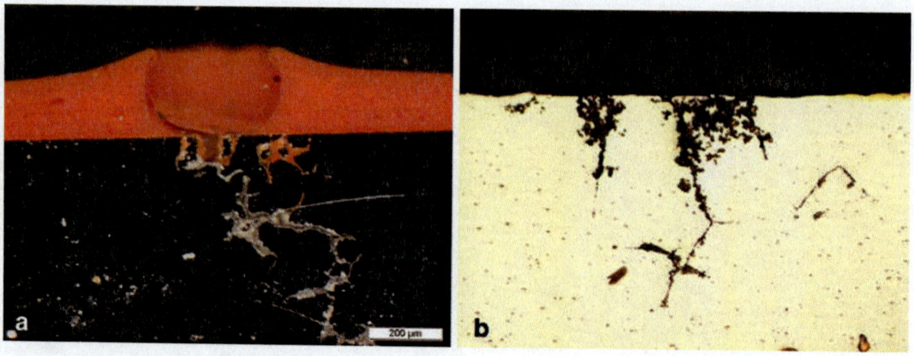

Fig. 8.4 Craters in the paint film due to "outgassing" from the base material; cause in this example: pre-corrosion of the base material

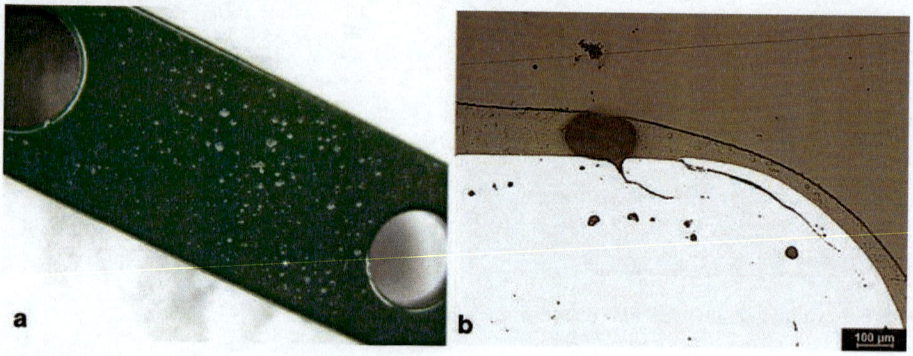

Fig. 8.5 Craters in the paint film due to "outgassing" from the base material. Example here: Zinc die casting

- The paint booth: lubricants from moving parts, release agents from plastic attachments, hoses and sealing compounds, insufficient cleaning of oil and waste water separators, unclean ceiling and floor filters
- The coating material: impurities in the coating due to transport and storage, recovery powder
- The workpiece: release agent residues on castings, drawing and lubricant residues, soldering pastes, residues from packaging or insulating materials
- The painting process/auxiliary materials: unsuitable cleaning agents and cleaning cloths, adhesive tape residues, contaminated conveying and dosing air
- The environment: Suction of contaminated air from other parts of the plant into the cabin.

Prevention of Craters

The above points show many sources of error that can cause craters in a paint film. In order to avoid these, precautionary measures can be derived from this. In principle, silicone-free products should be used, care should be taken to clean the surfaces thoroughly and a final rinse with deionised water should always be used in the pre-treatment. Using pre-corroded surfaces as an adhesion primer for lacquers will lead to problems in any case. If there are no disturbances in the paint film in the form of craters, you will get infiltrations and bubbles after a short time when using the component outdoors.

Repair

There are only two possibilities: either remove the complete lacquer layer and recoat or sand the lacquer until the defects are removed and the layer appears smooth and even. The recoating is applied with suitable lacquer materials.

8.1.2 Blisters in the Coating Film

On coatings, the formation of bubbles is a result of delamination and volume increase under the coating layer. The content of the bubbles can be gases or liquids that have either penetrated the coating from the outside by osmosis or have escaped from the substrate (e.g. wood). In the case of metal surfaces (e.g. steel), corrosion products of the metal usually form under the bubbles.

Paint films absorb a certain amount of water vapour when offered (humid weather) and the water evaporates again in dry weather. This process is common and normally does not damage the paint structure. If the substrate is not properly prepared, hygroscopic, i.e. water-attracting substances (salts) may remain as impurities. These cause local concentrations of moisture which lift the paint film as bubbles. Bubbles can appear as elevations of different size, arrangement and frequency. They can occur between individual layers, but also under the entire resist film. If the paint film dries out (dry weather), most of the bubbles form again (Fig. 8.6).

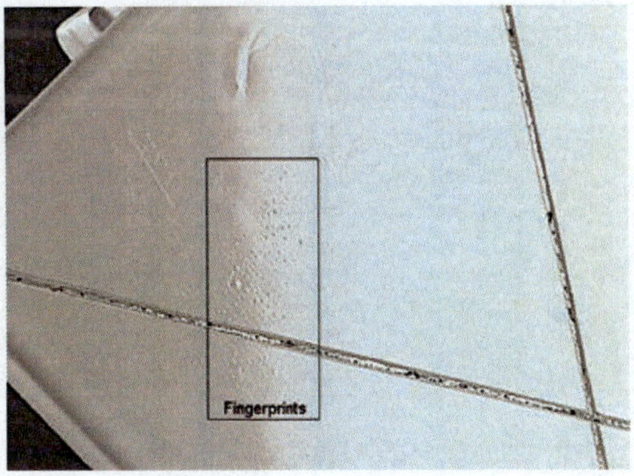

Fig. 8.6 Bubbles in the paint film caused by hand perspiration (fingerprint on the base material)

Reasons of Blistering

- The surfaces to be painted have not been sufficiently cleaned. Contamination due to salt residues (from pre-treatment, from hand perspiration) remains on the surface. The arrangement of the bubbles allows conclusions to be drawn about the cause, e.g. bead-like = wipe marks or run-off marks, nest-shaped = fingerprints.
- Outgassing from the base material, mostly single bubbles.

Prevention of Blisters
The surfaces to be painted should be intensively cleaned. Residues of mechanical processing and chemical pre-treatment should be avoided. To prevent drying or traces of run-off, a rinse in deionised water should be carried out at the end of the cleaning or conversion treatment.

To avoid fingerprints on the cleaned surfaces, the parts should not be touched or only touched with suitable gloves (clean cotton gloves).

Repair
Paint stripping of the workpieces with subsequent cleaning of the surface, new build-up of the conversion coating with subsequent recoating. The necessity of a new build-up of the conversion coating depends on the requirements placed on the component (Figs. 8.7 and 8.8).

8.1.3 Pinholes in the Coating Film

Pinholes are small, fine surface defects (pores, holes) in the paint surface. They are basically a variant of cratering, but with the special feature that the crater-shaped appearance continues in the form of very fine pores in the coating layer. Pinholes are caused by factors

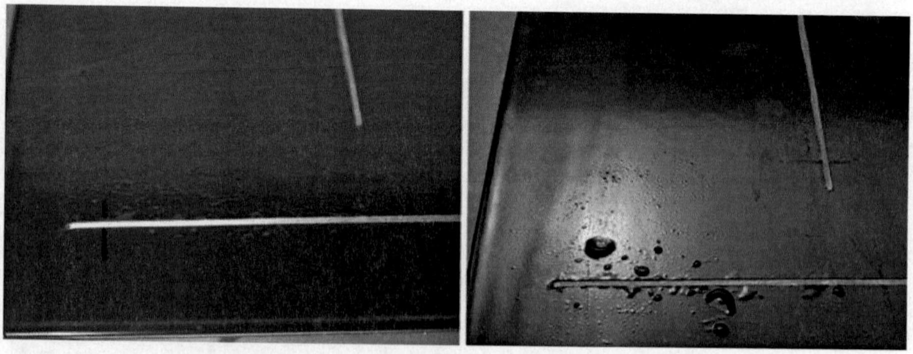

Fig. 8.7 *Left:* Bubbles in the paint film, seen after the corrosion test; *right:* after the boil test. Cause: Hand perspiration – touching the parts after conversion treatment without gloves

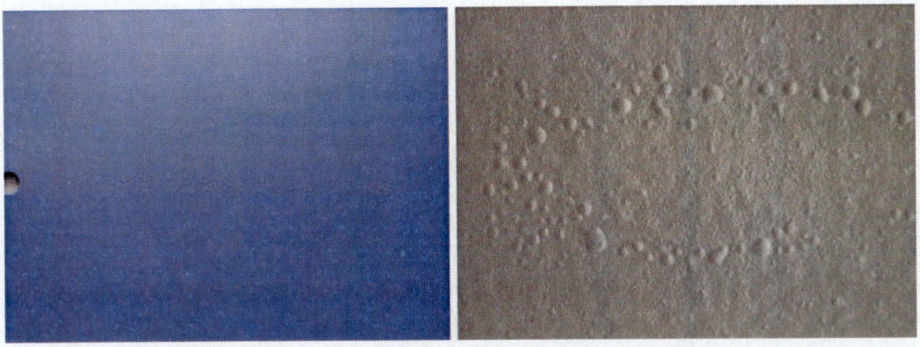

Fig. 8.8 *Left:* Bubbles in the paint film caused by impurities on the base material; *right:* pearl cord-like run-off traces (insufficient rinsing time or quality of the rinsing water)

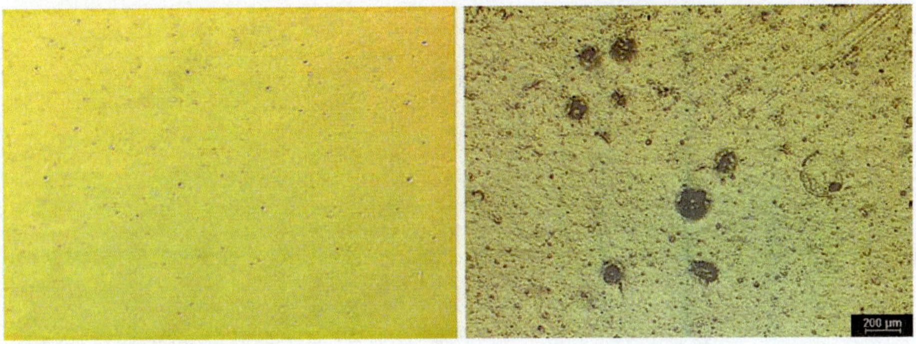

Fig. 8.9 Pinholes in a polyester powder coating caused by volatile components

that can be attributed to the substrate or the deeper layers of the coating. This is the case, for example, when residual air, moisture or reaction products escape and the already dried (cross-linked) coating layer can no longer flow together. with most powder coatings no or only very small quantities of volatile components are released during the baking process. Disturbances in the form of pinholes can be caused by moisture, split-off products and/or air pockets during baking. Powder coatings should be formulated in such a way that they allow these volatile components to leave the coating film without leaving permanent disturbances (Fig. 8.9).

If there is a high number of pinholes in the coating film, this can also lead to a change in gloss (matting), in glossy systems.

Reasons for Pinholes

- Moisture content of the powder too high (e.g. due to incorrect storage)
- Layer thickness too high; during the stoving process the volatile components can no longer escape completely

- Incompatibility with other powder coatings
- Fast heating curve; this does not allow sufficient time for air pockets or moisture to escape from the coating film.
- Pores, cavities and holes in the substrate.

Prevention of Pinholes

Powder coatings should always be stored in a dry place and not above 25 °C. In order to avoid pinholes caused by too high coating thicknesses or too rapid heating or too high temperatures, the powder manufacturer's instructions must be observed. The booth and guns must be carefully cleaned each time the powder is changed. The compatibility of the powder coatings has to be checked in advance. In the case of porous base material, special additives for degassing or special powders can be used.

Repair

In order to obtain a lacquer surface without disturbances, either the workpiece can be stripped and recoated or the lacquer is sanded down until the disturbances are removed and the layer appears smooth and even. The new coating is applied with suitable lacquer materials.

8.1.4 Bits, Spots and Specks

Bits, spots or specks in the paint film are elevations or inclusions of foreign particles and/or contamination. The causes of specks can be very different: dirt particles of any kind, sanding dust, metal particles, pigment agglomerates, gel particles, etc. In the case of spots or specks, the size of the foreign particle is in the range of the resist film thickness or above (Figs. 8.10, 8.11, 8.12 and 8.13).

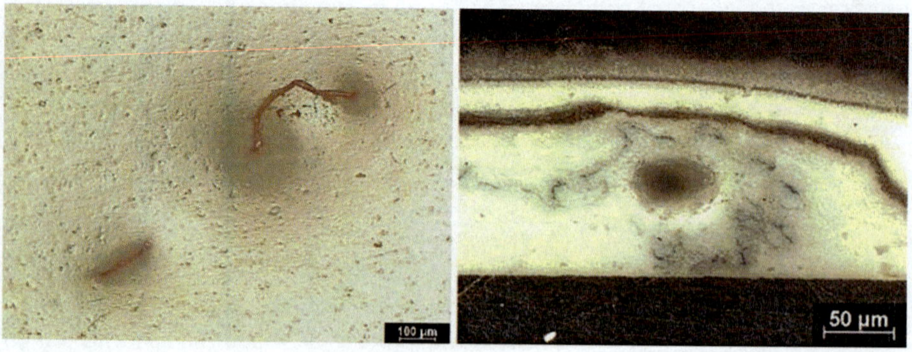

Fig. 8.10 Spots in the paint film caused by fabric fibres

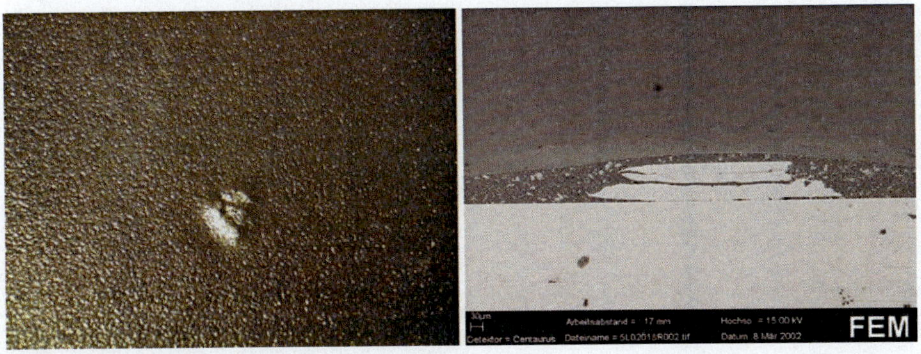

Fig. 8.11 Bits in the paint film due to metal chips

Fig. 8.12 Bits caused by agglomerate of filler material

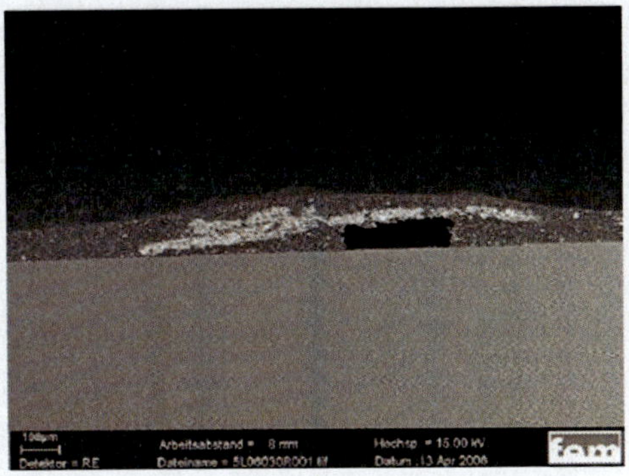

Reasons for Bits/Spots

- Fibres, fluff from cleaning tissues or clothing
- Metal chips, press defect-a die-pick-up, base material (overlaps), welds
- Sintering of contamination from the furnace
- Contamination from the conveyor or transport chain
- insertion of dirt from plant cleaningcontamination in the powder itself, e.g. caused by improper storage
- Particles of a different colour in the coating film due to inadequate cleaning during colour change, in the case of adjacent booths due to paint particle transfer or contamination at the powder manufacturer.
- Powder coating: insufficient dispersion – metallic pigments too large – agglomerates of e.g. filler.

Fig. 8.13 Bits due to large aluminium pigments

Prevention from Bits, Spots, Specks

The mentioned sources of defects contain the corresponding prevention measure. Essentially a coating company should make an incoming goods inspection. Thus, contaminations from the powder manufacturer or the supplier can easily recognized. An additional, very helpful measure is to sieve the powder coating over an ultrasonic sieve before application. The mesh size should not be larger than 120 μm.

Repair

Depending on the cause of the fault and the number of defects, the coating must be removed and the workpiece recoated. If die-pick-ups and impressed metal particle appear, the surface has to be mechanically reworked again.

8.1.5 Point-Like Corrosion Phenomena on the Paint Surface

Flying sparks and industrial dust (rust film) cause spot corrosion on the surface of the paint. The metal particles oxidize at high humidity and then "eat" into the paint surface. Hot, partly glowing particles can also be hurled onto the paint surfaces by mechanical processing or brake abrasion (railways, trams) and burn into the surface there (Fig. 8.14).

Reason for Corrosion Spots on the Paint Surface

- Industrial dust: precipitation from chimneys of foundries and iron processing plants, rail abrasion and road grit
- Flying sparks: Burning of overhead lines of tramways and railways. Flying sparks from flexing, welding or grinding; the hot, partly glowing particles burn into the paint surfaces.

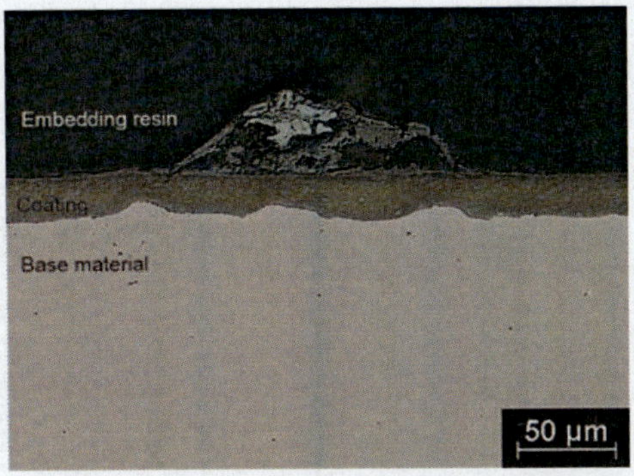

Fig. 8.14 Iron particles pressed into the paint surface

Fig. 8.15 Marking of the intermediate layer of cardboard during humidity storage from a powder-coated paint surface, RAL 8017

Prevention of Corrosion Spots
Immediate removal of the metal particles and regular care of the coating process contributes to the prevention. Cover surrounding paint surfaces during flexing and welding work.

Repair
Use commercial metal dust remover and repolish if necessary. If the metal particles lie deeper, use fine sandpaper to sand down the defected areas and then polish them up.

8.1.6 Staining Due to External Influences

This means etching or discolorations on the paint surface by different causes. The defect appears spot-like in various shapes, colours and sizes (Figs. 8.15, 8.16, 8.17, 8.18 and 8.19).

Causes

- Industrial fumes, the defect is indicated by matting or loss of gloss
- Tar

Fig. 8.16 Water absorption of the paint film in damp packaging, due to improper outdoor storage of the components

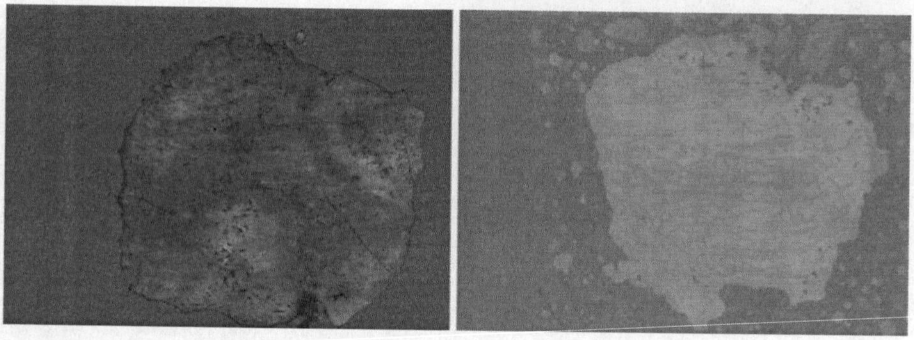

Fig. 8.17 Colour change of metallic coatings due to mortar

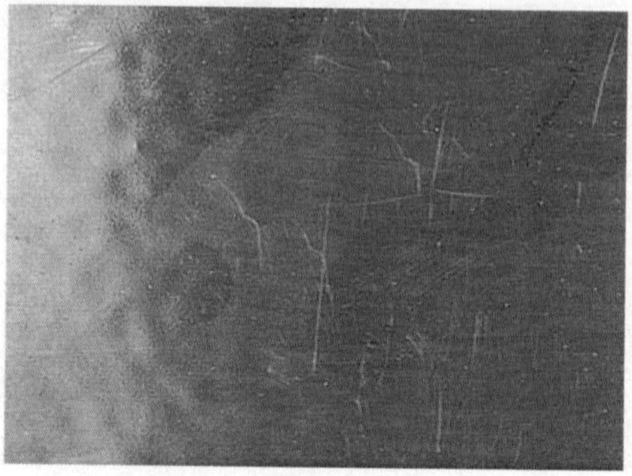

Fig. 8.18 Dulling in the paint surface caused by film packaging

- Insect secretions
- Bird droppings
- Acid or alkaline reacting substances, e.g. mortar
- Wetness, water absorption of the paintfilm in closed packages with condensation water formation. the defect appears as a pale spot of different colours, size and shape.
- Plasticizer diffusion out of packaging foils. This defect appears as matting of the paint film.

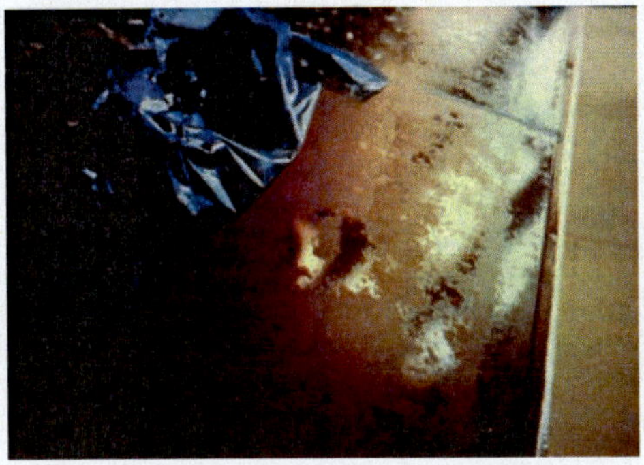

Fig. 8.19 Dulling due to adhesive of the cover foil

Prevention
Always remove foreign substances from the surface immediately. Regular care of the painted surfaces. Suitable materials must be used to avoid staining by packaging material. The packaging must allow air circulation and should not remain on the painted surface for an excessively long time.

Repair
The repair is based on the damage. For small areas, you can try to sand the defect and polish the paint. In the case of large-scale defects, the area must be sanded and repainted.

Staining caused by water absorption can partly be removed by heating the paint layer again to over 150 °C. This is only possible, however, with components that can withstand this heat load or are not too large for the curing ovens. However, this is only possible for components that can withstand this heat load or are not too large for the curing ovens. In the case of locally limited defects, drying with a hair dryer/hot air can lead to success.

8.1.7 Colour Variations: Poor Hiding Power

A translucent substrate is the consequence of a poor hiding power of the paint. In the case of powder coatings, this can mainly have two reasons.

Causes

- Poor hiding hue, e.g. heavy metal free pigment coatings, especially with yellow- or red hues
- strong fluctuations in film thickness or insufficient coating film thickness
- Substrate not matched in colour, large colour difference between substrate and paint colour.

Prevention
Application of a sufficient film thickness, according to the specifications of the paint manufacturer. If necessary, carry out preliminary tests with black and white sheets or tests with different film thicknesses to determine the necessary film thickness.

Repair
The paint must be slightly ground and re-coated with a sufficient and uniform film thickness (Figs. 8.20 and 8.21).

8.1.8 Loss of Adhesion

The loss of adhesion of paint layers appears in two different ways. This depends on the fact if the disturbance of adhesion concerns the whole paint up to the substrate, or the insufficient intercoat adhesion of single paint layers.

Fig. 8.20 Translucent substrate with lettering; insufficient hiding power

Fig. 8.21 Colour differences due to layer thickness differences

Causes

- Adhesive substances that remain on the substrate to be painted, e.g. silicone, oil, grease, wax, dirt, rust, grinding residues and passivation or oxide layers.
- An unsuitable or insufficient primer or pre-treatment (conversion coating)
- Insufficient or incorrect preparation of the substrate – insufficient cleaning
- Defects during the hardening of the coating.

Prevention

In order to avoid adhesion losses, the appropriate and sufficient cleanings must be carried out for the various base materials and adhesion promoting layers must be used. The application of chemical conversion coatings or primers always also depends on the requirements placed on the component or workpiece.

Repair

All layers with disturbed adhesion must be removed. Thorough sanding and cleaning of the substrate is necessary. Afterwards, the new coat is applied with suitable primer and top coat systems (Figs. 8.22 and 8.23).

Fig. 8.22 Paint peeling over a large area. Cause: improper pretreatment of the aluminium panels

Fig. 8.23 Insufficient pre-treatment of the aluminium substrate for outdoor use (here phosphating of the aluminium surface)

8.1.9 Chalking of the Paint Surface: Color Change

Chalking means the “liberation” of pigments, due to a binder degradation on the paint surface.

Causes

- Weathering influences such as aggressive environmental burdens with sulphur dioxide and nitrogen oxides in combination with werness and UV exposure
- Binders not suitable for outdoor use
- Insufficient hardening of the powder coating layer
- Poor care of the paint surface.

Prevention

- The curing conditions of the powder manufacturer must be nocticed and kept.
- Use only systems suitable for outdoor application, e.g. polyester, polyurethane or acrylate systems.
- Regular care of the painted surfaces.

Repair
Repolish and preserving of the surface with appropriate abrasive-acting cleaning agents, which are recommended by the manufacturer. If the chalked surface could not repaired, the damage area must be ground and recoated.

The GRM guidelines [1] provide information on the cleaning of paint surfaces (Fig. 8.24).

Fig. 8.24 Powder coating after nine years of weathering. Light areas: Chalking of the paint surface; dark, colour-intensive area after cleaning

8.1.10 Gloss-Differences Haze Formation: Blooming Effect

Under certain conditions a few resin combinations, which also find application for powder coatings, show a kind of haze formation on the paint surface. This effect is called blooming [2].

This effect occurs, if a powder coated component, whose resin-combination tends to "blooming", is tempered additionally, after curing and cross-linking, for a sufficient long period above the glass transition temperature. The effect appears as a whitish haze or patina.

Haze formation also occurs, if specific powder-formulations are cured at the lowest temperatures and shortest dwell times, prescribed in the manufacturer's guidelines.

Causes/Requirements

- Powder coating which tends to haze
- Inappropriate resins used in the powder formulation
- Thick and thin-walled parts cured at the same temperatures → Thick-walled parts are cured at lower temperaturesSlow cool down of the workpieces
- Longer storage of the parts in the oven, e.g. leaving the parts in the oven after switch-off

Prevention

- Observe the manufacturer's curing conditions
- Do not store coated and cured films for a long time above the glass transition temperature
- Use powder coatings which do not show this effect

Repair

By cleaning of the paint surface, the haze can easily be removed (Fig. 8.25).

8.2 Corrosion of the Metal Surface

8.2.1 Description of the Different Types of Corrosion

According to DIN EN ISO 8044-1999, corrosion is the reaction of a metallic material with its environment, which causes a measurable change in the material and can lead to limitations of the function of a metallic component or an entire system. In most cases, corrosion is electrochemical, but it can also be chemical or metal-physical in nature. Corrosion stands for the reaction, while corrosion phenomena are related to the type of corrosion and

Fig. 8.25 Two examples for haze formation of a powder coating. The lowest temperature and shortest dwell time, recommedded by the manufacturer, were chosen for curing. Both coating surfaces show a white haze, which was polished with a rag, on the part down left.

mean measurable changes in the material or component. Corrosion damage is the impairment of the function of the metallic component or system [3].

A distinction is made between types of corrosion without and those with additional mechanical stress [4, 5].

8.2.1.1 Types of Corrosion Without Mechanical Stresses

- *Surface corrosion:* Uniform corrosion with almost the same removal rate on the entire surface.
- *Pit corrosion:* Corrosion with locally different removal rates. The cause is based on the effect of corrosion elements. The manifestation is characterized by troughs whose diameter is much greater than their depth.
- *Pitting corrosion:* Corrosion in which the electrolytic metal removal only takes place on small surface areas and crater-like depressions are formed which undermine the surface or are like pinholes. Virtually no surface removal takes place outside the pitting areas. The depth of the pitting is usually deeper than its lateral spread. As with pitting corrosion, pitting corrosion is tied to the presence of corrosive elements.
- Crevice *corrosion:* Locally accelerated corrosion in crevices caused by concentration differences in the corrosion medium and the resulting formation of corrosion elements.
- *Contact corrosion:* Accelerated corrosion of a metallic area; it is due to a corrosion element formed by a metal/metal or metal/electron-conducting solid pairing with different free corrosion potentials. Here, the accelerated corroding metallic region is the anode of the (galvanic) corrosion element. It is also called as galvanic corrosion.
- *Corrosion due to different aeration:* locally increased corrosion due to different oxygen influx in the medium. Formation of a corrosion element, whereby the less aerated areas are removed more quickly.
- *Corrosion under deposits (contact corrosion):* Locally increased corrosion when in contact with impurities. This type of corrosion can be either crevice corrosion when in

contact with electrically non-conductive solids or contact corrosion when in contact with electron-conductive solids.

- *Selective corrosion:* Type of corrosion in which structural components, areas near grain boundaries or alloy components corrode preferentially. These include: Intergranular corrosion (selective attack of grain boundary regions); Interdentritic corrosion (selective corrosion along dentritic segregations in the cast structure); Transgranular corrosion (selective corrosion that is approximately parallel to the direction of deformation through the interior of the grains); dezincification (selective dissolution of zinc in copper-zinc alloys with the formation of drop-shaped or layered copper sponge); spongiosis (selective dissolution of the structural constituents ferrite and pearlite of cast iron in cases of insufficient protective layer formation); layered corrosion attack (selective attack of segregation zones leading to flaking and/or bulging of the material).
- *Acid condensation corrosion (dew point corrosion):* Corrosion caused by acid condensing below the dew point.
- *Condensation water corrosion (humidity corrosion):* Corrosion caused by water precipitating on metal surfaces as a result of falling below the dew point.
- *idle time corrosion:* Corrosion that only occurs during the operational *down time* of a plant.
- *Microbiological corrosion:* Corrosion of metallic materials with the participation of microorganisms and their metabolic products.
- *Tarnishing:* Reaction of metals with gases to form thin films that cause interference colors or reduce gloss.
- *Scaling:* chemical corrosion of metals in oxygen-containing gases at high temperatures.

8.2.1.2 Types of Corrosion with Simultaneous Effect of Corrosion and Mechanical Stress [4, 5]

A distinction is made depending on the type of mechanical stress:

- *Stress corrosion*: temporally constant tensile stresses, resulting from load and/or residual stresses
- *Vibration corrosion*: time-varying stresses, mostly cyclic. The stresses can be in the form of tensile, compressive, tensile swell, bending or torsional stresses, as well as combinations of these stress types.

8.2.2 Filiform Corrosion

Filiform corrosion on coated steel parts has been known for a long time and has also been studied in detail. It was described 1944 by Sharmann as "the growth of hair-like traces of corrosion". He referred to this as corrosion under the "film" [6]. This appearance was then later found on other materials, such as aluminum and magnesium. In aircraft construction, filiform corrosion of paintedaluminum has been known for a long time, while in building industry it was only noticed in the 1980s, especially in coastal areas. Particularly affected are areas near the sea and especially where the air is polluted with maritime (salts and high humidity) and industrial particles (dust, acid rain, SO_2).

8.2.2.1 Causes of Filiform Corrosion

For the development of filiform corrosion, certain conditions must be met:

- Defects in the coating layer must exist like, cracks, scratches, pores, microcracks, crevices, undercoated edges, trapped salt crystals or dust particles.
- The presence of certain starter salts, so-called initiators, is necessary. Such initiator salts are mainly chlorides, but also sulfates and nitrates.
- During the formation and development of the filiform corrosion a higher relative humidity must be present.

If these three factors coincide, the occurrence of filiform corrosion is very likely. These conditions are most likely to be present in coastal areas with their saline atmosphere and in industrial areas.

8.2.2.2 Filiform Corrosion on Aluminium and Aluminium Alloys

Aluminium is only affected by filiform corrosion if the metal has been coated with an organic coating (paint) or if the oxide film is the coating.

An active corrosion element forms under the coating, which travels at a low speed and leaves a trail of corrosion products. The attack does not show itself here in the usual formation of infiltrations or blisters, but in the form of thin, sharply defined threads, see Figs. 8.26, 8.27, 8.28 and 8.29.

The characteristics of filiform corrosion have been described by different authors. These essential characteristics are observed in the filiform corrosion of steel, aluminium and magnesium:

Fig. 8.26 Filiform corrosion in the mitre area of a window frame

Fig. 8.27 Filiform corrosion on a milled water slot after 10 years of outdoor weathering in Italy

Fig. 8.28 Result after filiform corrosion test according to DIN EN ISO 3665, test duration 1000 h

Fig. 8.29 Result after 28 days filiform corrosion test according to PPAP 3002 and so-called drawer corrosion after storage in a room climate

- The filaments run on irregular paths, but also directed over longer distances [7, 8–9].
- The threads do not cross each other [7, 8–9, 10].
- Oxygen is necessary for the growth of the filaments [8, 9, 11].
- Increased relative humidity is necessary for filament growth (optimal at about 80% relative humidity) [8, 9–11].
- The filament head contains a concentrated salt solution with low pH values [7–10, 12].
- Dry corrosion products are contained in the filament body (trunk) [7, 8, 12].

- Of the starter salts, the anions remain in the filament head, small amounts are also found in the filament root, while the cations are not found [8].

To start filiform corrosion, a number of factors must be met. This concerns above all the environmental and site-related parameters.

The Environmental and Site-Related Influences Relate to

- the location of the object,
- exposure of the components to salt containing humid air, (so-called starter salts such as chloride),
- a certain time for the corrosion-causing substances to act,
- the dissolve point of the starter salts must be reached.

Every salt has a so-called flow moisture at a given temperature. Due to the hygroscopicity of the salts, they can extract so much moisture from the atmosphere at certain relative humidities that they begin to dissolve, i.e. go into solution. For many salts or salt mixtures, this range is approximately between 60 and 95% relative humidity at a temperature of 20 °C. This does not mean, however, that below 60% relative humidity the salts are dissolved. However, this does not mean that filiform corrosion cannot occur below 60% relative humidity. The higher the humidity, the more certain it is that the deliquescence moisture content of the starters will be reached.

If, the part is constantly covered by a film of water, no filiform corrosion will occur. Coastal areas or areas with a high annual average of relative humidity are more predestined for filiform corrosion than areas with a low annual average of humidity [13].

The Following Parameters Are Considered to be Component-Related

- Defects or pores in the paint film
- Mechanical damage
- Coated areas with low coating thickness
- Construction-related weak points, e.g. edges, mitres and drill holes
- Incorrect chemical pre-treatment or lack of chemical pre-treatment

8.2.2.3 Formation and Process of the Corrosion

In general, filiform corrosion will follow the scheme described below.

It is triggered by the effect of starter salts, especially chlorides. The attack occurs at imperfection of the coating, e.g. at pores, paint spalling, defects on edges with low paint film thickness.

The starter salt fix on these weak points. At a certain relative humidity, the salt absorbs water until the yield point is reached and a highly concentrated electrolyte droplet has formed. At this place a corrosion cell developes due to the oxygen of the atmosphere.

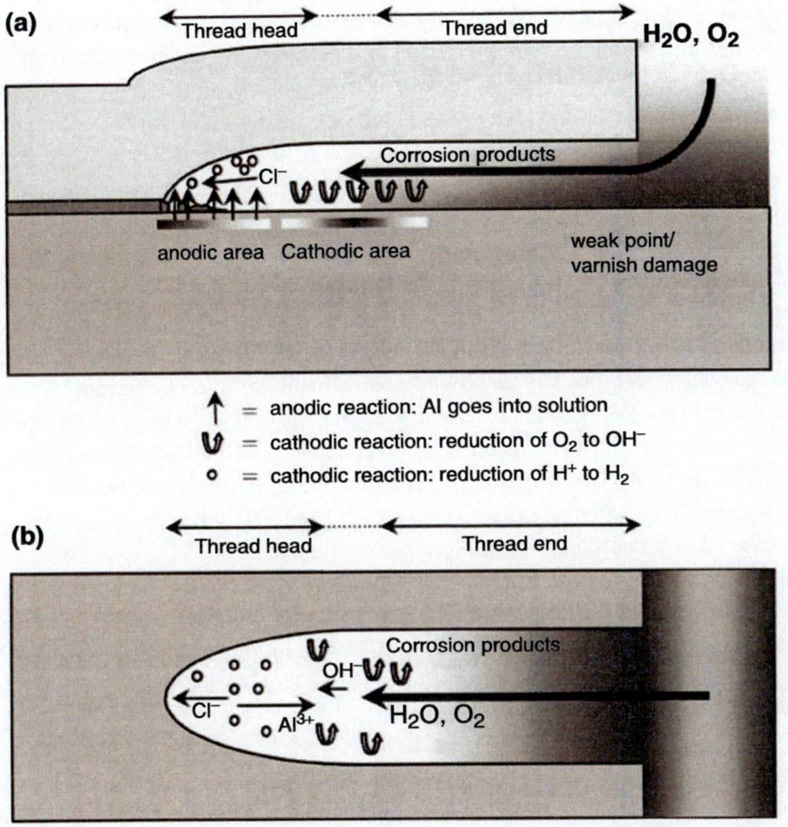

Fig. 8.30 Schematic figure of the mechanism of filiform corrosion. **(a)** shows a longitudinal section through a filiform threat, **(b)** a top view [14]

The cathode is formed at the oxygen-rich site (damaged spot) and the filament head, the anode, is formed at the oxygen-poor site. Aluminium is dissolved at the anode and oxygen is reduced at the cathode with the formation of aluminium corrosion products.

The filament head migrates away from the damaged spot, into the coating. The driving force for filament growth is the oxygen concentration differences between the anode and cathode. Figure 8.30 shows the mechanism of filiform corrosion on aluminium substrates.

8.2.2.4 Appearance of Filiform Corrosion

Filiform corrosion appears as capillary or filiform infiltrations of different widths. They expand between the metal surface and the paint layer starting from the damaged spot in straight or windet paths and cause the lifting up of the paint. Responsible for this is the active corrosion cell in the head if the filament, where the real corrosion process take place. The rest of the filament is inactive and filled with dry, stable corrosion products. The filaments can achieve high length but only a relatively narrow width. Filiform corrosion is related to pitting and other aggressive types of corrosion, but since the filament head is always moving, the depth of corrosion is relatively shallow, so that there is generally no danger to load- carrying components. The depth of corrosion is strongly dependent on the alloy composition. For alloys such as AlMg1 and AlMgSi0.5, it is relatively small; for alloys containing Cu, Zn or higher Si contents, it can be considerably greater (>100 µm) [7, 9, 10–16, 19].

8.2.2.5 Ways of Preventing Filiform Corrosion

The causes and mechanism of filiform corrosion were clarified in work by Hoch, Tobias and Slabough [9–11]. The influences of the base material, cleaning, pretreatment and paint systems have now been investigated by numerous institutes and aluminium industry [16, 19, 20]. Extensive investigations on the influence of anodic oxidation as a pretreatment method to prevent filiform corrosion have been carried out by Pietschmann and Pfeifer [16]. The results of this investigation showed that pre-anodization can be used to prevent filiform corrosion. Outdoor weathering tests by [21] and [22] showed very good resistance of the pre-anodized samples with different paint systems.

Today, in the field of architecture at locations with a special corrosive load and environmental conditions for a possible occurrence of filiform corrosion, the so-called pre-anodization is used as a bonding and corrosion protection layer. In the regulations of the quality label Al 631 [26], the requirements regarding execution and coating properties have been defined.

8.2.2.6 Repair

If a workpiece is affected by filiform corrosion, it usually is the only way to remove the workpiece and the paint and recoat again. For the case that the affected workpiece can not be removed or thereby incurred costs are too high, the part must be repaired on-site. Therefore, this are must be ground up to the substrate and cleaned from grindings and other contaminations. To achieve a permanent protection against corrosion the metal surface of aluminium parts should be pre-treated with brush anodizing. This builds a 3 to 8 µm thick oxide layer, which permanently protects the workpiece against corrosion. The area must be rinsed afterwards with enough water, dried and re-coated with appropriate liquid paint. The execution of tampon anodization is described in detail by [23].

8.3 Examples of Damage Cases from Practice

Various examples from practice will be used to explain how and where the errors described can arise and how they show themselves to the observer.

The result and the quality of the powder coating is influenced by the workpiece, the pre-treatment, the application, the oven, the powder coating and the environment [24]. All of these factors can lead to damages, which can then manifest themselves in the errors described (Sect. 8.1).

In many cases it is possible to localize the cause. A microscope and scalpel can often be very helpful. With modern analysis methods, such as the scanning electron microscope (SEM) with energy dispersive X-ray microanalysis (EDX), IR spectroscopy, but also the classic methods for determining the coating properties, the causes of defects can be identified. It is helpful to set up regular production and product control in order to detect defects at an early stage.

Testing of coatings always has a direct or indirect relation to the failures or damages that may occur. Production control and testing helps to determine applicability, to define formulations, to maintain quality assurance during production, to investigate the development of application and storage parameters and, last but not least, to detect and avoid possible defects in advance.

Testing after the occurrence of an error supports the search and determination of the causes, not least in order to avoid later errors of a similar nature.

8.3.1 Causes of Defects

Defects often only become apparent after coating and curing, but often also after installation and a certain period of use.

In spite of a working company in-house control, disturbing defects, that means decorative and functional defects of the finished product can not avoided every time. The cause of a defect can be searched for in various process steps. Thus, the search for the cause of defect should be during the production of the coating powder, during the coating process itself, during transport and storage of the work pieces or on the substrate.

8.3.1.1 Defects During Production of the Powder Coating

Mistakes in paint formulation can be:

- the dispersion and homogenization of the powder material in the extruder
- the particle size distribution of the powder
- unsuitable selection of pigments/binders
- disturbances in the paint film caused by gel particles in the binder (resin)
- faults caused by additives

Hardener/Binder

If, for example in the production of powder coatings, the correct mixing ratios of hardener and binder are not maintained, the cured powder coating film may not adhere properly and the sensitivity of the coating film to solvents and cleaning agents is higher.

Pigments

The use of thermally unstable substances, additives with low UV resistance or pigments that are not lightfast can lead to clearly visible colour changes even within the curing window specified by the manufacturer. The type of oven heating also plays a major role in this context. If the crosslinking of powder coatings takes place in a curing oven directly heated by gas, the nitrogen oxides formed can change unstabilized pigments and resins and thus cause color variations. Often, these colour variations only become apparent when the components are joined together, e.g. in the case of window frames. If UV-unstable pigments have been used, colour changes will occur within a short period of time on the surfaces exposed to the sun without any associated matting (weathering).

Additive

Additives, i.e. waxes and low-molecular polymers, which float to the surface during melting and influence the flow of the coating film by changing the surface tension, are used to improve the coating film flow. Waxes are also used to matt powder coatings.

When using certain additives and/or an incorrect metering of them, a haze may form on the lacquer surface (see Sect. 8.1.10). Here, a film forms on the surface during the crosslinking of the lacquer film in the area of the lower baking conditions, which can already be seen when handling the parts after coating. Already when handling the parts, wipe marks and fine scratches are formed, which cannot be removed easily during installation. If one rubs off these surfaces with an appropriate cleaning agent, the overall gloss changes. This haze formation can also occur when thick-walled and thin-walled parts are run through the baking oven at the same time. In the case of thin-walled parts, the paint film will be without a haze due to the faster attainment of the curing temperature and due to higher object temperatures, whereas in the case of thick-walled parts, this haze will show on the paint surface. The additives can not only float to the surface, they can also deposit on the base material or the conversion layer, which leads to problems in the resist film adhesion. This can become noticeable when processing the parts, e.g. when cutting them by slight flaking on the saw cut.

Problems with the dispersion and homogenization of the coating raw materials lead to coating flaws, such as pinholes in the melted and cured film. Not completely dissolved resin particles can also lead to flaws (craters, specks) in the cured film. Pigments or pigment agglomerates that are not completely dispersed appear in the paint film as dot-shaped impurities of the corresponding pigment colors.

Table 8.1 shows the various failure modes in the production of the powder coating and explains the causes and manifestations of the resulting failure.

8.3.1.2 Material Defects

Defects of the substrate (workpiece, raw part):

Table 8.1 Possible errors in paint production

Defect possibilities	Problem	Appearance
Binder/hardener ratio	Insufficient cross-linking	Poor weather resistance Insufficient mechanical-technological properties Lack of paint adhesion
Pigmentation	Use of non lightfast pigments	Insufficient weather resistance obvious colour changes on sun exposed surface within a short time
Homogenization/ dispersing	Inhomogenous coating powder	Craters, specks or differently coloured particles or unpigmented paint particles (gel particles) in the film
Binder	Binder system not suitable for the purpose	Chalking during weathering temperature-sensitive colour differences
Additive	Incorrect metering of additives, e.g. waxes or low-molecular polymers	Floating or settling of the additives → hazing or disturbance of the paint adhesion

- Residues on the surface (anti-corrosion oils, adhesive residues, marking with waterproof markers or others)
- Inclusions and cavities (on cast parts)
- Graphite residues on aluminium profile material
- Lamination or compression defects (die-pick-up)
- Impression of metal cuttings from mechanical treatment
- Pre-corrosion.

Markings, Transport Protection of the Blanks

Thick layers of corrosion protection oil or corrosion protection oils already resinified by a reaction with the air and UV radiation cannot be reliably cleaned off in the pretreatment. After stoving, this become visible by crater-like, accumulating depressions, a missing conversion layer in these areas and, as a result, insufficient paint film adhesion.

Adhesive residues as well as markings of blanks with waterproof felt pens also show these defect patterns. These residues must be removed from the visible surfaces before pretreatment, as they cannot be reliably removed. This leads to a not completely covering or missing conversion layer as well as clearly visible level differences in the paint film. The defect become visible itself in insufficient resist film adhesion.

Manufacturing

Another problem of the base material is graphite residues from the production of the aluminium profiles. These residues cannot be removed with the usual cleaning steps. Graphite is used directly after pressing the aluminium alloy as a support and sliding block for the profile. Residues of graphite on the profiles can only be removed with the help of an intensive alkaline pickling. The defect is visible by a missing conversion layer and thus a lack of paint adhesion and possibly wetting disturbances or elevations in the paint film.

Pinholes or blisters in the hardened paint film are caused by cast defects. Pressed-in chips from machining also lead to this appearance.

Cavities, pores and oxides on the surface of castings lead to craters (outgassing during the baking process), problems in paint adhesion and corrosion problems due to a incorrect conversion coating.

Transport and Storage of Raw Parts

The coating of pre-corroded surfaces is always critical. Depending on the depth of the corrosion, the pre-treatment must be carried out or adjusted accordingly in order to obtain a flawless and clean surface. If the corrosion cannot be completely removed, paint disturbances, such as craters, or paint adhesion problems will result. Often a prolonged pickling time results in a scarred surface that cannot be fully compensated for by painting.

Table 8.2 shows the various failure modes already inherent in the basic material and explains the possible causes and appearance of the resulting failure.

8.3.1.3 Failures in Coating Operation

Failures in coating operation can be:

- Pre-treatment – cleaning, conversion layer
- Layer (coating) thickness
- Cleanliness in the application – Cleaning of the spray booths
- Problems during the application
- Over- or underbaking of the coating film

Table 8.2 Possible defects in the base material

Defect possibilities	Problem	Appearance
Pressed-in chips or dirt	Not removeable during pre-treatment process	Pinhole or blisters in the paint film
Pre-corrosion	Different pickling attack on the surface → scarred surface	irregularities in the paint film
	Incomplete removal of corrosion products	Less paint adhesions – craters and bubbles
Graphite strips during aluminium extrusion molding	Residues cannot completely be removed	Problems with paint adhesion and corrosion caused by a missing conversion layer for these sectors
Inclusions, cast defects of the substrate	Accumulation of pre-treatment solutions in the cavities → sudden discharge of these substances or also of air in the curing oven	Outgassing, craters, blisters
Residues of markers, anti-corrosion waxes or oils	Cannot always be removed completely in the chemical pretreatment	Delamination of the coating, craters or blisters

Pre-treatment

The chemical pre-treatment offers a large number of error possibilities. The introduction of grease residues, used and dirty active baths, carry-over and insufficient rinsing lead to a failures in the coating. Paint flaking during machining (mouse teeth), blistering and paint peeling during use or after weathering. Also, touching pre-treated, conversion-coated workpieces with unprotected hands can lead to fine blisters in the paint film within a short period of time outdoors, which correspond to skin lines like a fingerprint (salt residue on the surface due to hand perspiration). Defects in the pretreatment can be detected very quickly and easily with the so-called boiling test. For this purpose, the sample is immersed in deionized boiling water for two hours. Defects in the conversion layer are indicated by the formation of bubbles in the coating film. The fingerprints already mentioned can also be clearly shown here.

A crucial step in pretreatment is the rinsing of the parts. Adhesion problems and paint delamination can result from insufficient rinsing and/or unsuitable water quality. Residues of pre-treatment chemicals, salts or biological impurities lead to bubbles, delamination or corrosion during use or after testing with appropriate test methods. A higherlayer weight, caused by drying off of the pre-treatment solution on drain edges, also leads to the mentioned failures.

In the no-rinse process, the workpieces are only blown off with compressed air or a fine water mist after the conversion treatment. Drying on the drain edges and drain traces in the case of insufficient spraying or blowing off lead to a higher coating thicknesses. This results in adhesion problems of the coating layers.

Cleaning/Rinsing

Another problem can relate to the cleaning of the parts. Shorter cycle times in degreasing and pickling or shorter cleaning tunnels can contribute to the tendency of a faster infiltration and corrosion of the parts. For example, tests with aluminium profile sections of the same alloy, with yellow chromate and approximately the same coating weight, but with different cleaning cycles and times, showed very significant differences in filiform corrosion resistance. Specimens with a higher pickling removal rate show significantly better corrosion resistance in outdoor weathering [23] and also in laboratory tests [27] than those with a low removal rate.

Another important criterion for good paint adhesion is sufficient rinsing of the surfaces. To avoid residues, e.g. salt residues on the surface, the final rinse should be operated with fully demineralized water. With high requirements on corrosion protection and paint adhesion, the dripping water from the workpieces after the final rinse must not exceed a conductance <30 μS/cm [26]. In individual industries, conductance values of <10 μS/cm are required.

Contaminations

Cleanliness in the coating operation, especially in the area of the coating booth and recovery, is one of the most important criteria within production. Even small amounts of remaining powder coating particles of a previously used powder coating can lead to matting of the coating film in the next paint film. Incompatibility reactions of the coatings are responsible for this effect. Residues of powders of different colors also lead to disturbances (color differences).

Dirt particles or metal chips that have not been removed during chemical pretreatment can also lead to imperfections in the paint film. Fine metal chips, produced during cutting or drilling, can be pressed into the surface during transport and can no longer be removed by chemical treatment. A visual sample inspection during workpiece feeding is therefore advisable in any case.

Part Geometry: Construction

A important role in surface treatment is the design and construction of the workpiece to be suitable for coating. This includes, the avoidance of blind holes and gaps in which pretreatment fluids can collect. These pretreatment residues then emerge explosively from the gaps during the curing process and lead to specks and crater-like depressions on the parts below.

Application

With non-bonded metallic paints a cloud-free painting on large flat surfaces is difficult.

Despite the possibility of applying relatively uniform layers with the existing automatic coating equipment, the use of recovery powder and/or an irregular addition of fresh powder results in a different deposition of metallic particles or metallic powder particles on the workpiece. The result is a cloudy surface which only becomes visible from a viewing distance of more than 20 m. In addition, the distance between the powder gun and the workpiece and the electrostatic properties of the powder coating film play an important role. Insufficient distances show up in streaking on large-area sheet metal parts. Manual recoating (by hand) on fouldings also leads to uneven coating and cloud formation. In this case, it is recommended to precoat structural fouldings, bends, etc. on workpieces manual before automatic coating and after to coat the entire surface automatically by applying a defined, uniform coating.

Table 8.3 shows the various defect possibilities that can occur in coating operation and explains the possible causes and appearance of the resulting failure.

Stoving Oven

If the temperature and heat distribution in the curing oven changes due to the failure of a burner or a so-called heating register, this can lead to a local lack of crosslinking of the coating film. The paint films then already show extremely brittle behavior during processing, and the adhesion is only moderate. These poor crosslinking are also associated with insufficient weathering behavior, which very quickly leads to chalking of the surfaces. A lack of crosslinking can only be clearly determined with a DTMA analysis (dynamic thermomechanical analysis) or a DSC (differential scanning calorimetry). A quick wipe test with solvents such as acetone, only shows a result if the paint film has a clearl lack of

Table 8.3 Defect possibilities which occur during coating operation and possible causes and appearance of the resulting failures

Defect possibilities	Problem	Appearance
Pre-treatment	Insufficient execution, rinse failures, residues from pre-treatment, washable conversion layer, too high or too low weight of the conversion layer	Poor paint adhesion – flakings, corrosion
Application	Insufficient cleaning of the installation after colour changes	Different coloured dots in the paint film. In case of incompatibility of the powders pinholes, craters, matting
	Layer thickness variations	Colour differences, uneven flow of the paint film
	Material mistakes or blending of two systems	Incompatibility reaction, matting, pinholes
Stoving oven	Over- or underbaking of the paint film	Yellowing of pale colours or colour changes. Worse mechanical properties with underbaking – brittle film, insufficient weather resistance, solvent sensitivity, worse paint adhesion, gloss differences
	Continuous infeed of coated parts into the oven, high air speed in the door area	Circling powder particles which settle down on other workpieces lead to a contamination of the paint film – differently coloured dots, spots, lumps

crosslinking. If the powder coating film is sufficiently cross-linked, wiping the surface approx. 5 times with gentle pressure with a cloth soaked in acetone will not dissolve the coating film. In the case of films that are not fully cross-linked, the paint already starts to dissolve, and in some cases it can be wiped off completely. This test is influenced by the amount of solvent, the time of wiping and the pressure applied to the coating film.

If there is a suspicion of a fully crosslinking of the paint film, the wipe test can be used in a parallel test. Parallel test means that the sample is divided, one part is post-crosslinked again at sufficient object temperature and holding time, the other remains in the initial state. The wipe test is now carried out on both samples. In a direct comparison, a statement can now be made about the behaviour of the paint films when exposed to solvents.

Another source of failure is the curing oven. Requirement for maintaining the so-called object temperature is precise knowledge of the oven characteristics and the heating behavior in the curing oven. Thick-walled parts of a large mass should not be cured together with thin-walled parts. While these thin-walled parts are usually sufficiently cross-linked, under-cross-linking occurs with the thick-walled elements with the consequences already mentioned. Excessively long dwell times or excessively high curing temperatures cause yellowing of the paint film, which can go as far as destruction. Especially with achromatic colors, such as white or gray, this yellowing is very noticeable. Even if this colour change is not objected to, these parts should not be exposed to weathering as they tend to chalk very quickly. Oven measurements should be carried out at regular intervals for production control and quality assurance.

8.3.1.4 Transport and Storage Defects

Failures during transport and storage:

- Mattings, marking and staining due to packaging material
- Release agent on rubber seals
- Airtight packaging → Condensation → Dulling, stains
- Soft paint film → Pressure marks, impressions

Packing

Some damages of the coating film occurs during transport, where the goods later showing abrasion and pressure marks due to improper packaging. High pressure loads also cause partial pressure points in intermediate layers of polystyrene or other air-filled foils, which only become visible when the protective film is removed. Removal by polishing is only possible to a limited extent.

Another point is the proper storage of the parts at the construction site until installation. Condensation can form due to airtight packed facade elements or in airtight packed transport pallets, which penetrates the paint film and can lead to matting due to swelling. It is therefore very important either to store the parts in a protected place until installation or to ensure adequate ventilation of the transport containers and to avoid water penetration. When using packaging films containing plasticizers, the effect of heat may cause matting in the paint film. The plasticizer contained in the foil diffuses into the paint film at higher temperatures and may cause similar phenomena as an organic solvent.

Transport and Storage

When storing and transporting raw material, care must be taken to ensure clean dry storage. The storage rooms should be dry and well ventilated. Semi-finished products which become wet should be dried before being placed in storage, and wet packaging should be removed. The formation of water condensation must be avoided. This can cause corrosion.

Table 8.4 shows the various possible failures which can occur during transport, storage and processing and explains the possible causes and appearance of the resulting defects.

8.4 Defects in the Coating Film

Various examples from practice will be used here to explain failures (error patterns) in the coating film, which have already been described in Sect. 8.1.

Table 8.4 Failures during transport, storage and handling

Defect possibilities	Problem	Appearance
Packing	"Soft" paint films	Pressure marks, visible support points, impressions
	Outdoor storage, long storage in packaging	Staining, lightening of the paint film, packaging material difficult to remove → Matting
Packaging material	Foils containing plasticizers	Discolourations, matting
Cleaning	Use of unsuitable cleaning agents	Solving the coating, discolouration, matting
Sealing material, adhesives	Materials not suitable for the application	No adhesion of the sealant or adhesive, discoloration, dissolution of the paint material

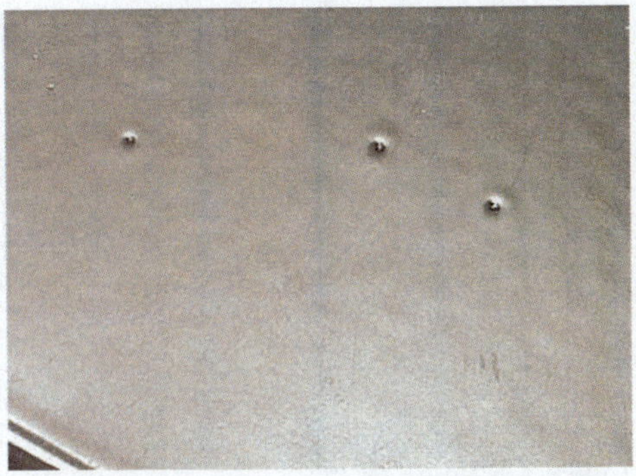

Fig. 8.31 Macro image of the paint defects. Cause of defect here: pressed-in metal chips

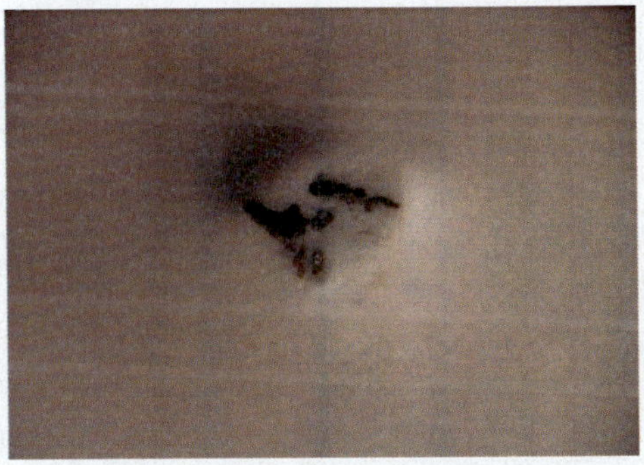

Fig. 8.32 Dark, metallic particles can be seen in the centre of the failure (chip, metal abrasion)

8.4.1 Spots in the Paint Film

Cause Metal Particles

The following pictures show examples of spots in the paint film caused by pressed-in metal chips or press fleas (Figs. 8.31, 8.32, 8.33, 8.34, 8.35, 8.36 and 8.37).

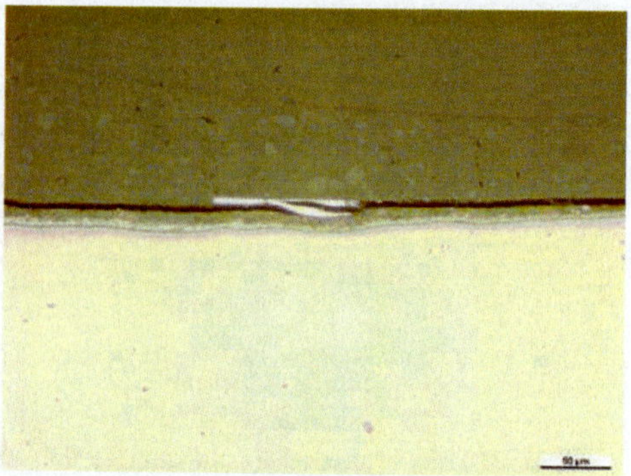

Fig. 8.33 Cross-section of the defect area. Metal particles are visible in the paint film itself and at the interface between paint and base material

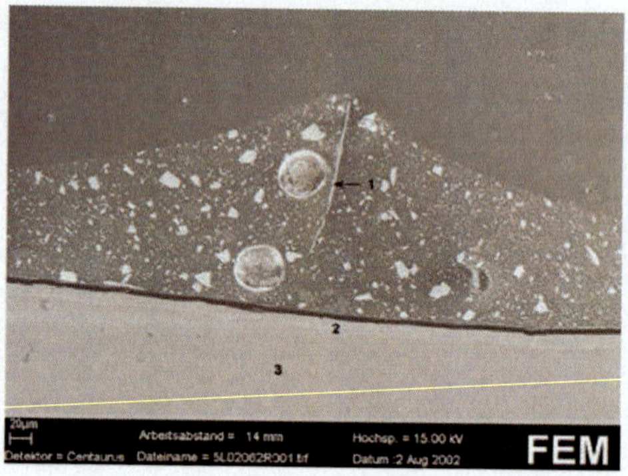

Fig. 8.34 SEM image of the paint defect in cross section. The cause of the defect are metal particles floating on top of the coating

Cause Contamination

Irregularly distributed spots of varying size were found in the paint film. Cross-sections through the defects resulted in foreign substances of dark color embedded in the paint film. With a EDX analysis, components of paint and contamination, e.g. fine iron particles and residues of grease, could be detected. The cause of this spot could be found in contaminated hangers. Figure 8.38 shows the cross-section through the defect area (Figs. 8.39 and 8.40).

Another example is contamination by fibres from clothes or contaminated cleaning cloths, see Fig. 8.41.

Cause Powder Coating

The following figures show various defects caused by the powder coating itself. These can be filler agglomerate (Figs. 8.42 and 8.43) or metallic pigments that are too large or "stuck together" (Figs. 8.44, 8.45, 8.46 and 8.47).

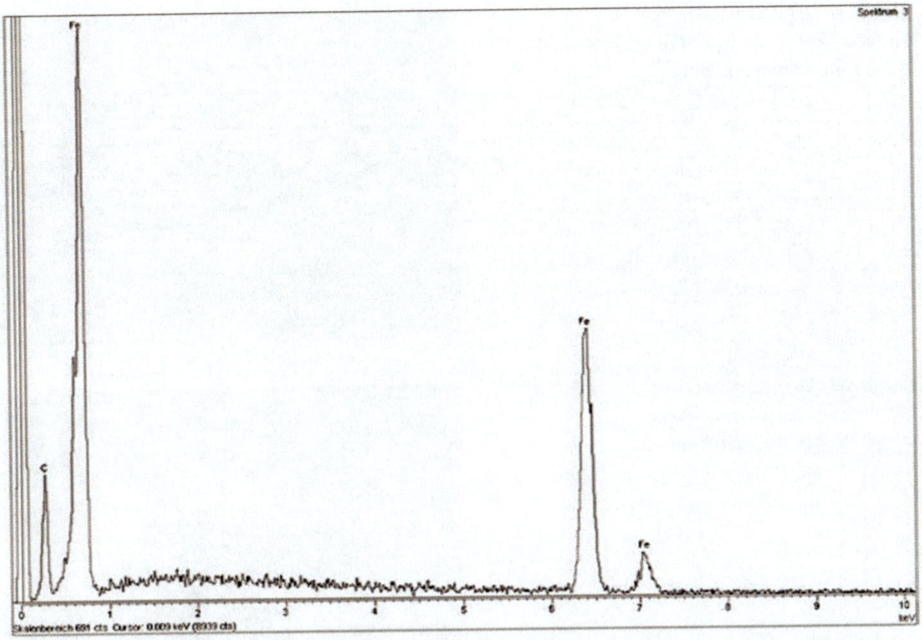

Fig. 8.35 Result of the EDX analysis of the metal chip; base material is aluminium, chip iron

Fig. 8.36 Lumps, pinholes in the paint film caused by a so-called press flea

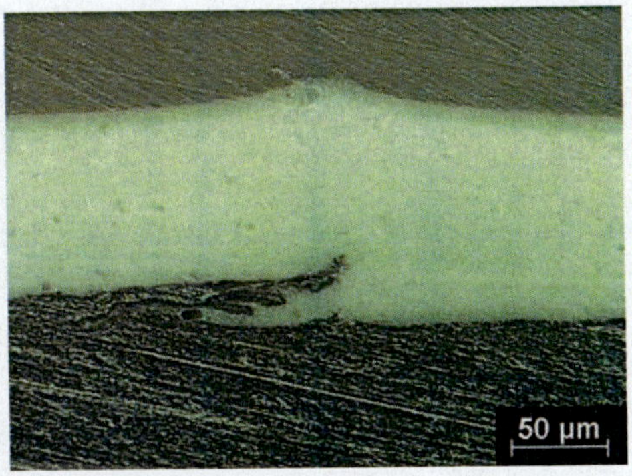

8.4.2 Craters in the Coating Film: Flow Disturbances

Crater in Coated Aluminium Profile

Craters and flow defects were found on aluminium hollow chamber profiles, chromated and powder coated. EDX analyses in the area of the defects, an okay area and on the base material show the presence of carbon and sulphur in the area of the craters. No impurities could be detected next to the defect. Causes here are oils or greases. Figures 8.48 and 8.49 show the area of defects on the paint surface. Figure 8.48, paint surface in the area of the

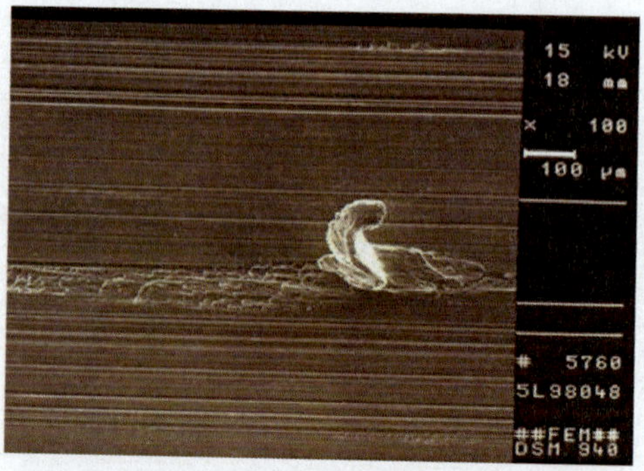

Fig. 8.37 SEM image of the surface after removal of the paint film – pressing defect

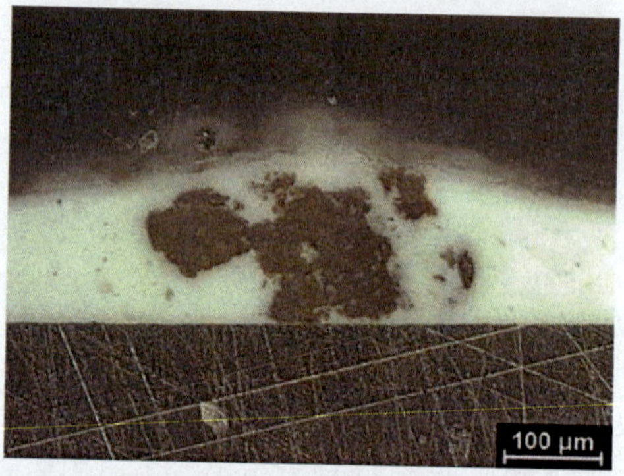

Fig. 8.38 Cross-section through the defect area – spot caused by residue/abrasion from the hanger

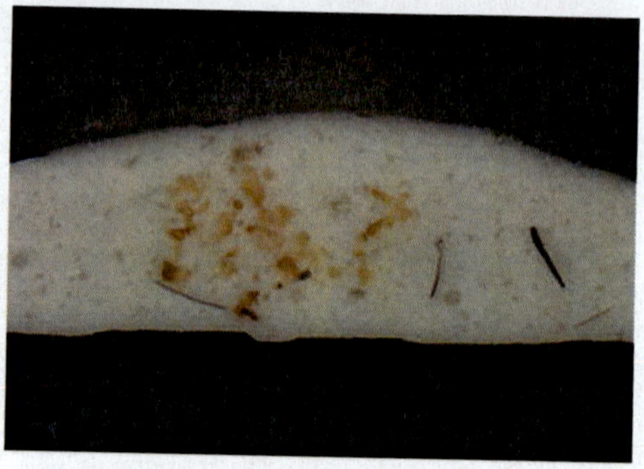

Fig. 8.39 Cross section through failure area. The cause was combustion residue from the oven

defect, Fig. 8.49 after removal of the paint film on the base material, Fig. 8.50 and 8.51 show the result of the EDX analyses in and next to the defect area. In the defect area, besides carbon (C), sulphur (S) and oxygen (O) can be analysed. Outside the defect, aluminium (Al) and chromium (Cr) from the pre-treatment can be found.

Fig. 8.40 Cross-section through the defect area. The EDX analysis showed barium and sulphur in addition to components of the coating. This is a contamination of another varnish (sticking powder particles which remove later from the hanger or in the oven)

Fig. 8.41 Contamination by fibres

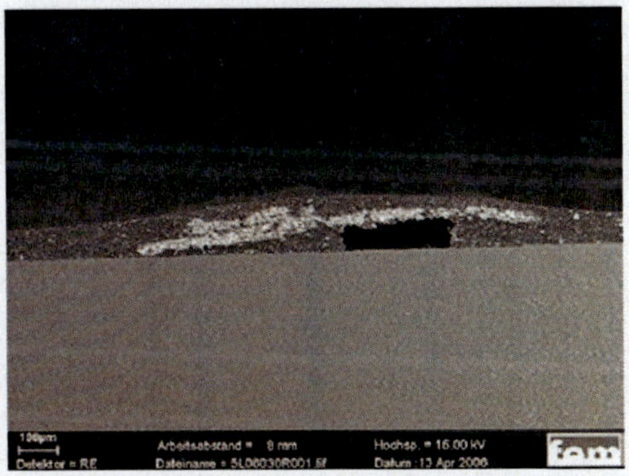

Fig. 8.42 SEM image of the cross-section through the defect area. Filler agglomerate in the resist film. This led to the "spot" defect in the resist film

8.4.3 Staining Due to Packaging

Coated components are packaged during transport and storage to protect the paint surface. If the parts are stored airtight packed in foils, colour change and staining may occur under certain circumstances.

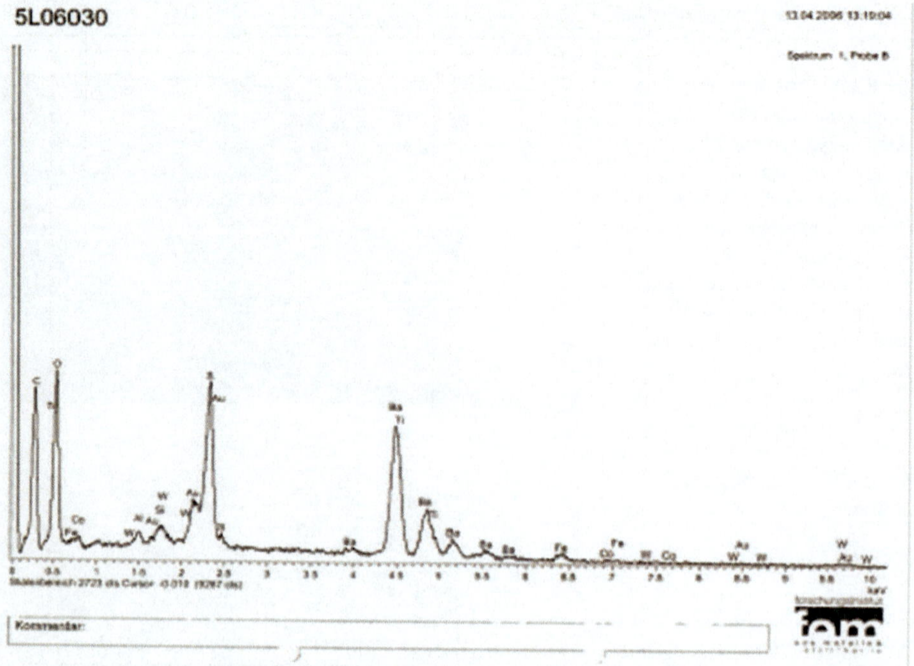

Fig. 8.43 EDX analysis of the defect area. Once bright area (filler) and as reference resist film. The elements Ba, S and O can be found in the defect area, from which BaSO4 can be concluded

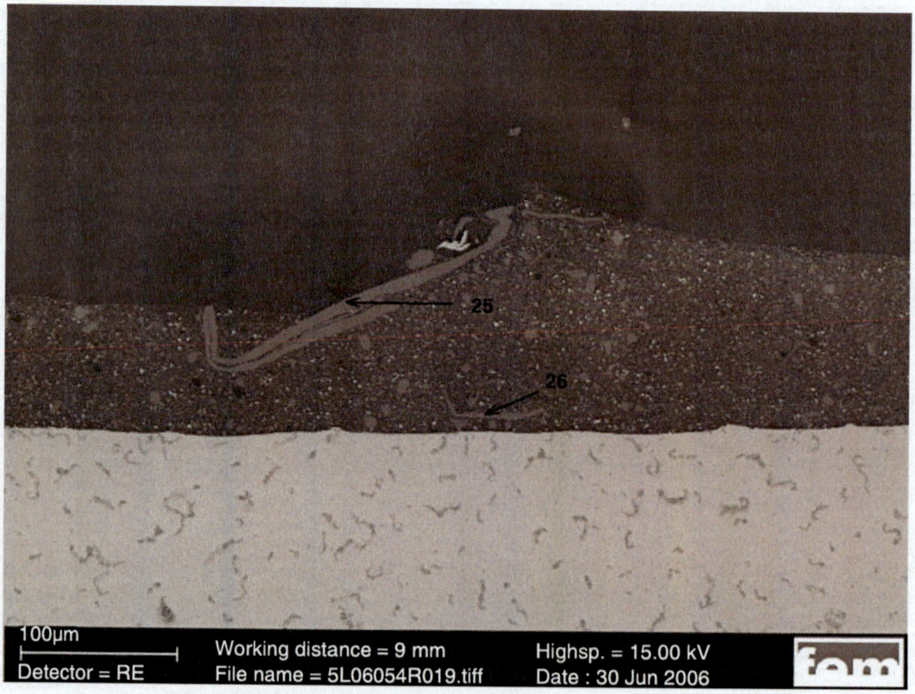

Fig. 8.44 Spots in the paint film caused by large aluminium pigments

Fig. 8.45 Reference resist film without defect

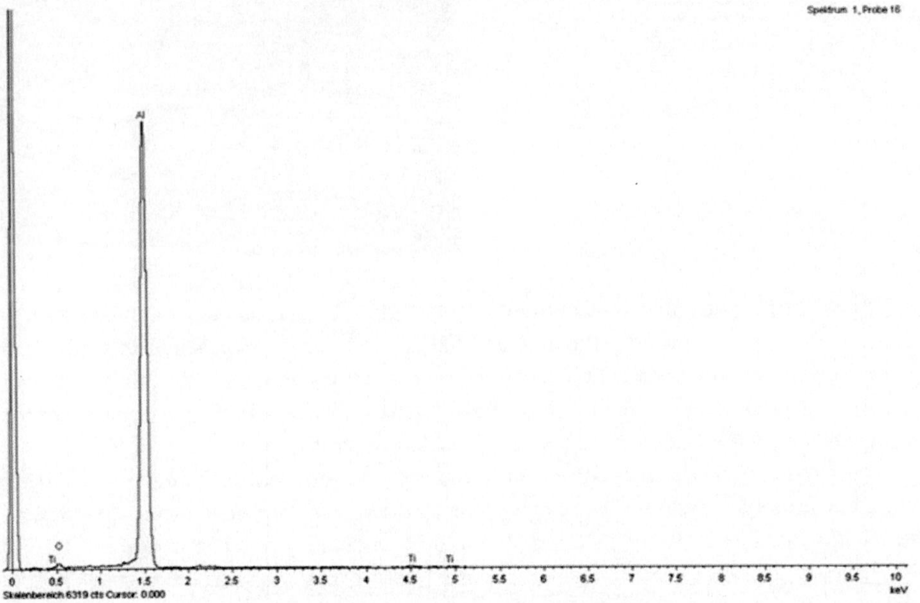

Fig. 8.46 EDX analysis of the interfering substance – Result: Al

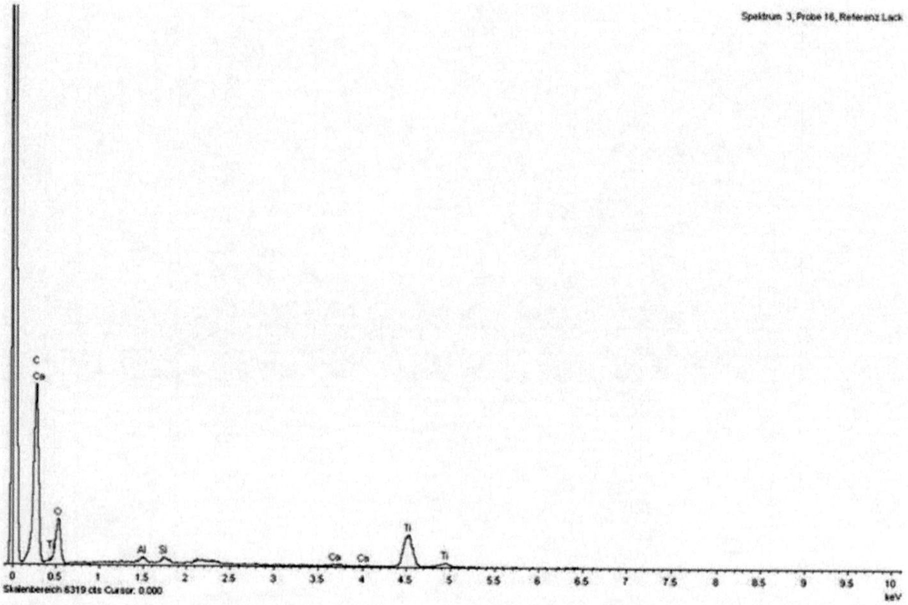

Fig. 8.47 EDX analysis of the reference range – resist film without interference

Fig. 8.48 SEM image of the paint surface in the area of the defect

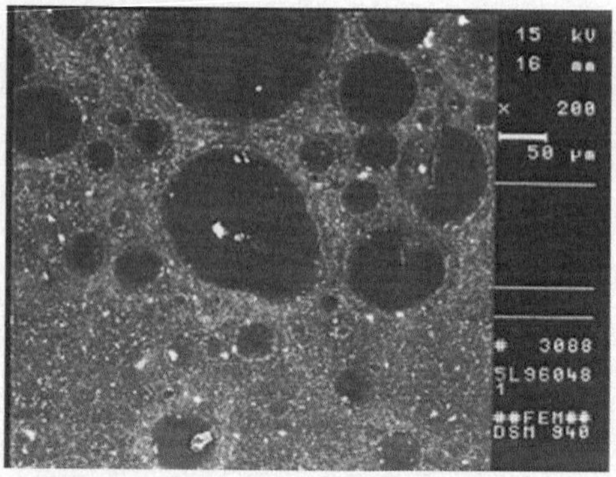

When airtight parts are stored outdoors, temperature fluctuations cause condensation to form. If the moisture cannot evaporate and the paint surface is exposed to this microclimate for a certain time, often at higher temperatures in the summer months, the paint layer absorbs water and bright, irregular spots form. This can be remedied by dry storage of the components and adequate ventilation of the packaged parts.

Once water stains have occurred, they can only be removed by removing the water stored in the polymer matrix. This can be done by drying the parts at an increased temperature in the oven, as far as the part size allows, or with the help of a hot air dryer.

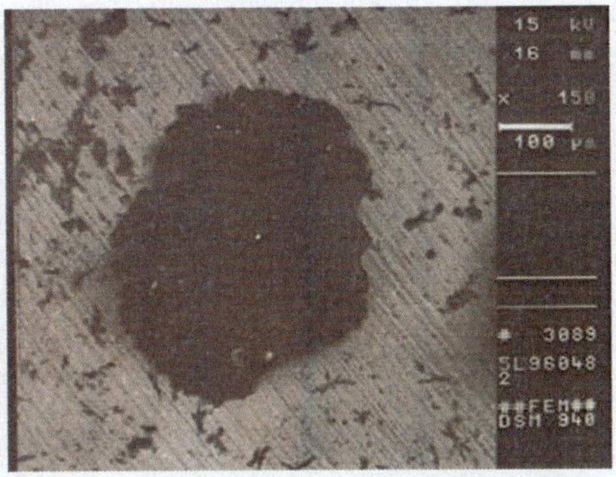

Fig. 8.49 SEM image of the metal surface after the paint film has been removed. Section of the defect area, dark spot = crater base, light area = base metal surface

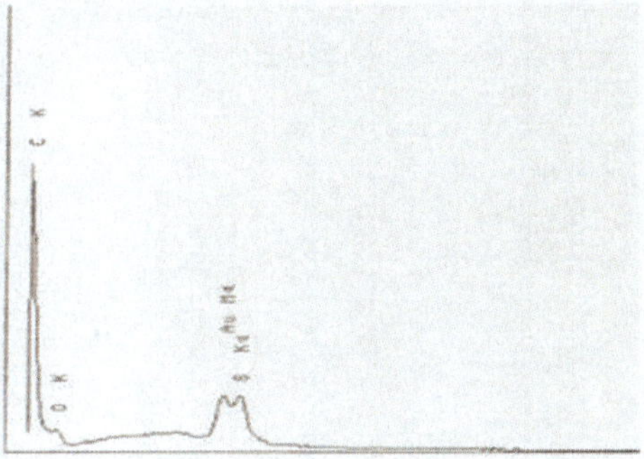

Fig. 8.50 Result of EDX analysis in the failure area, dark spot in Fig. 8.49

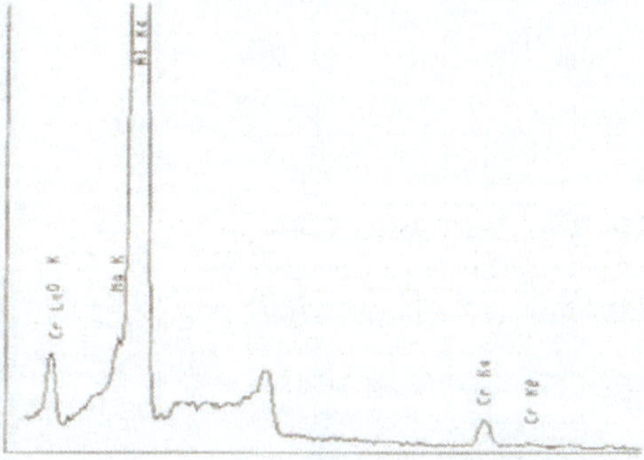

Fig. 8.51 Result of EDX analysis next to the defect area, bright area of Fig. 8.49

Figures 8.52 and 8.53 show changes in the paint surface caused by a foil containing plasticizers. During storage of the parts with packaging film, in addition to the described disturbances caused by moisture, matting of the paint surface also occurred due to components of the packaging film. The decisive factors here are the storage temperature and the time available for plasticizer diffusion (plasticizer migration) into the paint film (Figs. 8.54 and 8.55).

Fig. 8.52 Dark red painted facade with matt, irregular areas

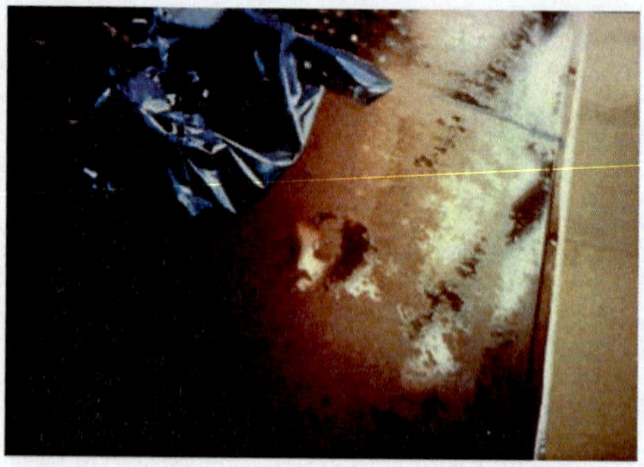

Fig. 8.53 Dulling due to plasticizer from the film

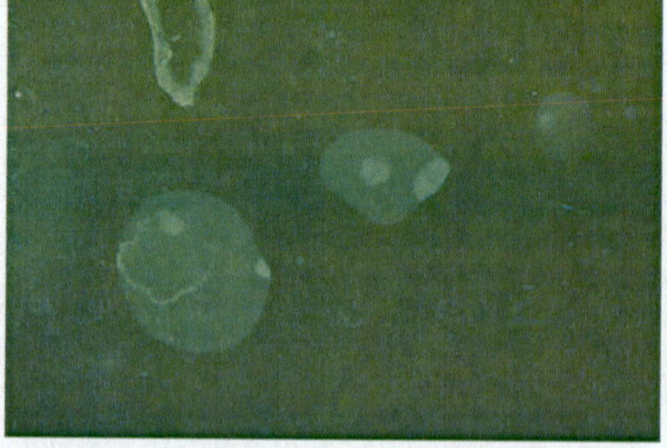

Fig. 8.54 Water stains in the coating after outdoor storage of the parts in airtight packaging

Fig. 8.55 Color change of the paint film due to water storage after the moisture storage test [26]. The color measurement of loaded compared to unloaded paint surface showed a color change of $\Delta L = 5$

Fig. 8.56 Colour differences on the sheet metal elements in the roof area due to a lack of crosslinking of the paint film

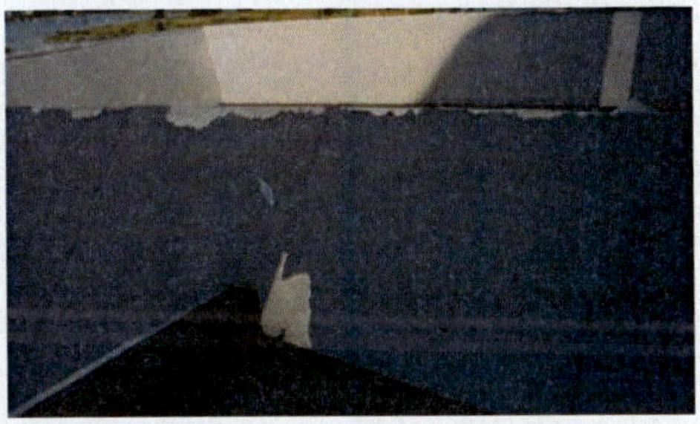

Fig. 8.57 Large area detachment of the paint film due to missing conversion layer on the base metal

8.4.4 Chalking of the Paint Film and Delamination

Facade Sheets with Colour Differences and Paint Delamination Within a Short Period of Time

A facade with large, powder-coated aluminium sheet elements showed extensive paint delamination and colour changes after a short time.

The cause of these defects was found to be, the lack of conversion treatment of the parts and, insufficient cross-linking of the paint film. Figures 8.56 and 8.57 show the colour differences of the sheets in the roof area and the extensive paint detachment from the base material (Fig. 8.58).

This defect can be avoided with a pre-treatment suitable for aluminium with a layer-forming conversion coating and sufficient cross-linking of the paint layer.

8.4.5 Detachement of the Paint Film

Aluminium Facade with Extensive Paint Detachment
(See Fig. 8.58).

Fire Extinguisher with Flaking Paint During Function Test
Powder coated hand-held fire extinguishers showed extensive paint spalling during the pressure test. The examination of areas with very poor paint adhesion and the direct comparison with areas of sufficient adhesion revealed insufficient pre-treatment of the surface. In the areas with insufficient paint adhesion, no components of a conversion layer could be found using EDX analysis. Figure 8.59 shows a fire extinguisher after the compression test with a chipped paint layer.

Profile Frame with Cracking and Detachments
For a glass roof, cassettes with gratings were used for light refraction. The frames, made of galvanized steel, were coated with a white powder coating. Of the total of 250 individual cassettes supplied, three showed cracks and partial paint peeling on the frames after

Fig. 8.58 Paint peeling on aluminium facade. Problem: Clear powder coating on a chrome-free, too thick conversion coating

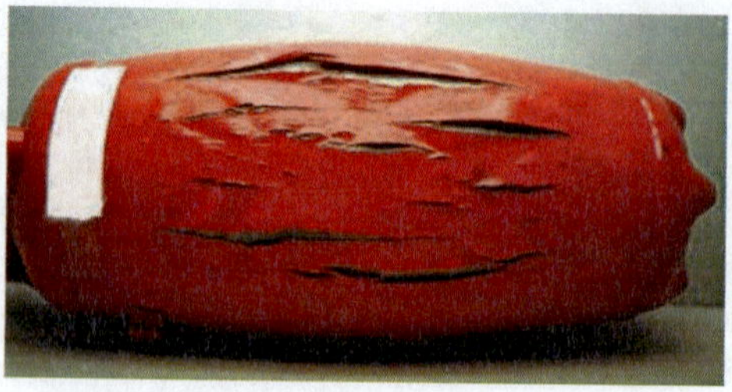

Fig. 8.59 Paint peeling during the pressure test of the fire extinguisher

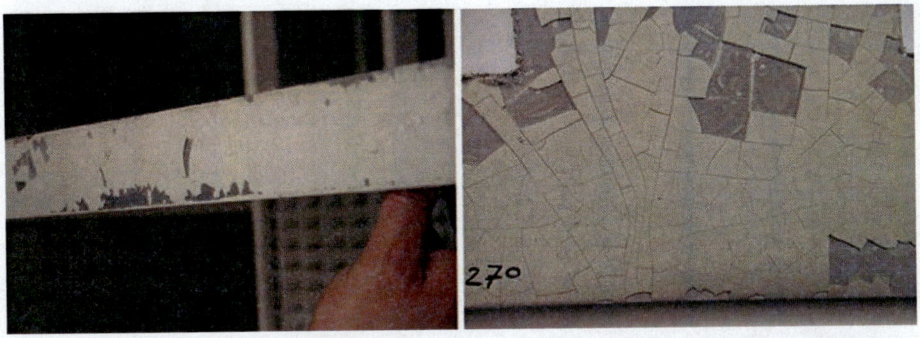

Fig. 8.60 Cracks and paint peeling on powder-coated steel frames. Cause: Lack of crosslinking of the paint film

only a short time. The examinations showed that these rejected parts were not sufficiently cross-linked. The paint film was damaged by weathering (UV and moisture) and this was shown by the severe cracking with paint peeling (Fig. 8.60).

8.4.6 Blisters and Craters in the Paint Film

Hot-Dip Galvanized Base Material with Problems in Paint Adhesion and Outgassing (Blisters in the Paint Film)

During the powder coating of hot-dip galvanised steel, blisters and craters occur in the hardened paint film, depending also on the thickness and quality of the zinc layer (Figs. 8.61 and 8.62). In the cross section Fig. 8.61 shows the defect originates from the porous pure zinc coating.

The galvanized container examined in Fig. 8.63 shows two defects. Outgassing, blisters in the paint film and huge adhesion problems of the coating. The blisters did not occur evenly distributed over the entire workpiece, but were particularly pronounced starting from the flanged edge and on partial surfaces (Fig. 8.63). With the help of cross-sections, SEM and EDX analysis, it was possible to determine that no mechanical or chemical cleaning and pre-treatment of the parts had taken place here. After removal of the paint film, residues of corrosion products could be analysed in the area of the particularly blistered surfaces (Fig. 8.64). The increased blistering in the edge area was due to white rust. The cause of the defects here was the lack of cleaning and pre-treatment of the metal surface.

Cavities Pores in the Casting Material

A die-cast aluminium housing coated with powder paint showed blisters after coating and baking of the paint layer.

A cross-section in the area of these blisters and craters shows pores and cavities in the base material. These "flaws" in the base material can cause outgassing of air, moisture or pretreatment residues during the stoving process. Blisters or craters can be found in these areas (Fig. 8.65).

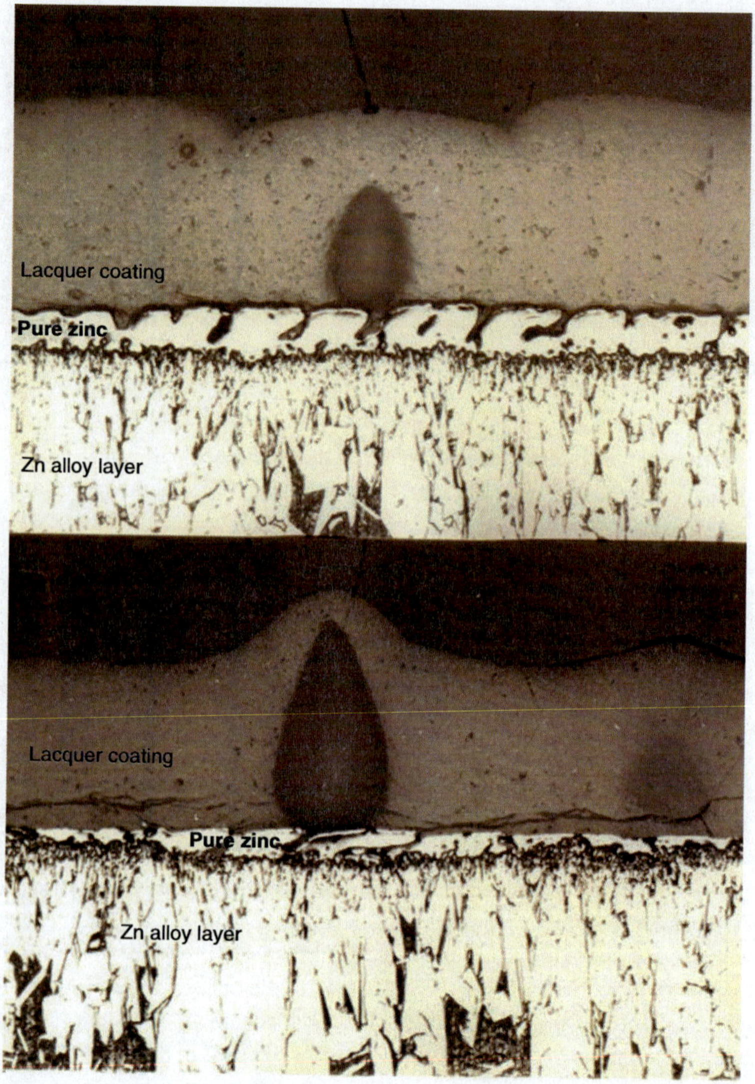

Fig. 8.61 Cross-section in the area of a defect. Base material hot-dip galvanized steel, layer thickness d > 120 μm

8.4.7 "Corrosion" of the Paint Surface

Staining and Matte Metallic Paint Surface

Shortly after completion, a façade in Hamburg showed run-off traces, matting and stains on the surface(Fig. 8.66). The coating is a single layer metallic powder coating with aluminum effect pigments without an protective clear coat. After finishing of the construction

Fig. 8.62 Craters and blisters in the paint layer on hot-dip galvanized steel substrate

Fig. 8.63 Container, hot-dip galvanised steel, powder-coated, with bubbles in the paint film

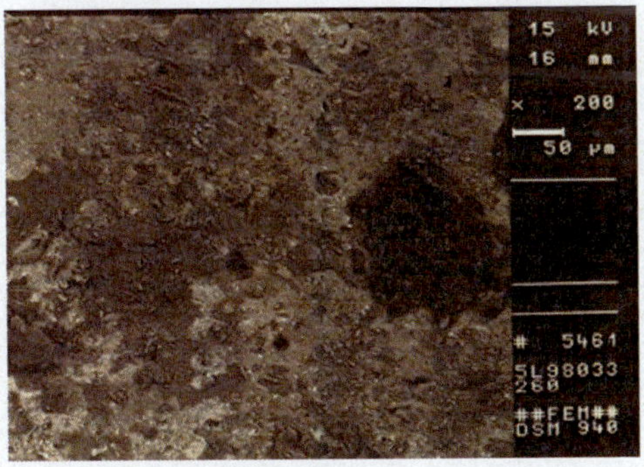

Fig. 8.64 SEM image of the surface in the area of the bubbles. Corrosion products can be seen on the surface

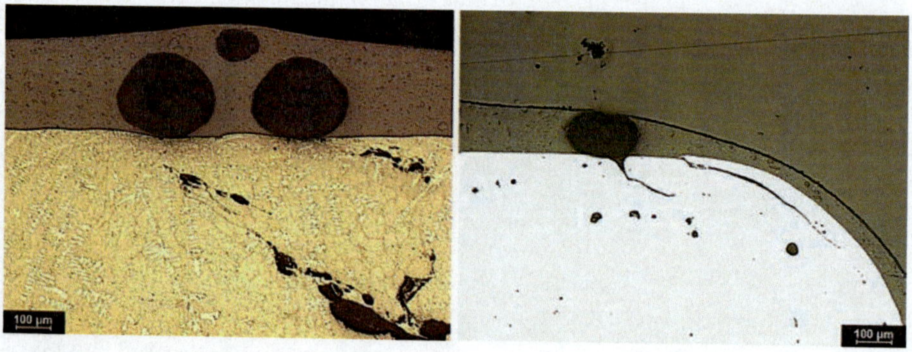

Fig. 8.65 Cross-section through the defect in the paint layer. Cause: Blisters due to a porous substrate

Fig. 8.66 Paint surface with run-off marks

an maintenance cleaning was done with alkaline cleaning agents. This led to a corrosion of aluminium pigments which are layering at the surface of the paint film. Examination of the paint surface in the area of the run-off marks revealed corrosion of the effect pigments. Figure 8.67 shows the SEM image of the paint surface with the aluminium corrosion products on the surface. The color changes, caused by cleaning, are due to chemical attack or alteration of the pigments by the cleaning agents. To avoid these defects, metallic coatings with a high amount of aluminium pigment should always be given an additional top coat of clear lacquer.

8.4.8 Metallic Coating Defects

Uneven, Cloudy Appearing Surface

A façade with front cladding of a building has been coated with a metallic powder system. After installation of the elements, the uneven appearance and the clearly visible colour differences were criticised.

Fig. 8.67 SEM image of the coated surface with corroded aluminium pigments

Fig. 8.68 Facade elements with dark edges and uneven and cloudy appearance

The error showed up in the form of:

1. dark areas at the edges of the sheet metal and
2. in an uneven and cloudy appearance, see Fig. 8.68.

The defect can be explained as follows: The sheets were first coated in an automatic booth and recoated again by manuell application in the area of the edges. The coating thickness on the edges was d ~ 130 µm and on the flat surfaces d ~ 80 µm.

Such high differences in coating thickness cause color differences in metallic painting. Also the charging of the powder, the distance gun – workpiece and the paint itself have a great influence on the result. The failure can be avoided by an even application of the paint and a change in the coating process. If the manual precoating of the edges is carried out before the automatic coating, a more uniform image can be achieved on the surfaces.

8.4.9 Failure in the Workpiece

Adhesion Problems at Profile Start and Ends

Aluminium insulation profiles from window construction, coated with a polyester powder coating, colour RAL 9010, showed adhesion problems of the coating at the beginning and end of the profile. In addition, blisters and craters were found in the coating.

The examination of the paint surface and base metal surface after removal of the paint film showed a high proportion of pre-treatment residues on the paint surface and also on the base material in the defect area. This is only possible if residues from the pretreatment emerge from cavities in the water dryer or in the curing oven and deposited on the base material and on the paint film.

During the analyse of the insulating webs, wrong stiffening could be found. Cavities with a diameter of approximately 1 mm were found in these areas.

Due to these capillary effect, pretreatment solutions can penetrate into the cavities, which can no longer be completely removed in the rinsing steps. Even very intensive rinsing cannot achieve a rinsing effect in these narrow cavities. In the water dryer or in the curing oven, these substances escape and, deposit on the surface and on the resist film. This leads to adhesion problems of the coating and to disturbances in the paint layer. The precipitation of these residues may also be influenced by the air flows in the water dryer or curing oven.

An EDX analysis of the coating in the area of the stiffener and the coating on the aluminium surface after the paint film had been removed showed the same elements. The elements sodium, phosphorus and oxygen were detected in the coating, see Figs. 8.69, 8.70 and 8.71.

8.4.10 Failures Due to Contamination

Contamination in the coating film can be caused by insufficient cleaning of the guns or the booth, but can also be present in the powder itself. See Diagram 8.15 for information on possible causes of this type of defect in the coating film (Figs. 8.72 and 8.73).

8.4.11 Weather Resistance

Colour changes, depending on the direction of ageing (location) and the weather resistance of the coating system, are shown in Figs. 8.74 and 8.75.

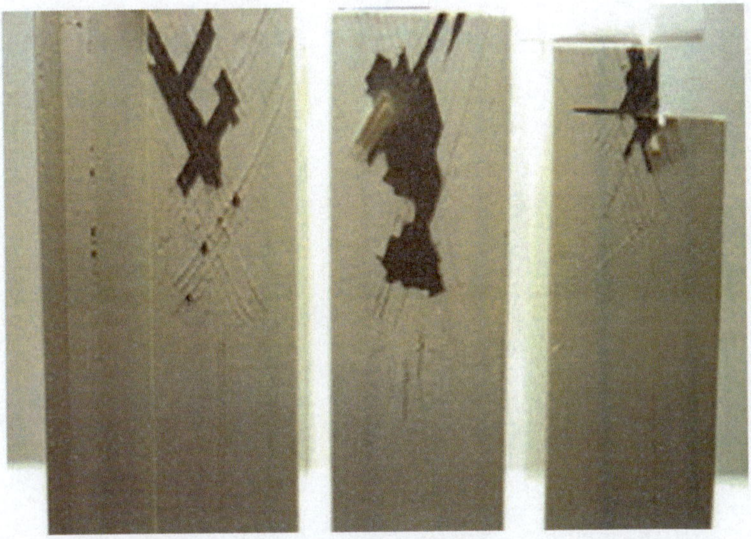

Fig. 8.69 Profile pieces with adhesion problems of the paint film

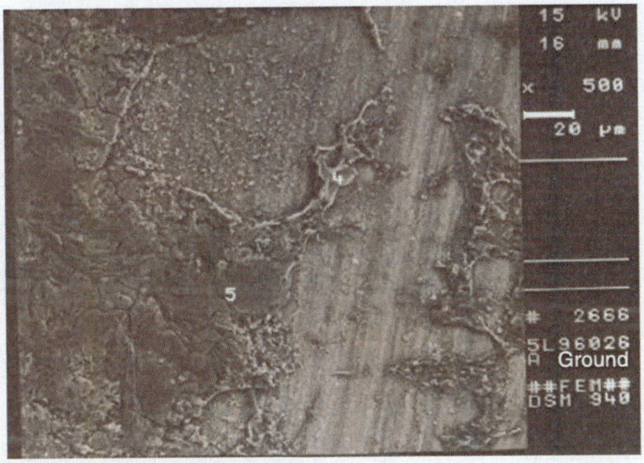

Fig. 8.70 SEM image of the coating on the base material

The pictures show a red painted door of a building. The profiles were coated with a powder coating, the filling is painted with a coil coating system. The colour tone was the same for both systems, approx. RAL 3016.

After ten years of outdoor weathering in the Netherlands, a clear colour difference can be seen between the powder coating and the coil coating paint, but the compass direction also plays a decisive role. The door facing south shows the greatest colour differences.

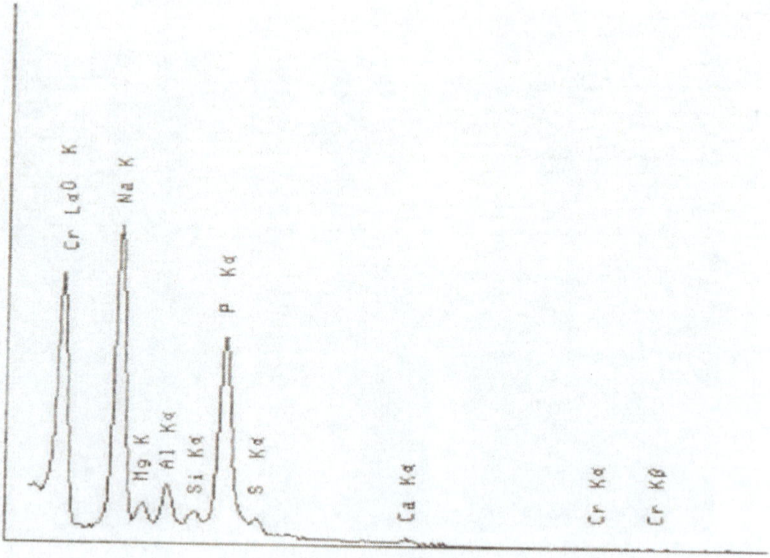

Fig. 8.71 Result of the EDX analysis of the coating on the base material

Fig. 8.72 Contamination by white powder coating: 1 g white to 10 kg black

Fig. 8.73 Impurities in a textured coating with differently coloured coating particles

Fig. 8.74 Door facing east

Fig. 8.75 Door facing south

Fig. 8.76 Casting surface with residues of cooling lubricant. It was contaminated with a high number of bacteria. Result after 240 h CASS test. The cooling lubricant could not be removed by the degreasing in the pre-treatment

8.4.12 Corrosion Caused by Biological Residues from Pre-treatment

If processing aids, such as cooling lubricants, remain on the surface for a long time, if the parts are stored incorrectly or for too long before coating, or if the lubricants are additionally contaminated with bacteria or fungi, this leads to problems in paint adhesion. During pretreatment, the coating of lubricant and bacteria cannot be reliably cleaned off. The result is an incomplete conversion layer and problems in paint adhesion and resistance (Fig. 8.76).

8.5 Troubleshooting: Diagrams and Tables

The following Diagrams 8.1, 8.2, 8.3, 8.4, 8.5, 8.6, 8.7, 8.8, 8.9, 8.10, 8.11, 8.12, 8.13, 8.14 and 8.15, describe the actions that can be taken when various faults occur to locate and eliminate the cause of the fault.

Problems in the coating can often not be attributed to one cause, but are the result of various errors, which makes a systematic search necessary. Only in this way can individual points be identified or excluded as the cause.

For this very reason, systematic troubleshooting requires a conscientious approach and a certain degree of understanding of the interrelationships.

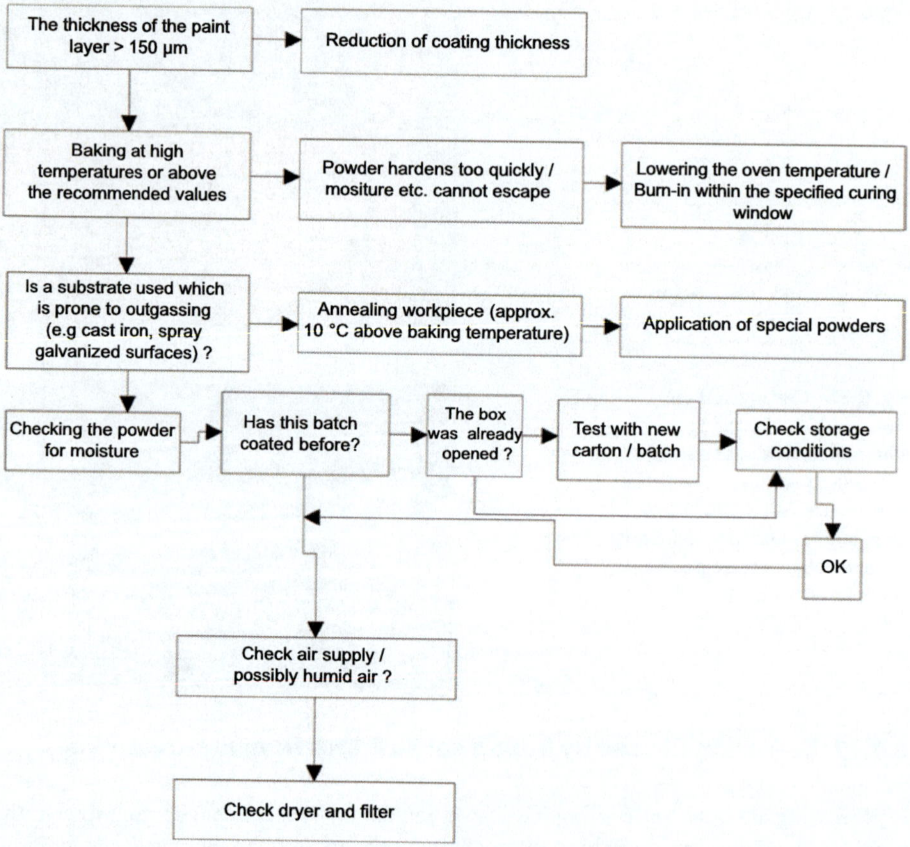

Diagram 8.1 Troubleshooting the occurrence of pinholes in the coating film

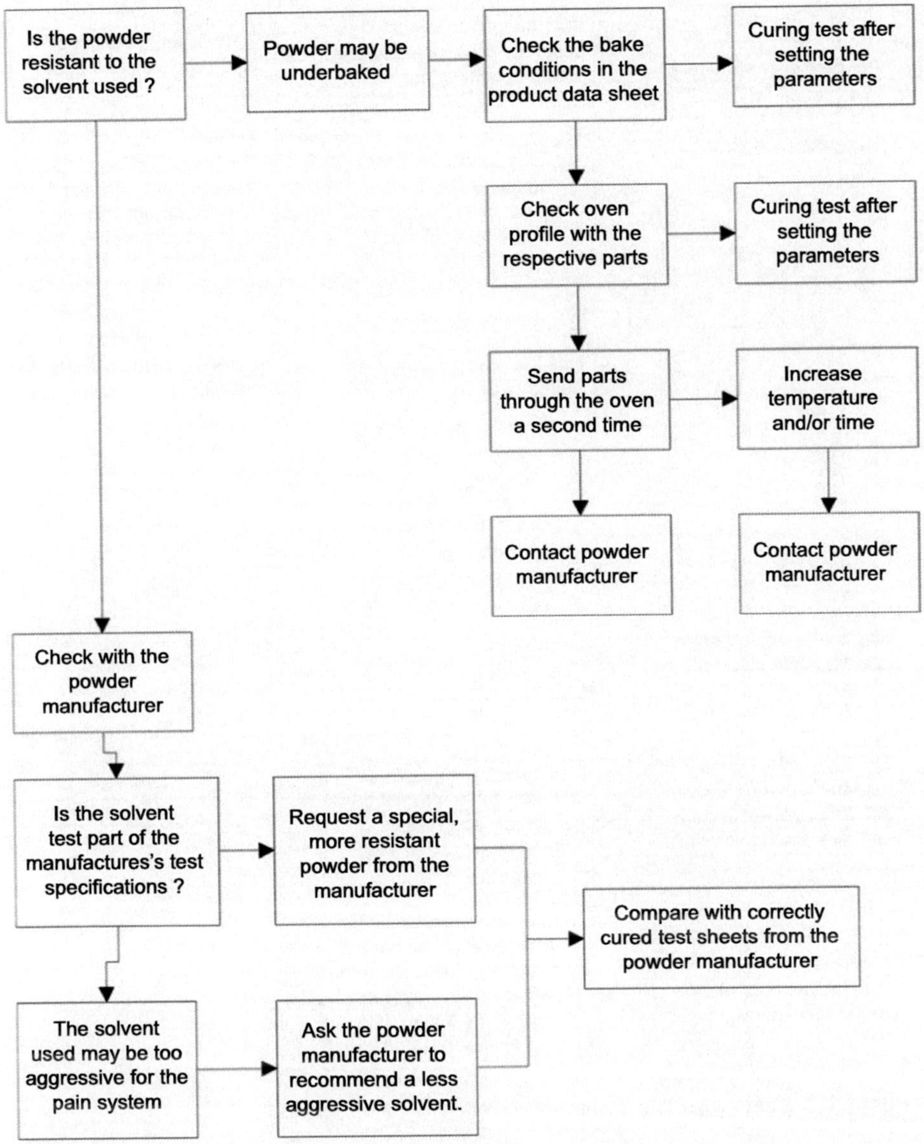

Diagram 8.2 Low chemical and solvent resistance of the coating

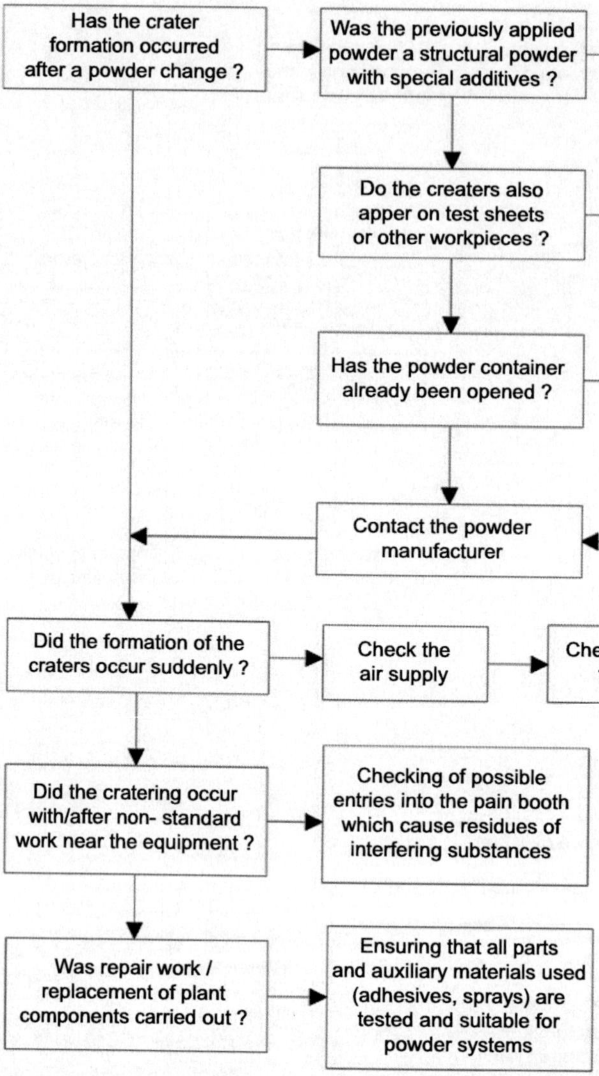

Diagram 8.3 Troubleshooting the occurrence of craters in the resist film

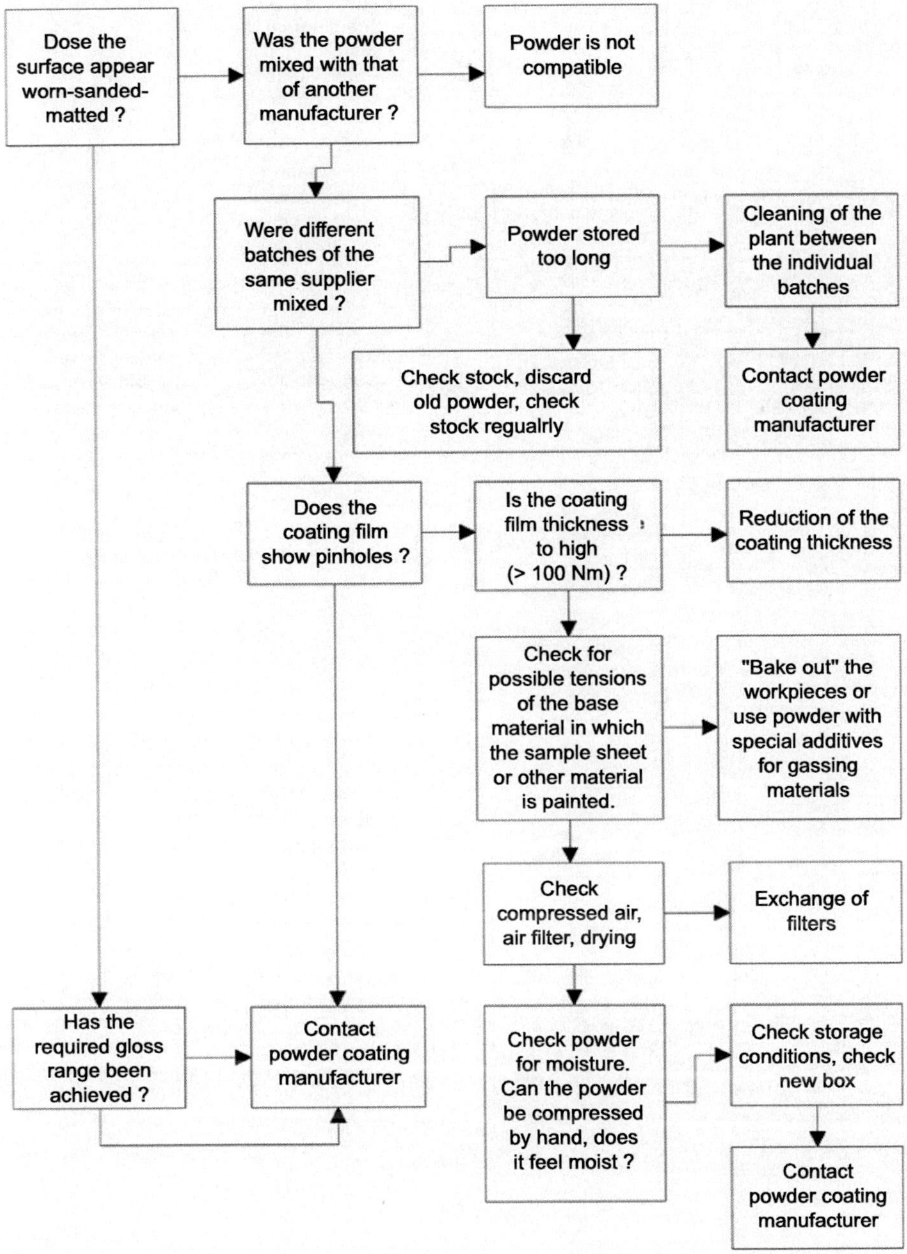

Diagram 8.4 Matting in the coating, lower gloss than specified

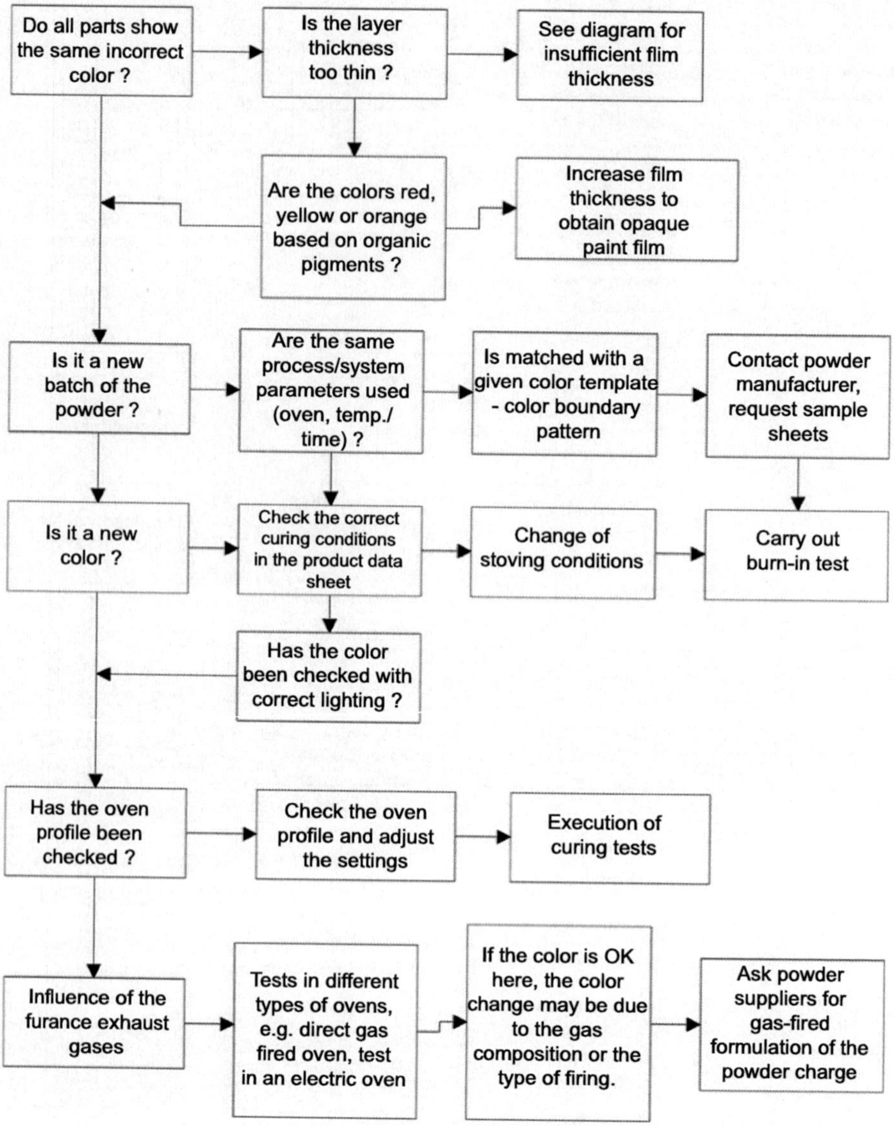

Diagram 8.5 Problem of incorrect color of the coating

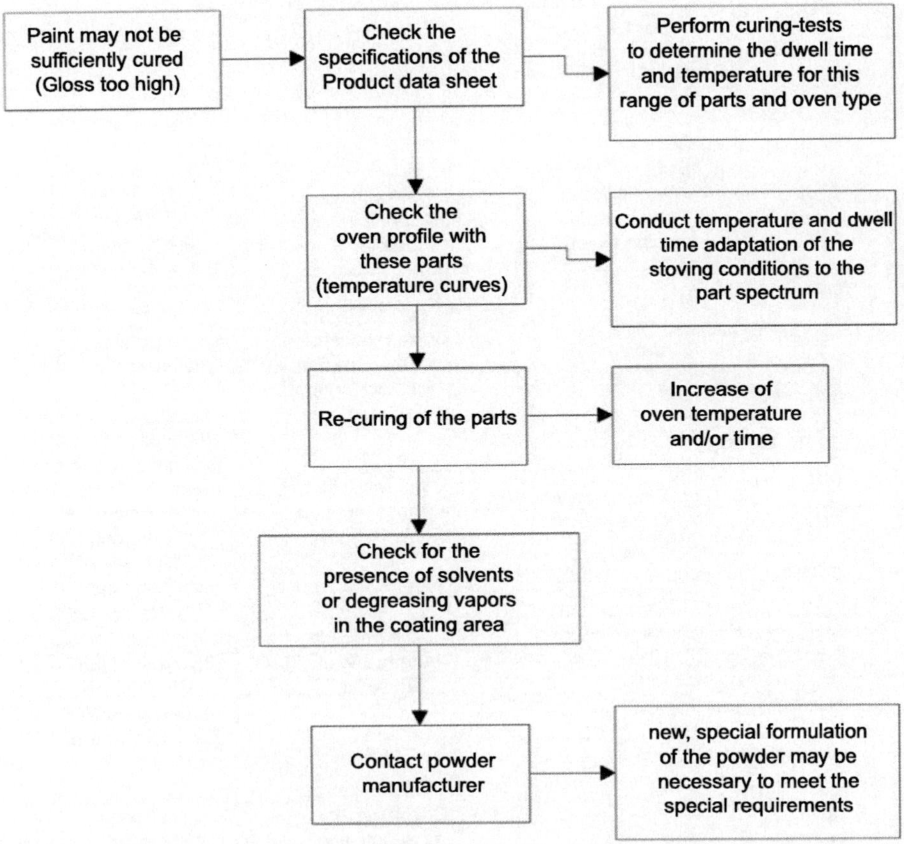

Diagram 8.6 Problem of insufficient coating hardness of the coating film

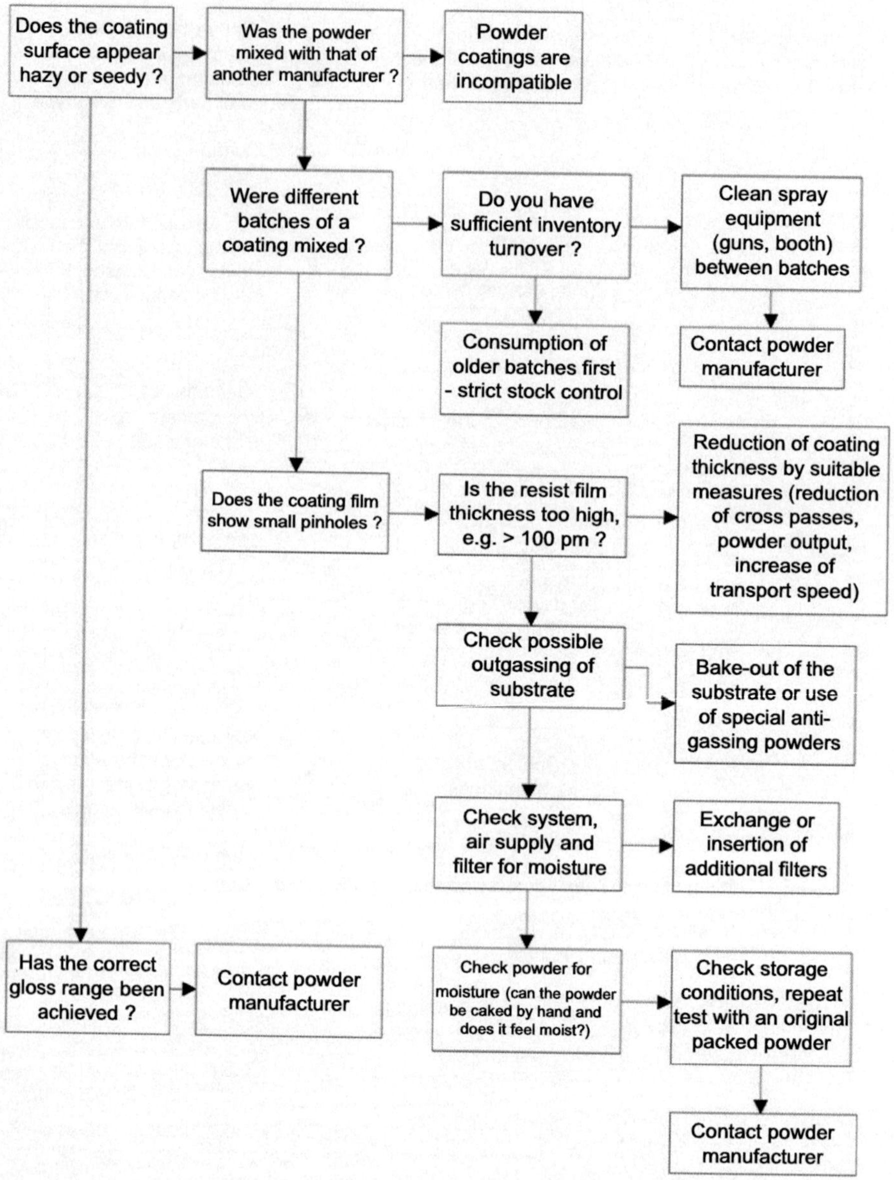

Diagram 8.7 Problem of low gloss areas in high gloss powders

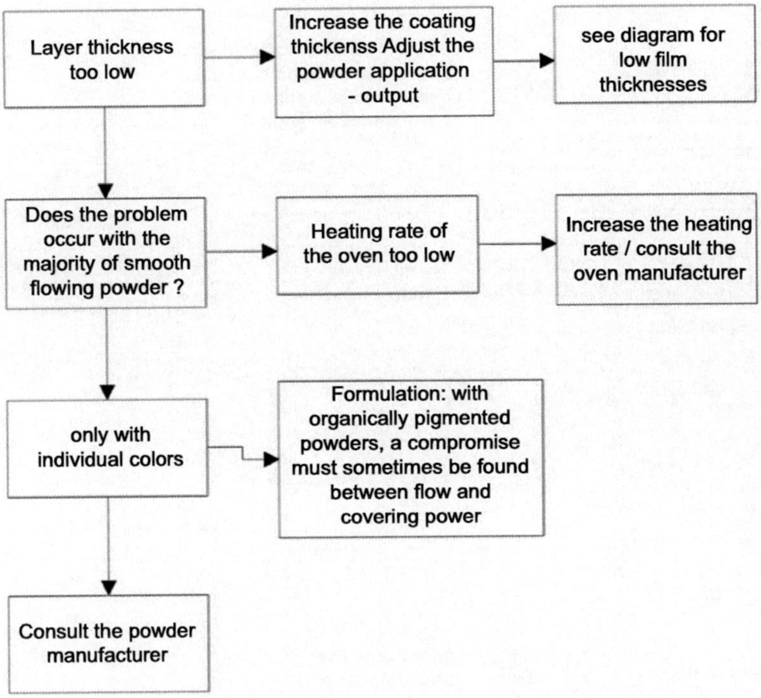

Diagram 8.8 Problems with poorly leveling paint films – Orange Peel

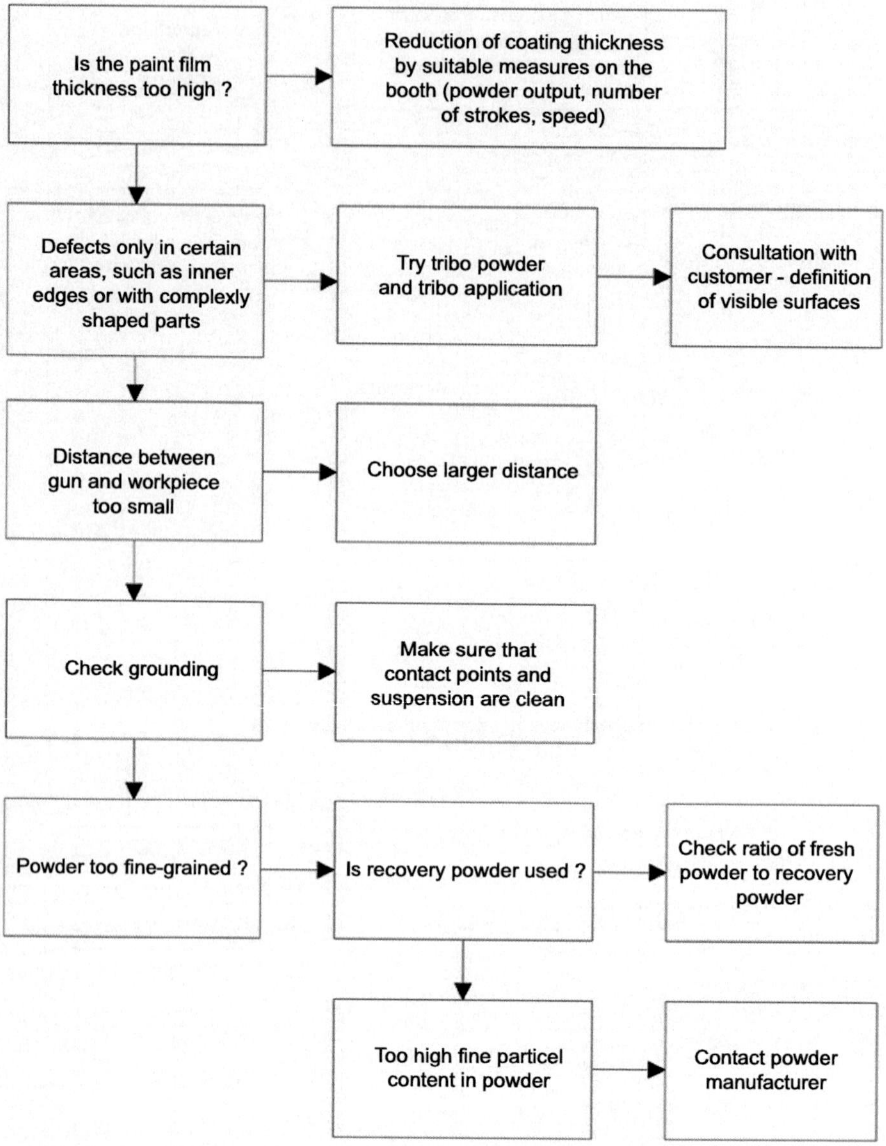

Diagram 8.9 Burnt-in paint film shows "open areas", backscatter effects

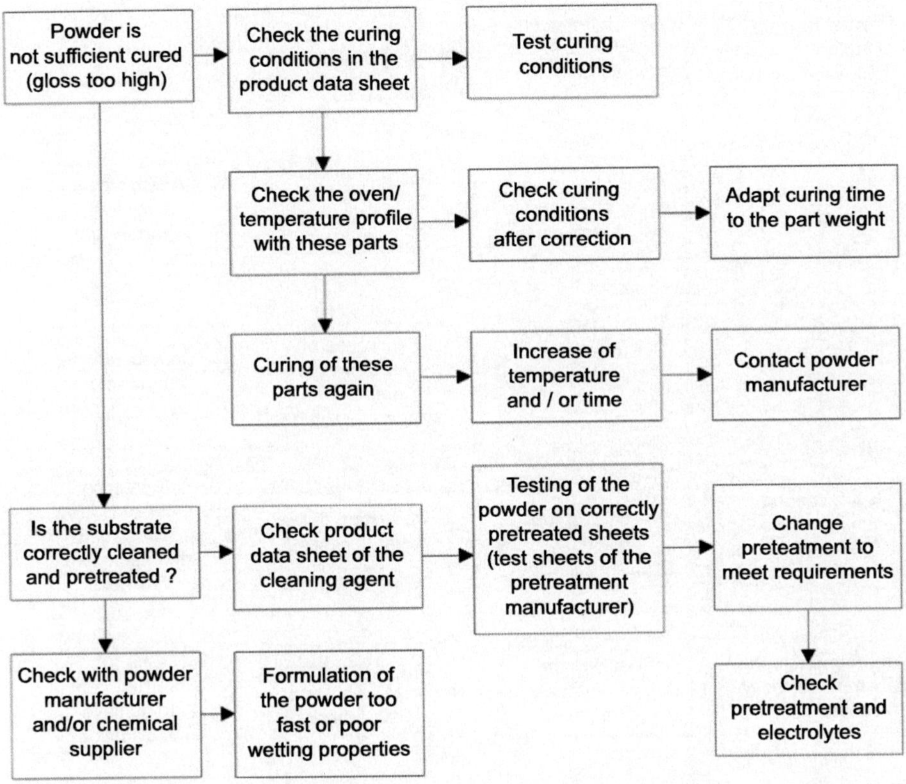

Diagram 8.10 Poor paint adhesion on the coated workpieces

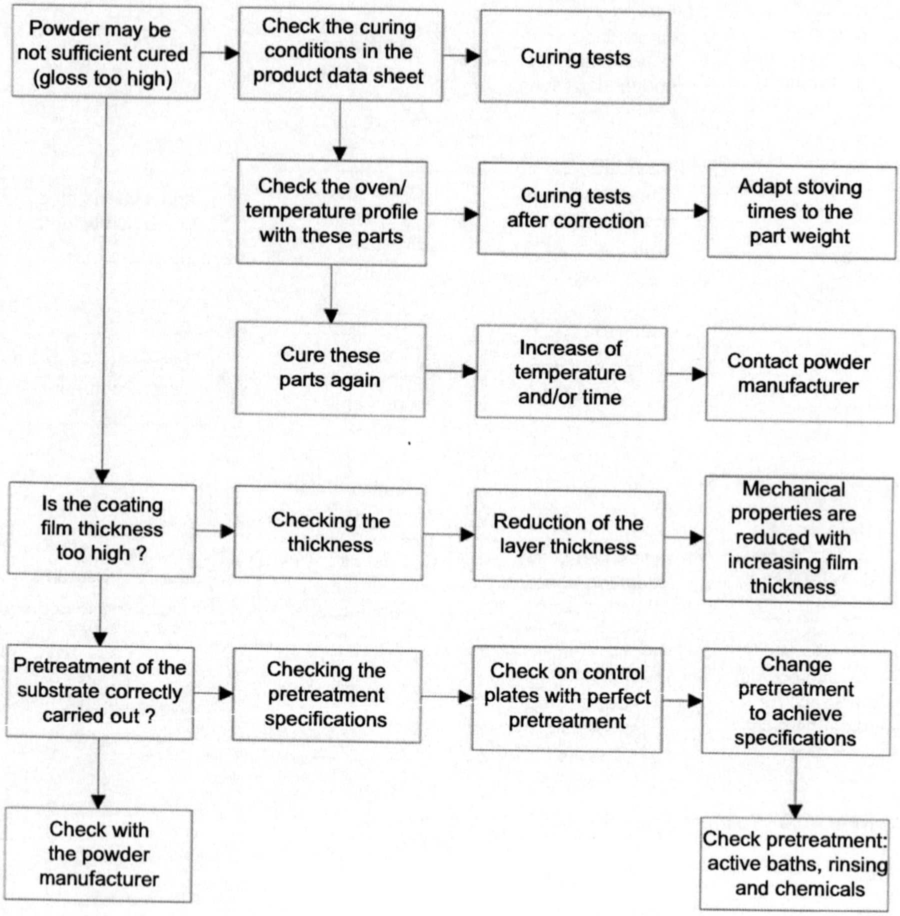

Diagram 8.11 Insufficient flexibility of the paint film, negative ball impact resistance

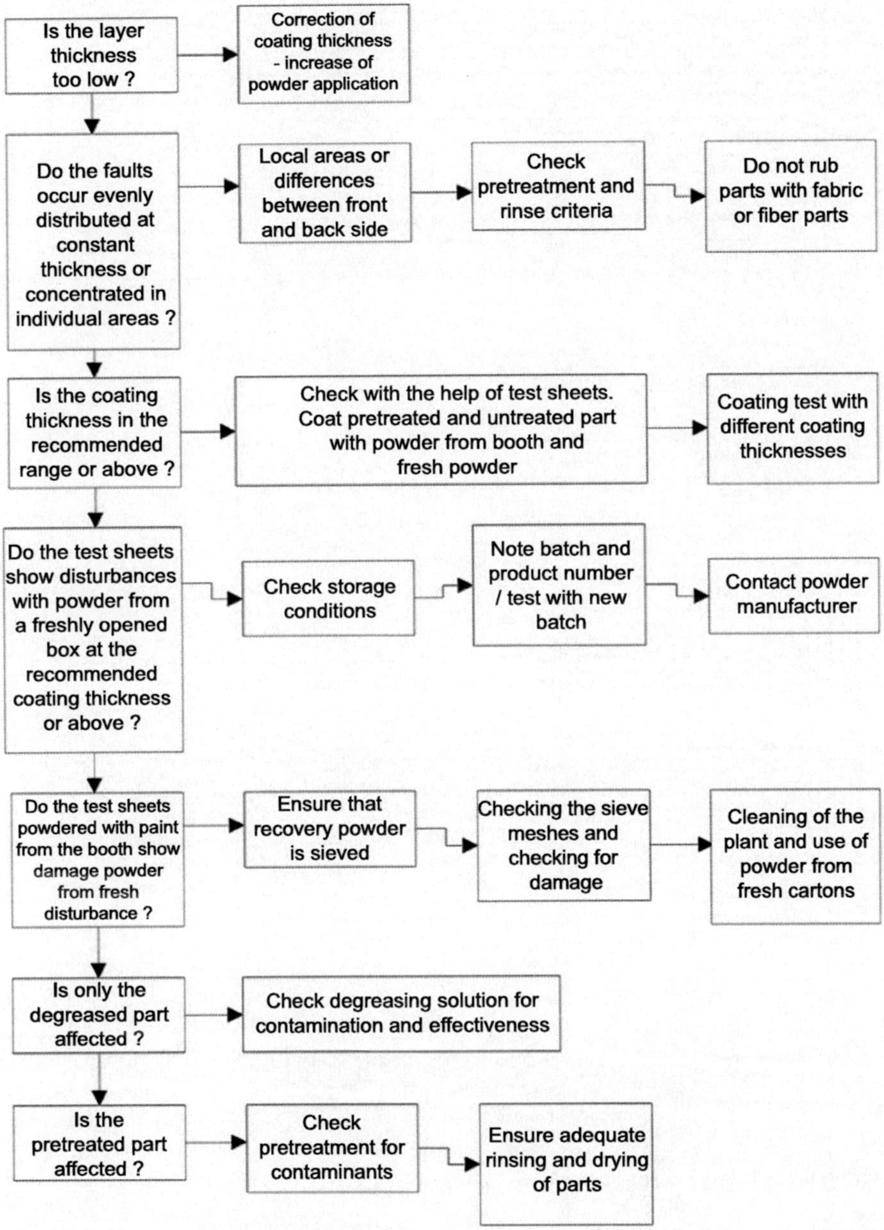

Diagram 8.12 Problem of faults or inclusions in the cured paint film

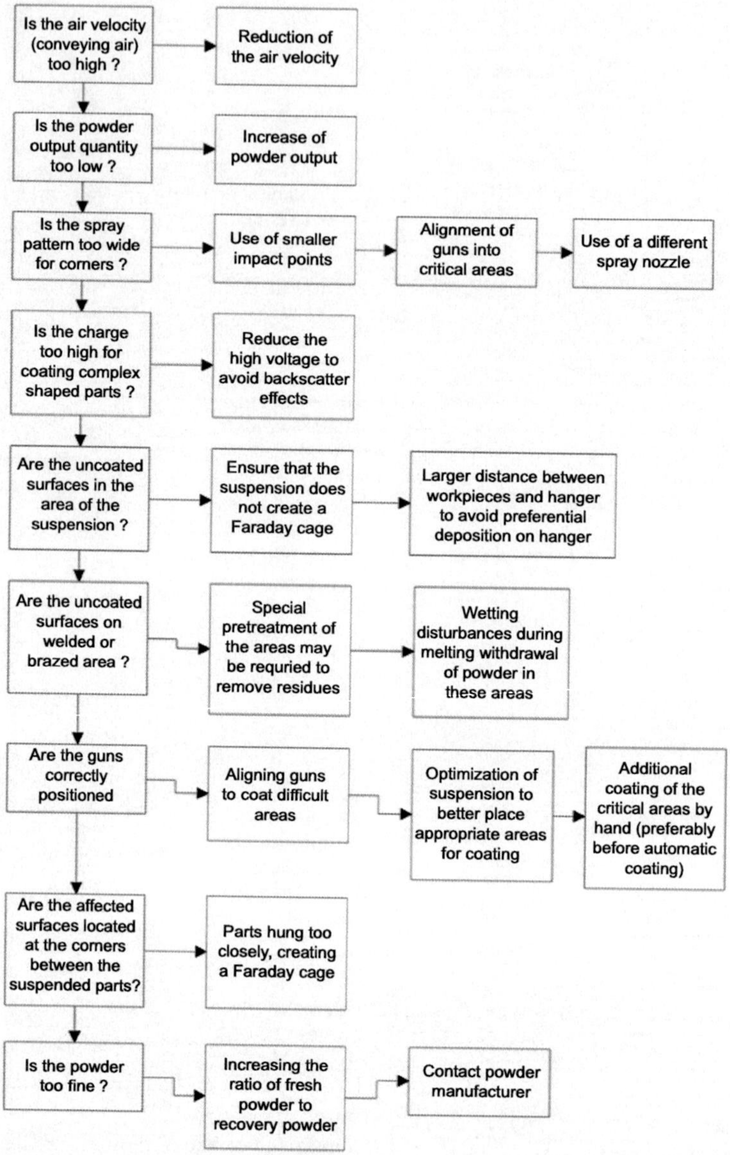

Diagram 8.13 Powder coating film shows uncoated areas

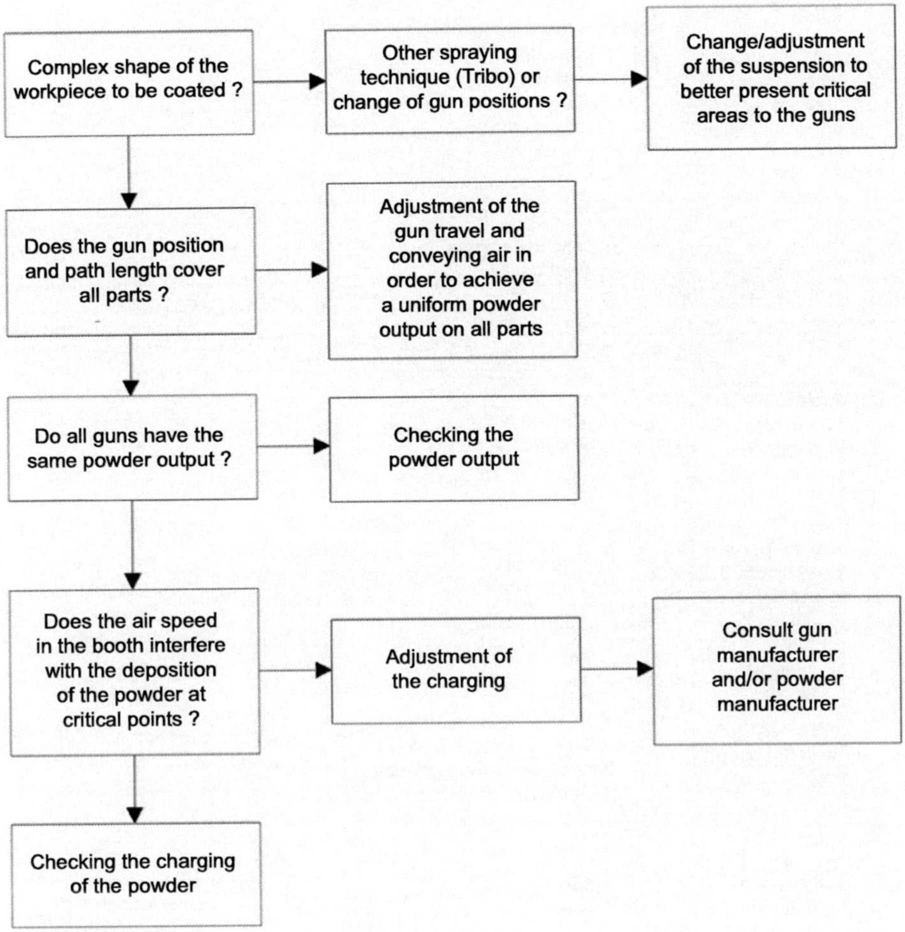

Diagram 8.14 Coated parts show different coating thicknesses

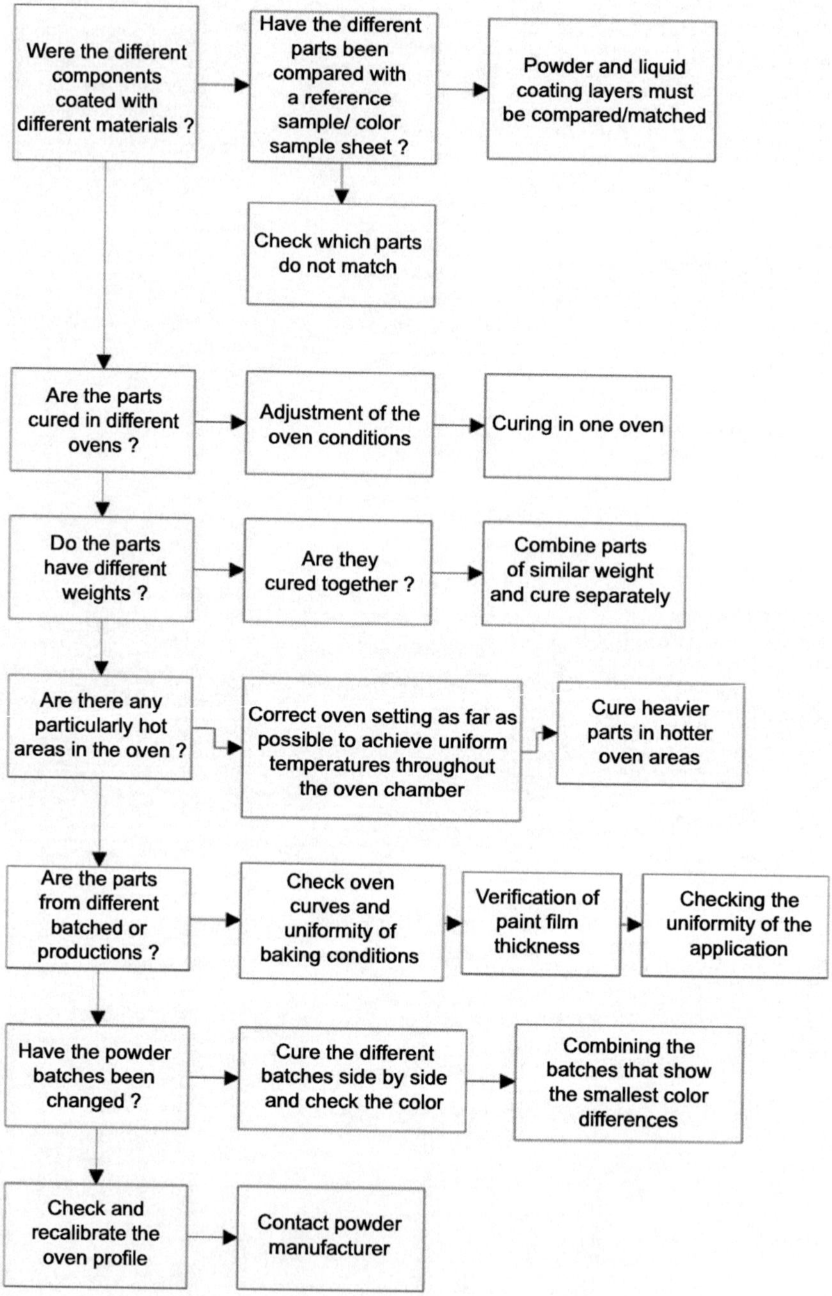

Diagram 8.15 The coated parts show colour and/or gloss differences between the individual workpieces

References

1. Güte- und Prüfvorschriften der GRM, Gütegemeinschaft für die Reinigung von Metallfassaden, RAL Gütezeichen 632
2. Martinazzo, F. et al.: Determination of the Chemical Nature of the „Blooming" Effect in Polyester Based Powder Coatings, 13th International Conference, Paint Research Association
3. Zorll, U. (Hrsg.): Lacke und Druckfarben, Römpp Lexikon. Thieme, Stuttgart (1998)
4. Kreysa, G. (Hrsg.): DECHEMA-WERKSTOFF-TABELLEN, Dechema, Deutsche Gesellschaft für Chemische Technik und Biotechnologie, D-60489 Frankfurt
5. Kunze, E. (Hrsg.): Korrosion und Korrosionsschutz, Bd. 1. Wiley-VCH, Weinheim (2001)
6. Sharmann, C.F.: Filiform underfilm corrosion of lacquered steel surfaces. Nature. **153**, 621 (1944)
7. Kaesche, H.: Untersuchungen über die Filigrankorrosion lackierter Stahlbleche. Mater. Corros. **10**(11), 668–681 (1959)
8. Scheck, K.: Doktorarbeit, Institut für Technische Chemie der Universität Stuttgart (1991)
9. Hoch, G.M.: A Review of Filiform Corrosion, Int. Corros. Conf. Ser. Volume Date 1971, NACE-3 134–142
10. Slabough, W.H., et al.: Filiform corrosion of aluminium. J. Paint Techn. **44**, 76 (1972)
11. Slabough, W.H., Chan, E.J.: Atmospheric components and filiform corrosion. J. Paint Techn. **38**(499), 8 (1966)
12. Funke, W.: Blistering of paint films and filiform corrosion. Prog. Org. Coat. **9**, 29 (1981)
13. Hagen, H., Rihm, K.H.: Filiformkorrosion auf organisch beschichtetem Aluminium. Farbe und Lack. **96**(7) (1990)
14. Mol, A.: Filiform Corrosion of Aluminium Alloys – The Effect of Microstructural Variations in the Substrate, ISBN 90-9014302-5
15. Hoch, G.M., Tobias, R.F.: Mechanismus of Filiform Corrosion, Paper 19, NACE Corrosion/71 (1971)
16. Pietschmann, J., Pfeifer, H.: Filiformkorrosion auf organisch bechichtetem Aluminium, Teil 1 bis 3. Z. Alu. **69**(70), 1019–1023, 1081–1084, 82–86 (1993/94)
17. CEN Working draft, Paints and varnishes – Specifications for organic coating of aluminium alloy extrusions, sheet and performed sections for architectural purposes – Part 1: Powder coatings, CEN/TC 139/WG 5 N 56
18. Hübner, W., Speiser, Ch.: Die Praxis der anodischen Oxidation des Aluminiums. Aluminium, Düsseldorf (1988)
19. Heinrich, M., et al.: Filiformkorrosion von Aluminium. Farbe und Lack. **100**(4), 177–180 (1994)
20. Kühne, S.: Filiformkorrosion auf Aluminium. Diplomarbeit 1993–1994, FH Farbe u. Druck, Stuttgart und FEM, Schwäbisch Gmünd
21. Heinrich, M., et al.: Korrosion von beschichtetem Aluminium. Metalloberfläche **49**(10), 802–806 (1995)
22. Unveröffentlichte Ergebnisse aus einem Forschungsprojekt des FEM in Zusammenarbeit mit Industriefirmen, gefördert von der EU
23. Pietschmann, J.; Freudenberger, J.: Filiformkorrosion bei beschichteten Aluminiumbauteilen, Teil 1. In: Galvanotechnik, **89**, 11 (1998); Teil 2 in: Galvanotechnik, **89**, 12 (1998)
24. Reiter, H., et al.: Tipps und Tricks der Drylac®-Beschichtung. Tigerwerk Wels
25. DIN EN ISO 105-B02: [2014–11], Farbechtheitsprüfung
26. Qualitätsrichtlinien der GSB-International, Al 631
27. Scamans, G.M., et al.: Control of the Filiform Corrosion Susceptibility of Aluminium Alloy Sheet. Vortrag, DFO Tagung Leichtmetall Anwendungen, März (2001)
28. DIN-EN 3665, 1997-08, Luft und Raumfahrt – Prüfverfahren für Anstrichstoffe – Prüfung der Beständigkeit gegen Filiformkorrosion von Aluminiumlegierungen

Further Literature

Pfeifer, H.: East report 1992
Laible, R., et al.: Oberflächentechnik, Bd. 271. expert
Pfeifer, H.: East Vortragsreihe, Beschichten von Aluminium
Pietschmann, J.: Schadensfälle in der Pulverbeschichtung. JOT 1996/4
FEM: Sammlung von Schadensfällen
Pietschmann, J.: Fehlerquellen und Fehleranalyse bei lackiertem Aluminium. Galvanotechnik. **89**(10), 3364–3373 (1998)
Pietschmann, J.: Seminarreihe Pulverlackierer unter sich. Vincentz, Hanover (1999)
Pietschmann, J.: Seminarreihe Pulverlackierer unter sich. Vincentz, Hanover (2000)
Freudenberger, J., Pietschmann, J.: Entwicklung eines Verfahrens zur Reparatur von – speziell durch Filiformkorrosion – korrodiertem beschichtetem Aluminium. Jahrbuch Oberflächentechnik, Bd. 55. Hüthig, Heidelberg (1999)
Pietschmann, J.: Schäden an Pulverlackschichten. Vincentz, Hanover (2003)

Images used in Chap. 8

FEM, Schwäbisch Gmünd: 8.1 bis 8.28, 8.30 bis 8.59, 8.61 bis 8.73
Schunter, Erich: 8.60
Ronal, Landau: 8.29 und 8.76
Atlas, Linsengericht: 8.74, 8.75

9 Paint Stripping

During powder coating, on the one hand, the hangers or brackets are also coated, and on the other hand, faulty coatings and disturbances in the coating layer occur time and again, which makes it necessary to remove them.

The task of paint stripping is generally to remove the metallic brightness from the workpieces or hanger parts, irrespective of the process used, so that they can be used again in the coating process. The paint stripping processes can be divided into nitrogen refrigeration processes, chemical processes and pyrolytic processes as shown in Fig. 9.1.

The nitrogen refrigeration process was developed for the stripping of thickly coated suspension parts. The parts are extremely cooled in liquid nitrogen at temperatures of minus 190 °C, which embrittles the powder coating so that it can be easily removed mechanically. Blasting with dry ice [2] can also be counted among the cold processes. This process was developed before 1960, but took a long time to become suitable for industrial use.

At the upper end of the temperature scale are the pyrolytic processes. They all work on the principle of heating the coating layer to over 400 °C. The binder is destroyed and pyrolysis or smouldering gases are formed, which have to be burnt off in an afterburner. The thermal processes can be divided into three groups:

- Pyrolysis or low temperature carbonization furnace
- Fluidized bed
- Laser beam or plasma

Under the collective term "pyrolysis oven" or "low temperature carbonization oven", very different designs are on the market. This ranges from simple devices of the type "baking oven with attached post-combustion chamber" to elaborate constructions in which the

J. Pietschmann, *Industrial powder coating*,
https://doi.org/10.1007/978-3-658-37592-8_9

Fig. 9.1 Methods of paint stripping [1]

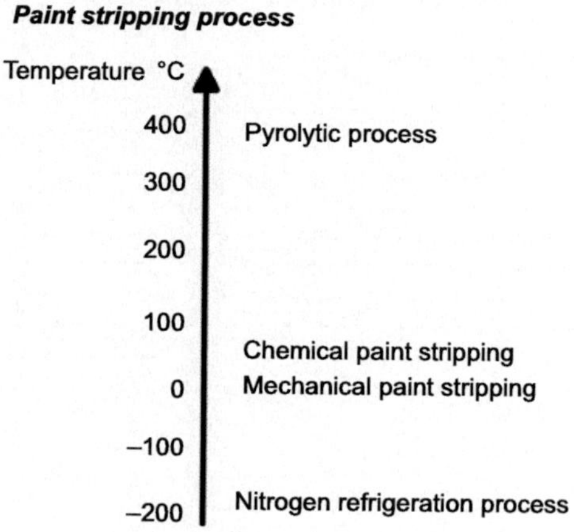

safety technology is often much more expensive than the actual carbonization chamber with combustion chamber.

The fluidized bed technology competes with these processes. Here, hot air flows through a retort filled with sand. A temperature of approximately 450 °C is maintained in this fluidized bed of sand. The parts to be decoated are immersed in this fluidized bed of sand, whereby the paint layer is destroyed by the high temperature. The swirling sand continuously removes the ash from the surfaces and thus supplies new heat. Thus, relatively short paint stripping times can be realized here, depending on the paint system and the thickness of the coating.

Destruction of the paint layer is also possible through the energy supplied by a laser beam or plasma. Particularly with laser beam technology, work can be very precise and targeted and it is therefore possible to remove paint only at certain points.

The thermal processes, in particular the fluidized bed and pyrolysis furnace processes, are used for the decoating of steel hangers. The prerequisite for these processes is in any case that the base material can withstand the temperature load without damage. These processes are critical for aluminium materials, for example.

Mechanical paint stripping processes include high-pressure water jetting, blasting with sand, dry ice or plant grains.

Chemical processes are among the most common processes in paint stripping. The paint stripping agents used can be divided into aqueous alkaline, water free organic or acidic processes.

9.1 Chemical Paint Stripping

9.1.1 Paint Stripper

A major strength of chemical paint stripping is its high flexibility and adaptability to special paint stripping tasks.

The paint stripping process itself consists of several process steps that are controlled by chemical and physical processes. With the most commonly used agents today, the paint stripping process consists of the following individual steps:

- Penetration and swelling of the lacquer layer,
- Chemical destruction of the binder and
- Removal of dissolved binder components from the surface.

The chemical reaction and the penetration process are essentially determined by the properties of the paint stripper, the paint binder and the temperature. In contrast, the removal of the dissolved binder components is a physically controlled process, which is determined by the flow conditions of the paint stripper on the surface to be stripped. The speed of these individual process steps varies greatly depending on the type of paint binder [4]. Table 9.1 provides an overview of today's chemical paint strippers.

The paint strippers for steel are aqueous-alkaline products with organic swelling agents which, in cooperation with the aqueous alkali, produce the desired effect. The organic swelling agents soften the paint film so that the aqueous lye can penetrate the film and the decomposition of the paint takes place.

Table 9.1 Classification of the paint strippers available today [1, 3]

Paint remover			
	Aqueous alkaline	Water free organic	Water free acid
Working temperature (°C)	50–90	20–80	20–50
Materials			
Iron/Steel	Suitable	Suitable	Suitable, problem: rinsing with water → attack of the base metal by acid
Zinc	Not suitable	Suitable	
Light metals	Not suitable	Suitable	
Non-ferrous metals	Not suitable	Suitable	
Coating type			
Polyester	Yes/p	Yes/p	Suitable, but not necessary
Hybrid- sytems – polyester/epoxy	Yes/p	Yes/p	Suitable, but not necessary
Epoxy powder	No	No	Yes
Acrylic powder	No	Yes	Yes
Polyamide powder	No	No	Yes
Liquid paints	Yes/p	Yes/p	Suitable, but not necessary
2 component lacquers	Yes/p + s	Yes/p + s	YES
Air-drying lacquers	Yes/p	Yes/p	YES
Electrocoating	Yes/f	Yes/p	YES

Detachment of the varnish: *p* pigment-shaped; *s flake-shaped*; *f* pancake-shaped

When the workpiece to be stripped is removed from the treatment bath, it is usually only covered with a thin film of stripping agent and some pigment sludge. Usually, the parts are sprayed off with water.

The aqueous alkaline paint strippers are not suitable for aluminium or galvanised parts. Water-free processes must be used here, but these are also slightly alkaline. The organic swelling agents contained soften the paint film, the alkaline component ensures the degradation of the binder.

9.1.2 Process Techniques for Paint Stripping

The simplest method for carrying out a paint stripping process is the immersion process, without movingthe workpiece or the paint stripping agent, whereby the physical mass transfer proceeds slowly.

Processes to increase the mass transfer or the mechanical pressure on the surface to be stripped are used today to increase the process speed.

The following procedures are used in practice [4]:

- Movement of the paint stripper and/or the materials to be stripped in the immersion bath
- Spraying the paint stripper
- Ultrasonic support of the paint removal in the immersion bath.

A simple method for accelerating the paint stripping process is the agitation of the paint stripping agent, which is achieved by a strong recirculation of the bath liquid. This creates an undirected flow in the dip tank and on the surface of the workpiece. The movement of the paint stripping material is better, as a directed flow can be achieved at the surface of the paint.

However, strongly agitated paint stripping agents not only increase the mass transfer on the surfaces to be stripped, but also in the already sedimented paint sludge. This leads again to a with the paint stripper and a grinding of the paint pigments. The consequences are hardly sedimentable paint stripper/coating pigment suspensions, whose paint stripping effect is severely limited. To avoid this, the paint stripping media should be cleaned of pigments more frequently.

Very efficient is the pointed application, where a high mass transfer with high mechanical pressure on the surface occurs. The requirement is a certain progress of the chemical destruction of the paint film. Depending on the coatings to be removed, it is possible to remove paint 4 to 5 times faster than with the immersion process. For small parts, completely independent automatic washing machines are available which, can be individually assembled as modular modules to form the actual paint stripping system.

A special feature is the removal of paint with ultrasound. The mechanical effect on free sonic surfaces is very high. The prerequisite is an advanced degradation of the paint film.

When using ultrasound, it should also be noted that liquids weaken the ultrasonic waves to varying degrees depending on their chemical composition and physical properties at the application temperature. Direct sonication of the surfaces is required, which is not guaranteed, for example, in the case of densely packed hangers.

9.2 Blasting with Dry Ice

Blasting with dry ice is an alternative in the field of decoating and cleaning processes. The "blasting tool" is based on CO_2 pellets which are accelerated at high pressure through a laval nozzle and then hit the surface to be decoated at very high speed and with corresponding force [2].

A significant difference compared to other blasting methods is that the blasting medium dry ice sublimates completely after blasting. If, for example, sandblasting produces 80 to 95% blasting material and 5 to 20% residues, only the residues remain unmixed when blasting with dry ice [5]. This considerably reduces disposal costs. Savings of up to 96% have been reported [6].

Due to the reduced waste volume and the chemically neutral behavior of CO_2, it is possible to clean machines or assemblies without having to remove them from the production line [5, 7]. However, the removed paint or dirt particles must be collected in a suitable manner.

The advantages of blasting with dry ice can be summarized as follows:

- No residual blasting medium: waste reduction
- Reduced or no masking required when cleaning or stripping paint from components.
- There is no water or chemicals to dispose of [8, 9].
- The process works "dry". This makes it possible to clean electrical systems without the risk of short circuits.
- The base material remains undamaged.

Against this are the following disadvantages:

- Only visible surface areas directly accessible to the beam can be processed.
- The pellets do not bounce off. Therefore, in complicated structures, the backsides cannot be machined, as is possible with sandblasting using the residual kinetic energy of the bouncing grains.
- Lower removal rates than sandblasting.
- High noise level during blasting, ear-protection is required.

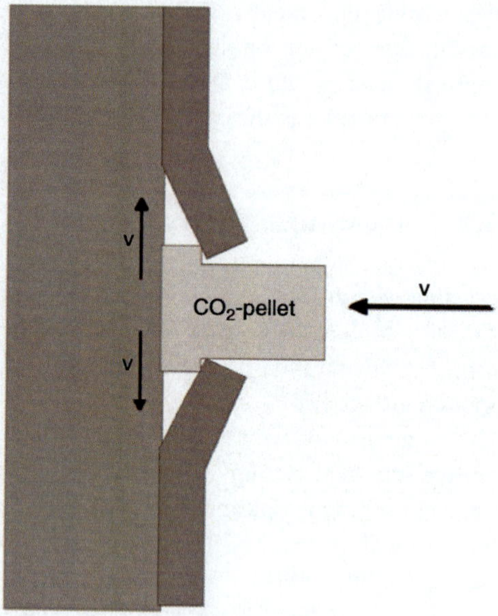

Fig. 9.2 Material removal effect during CO_2 blasting [2]

9.2.1 Operation Procedure

Figure 9.2 shows a model of the mode of action of dry ice blasting. According to Stratford [10], the removal effect of dry ice blasting is based on the fact that the pellets break apart after impact, rapidly increase in volume during sublimation and thus remove the coating from the inside.

According to current knowledge, the following effects overlap during dry ice blasting:

- "Abrasive effect" due to the kinetic energy of the pellets and the force of their impact.
- "Thermal shock" due to the large temperature difference between the component surface and the dry ice stream.
- "Expansion shock" due to sublimation of dry ice pellets on the surface.

According to [11], about 60% of the effect of the dry ice blast is due to the thermal effect and 40% to the kinetic energy of the pellets. The thinner the component and coating, the lower the thermal shock. The removal rate also decreases if the component has already been severely undercooled by previous blasting.

Investigations by [2, 12] have shown that changes in the surface structure of aluminium and brass base materials are to be expected, while no influence on the surface of stainless steel was found.

9.3 High Pressure Water Jet Technology

The term "high-pressure water jet" is not exactly defined. In literature and in practice, the terms high or ultra-high pressure water jet technology or jet cutting are often used.

The criteria of operating overpressure and blasting medium can serve as a basis for the subdivision of pressurized water jets. Typical pressure limits of water jetting are about 50 MPa (500 bar) and 200 MPa (2000 bar). The pressure range between 70 and 170 MPa is often referred to as the high pressure range. Pump pressures in excess of 200 MPa are considered ultra-high pressure water jetting [13].

Tap water is usually used as the abrasive. Under certain conditions, e.g. additives or inhibitors, but also insoluble additives can be added for additional abrasive support.

9.4 Paint Stripping with Laser Beam

Compared to the conventional methods of mechanical and chemical paint stripping, laser paint stripping offers the following advantages in principle [14]:

- non-contact and therefore base material friendly process
- highest precision during machining, depth of cut controllable in the μm range
- low heat influence on the base material
- free of abrasives and chemicals
- suitability for online process control and regulation due to the good controllability of the laser beam sources
- good automation capability.

Two different classes of lasers can currently be used as laser beam sources for industrial decoating [15]:

- Solid-state laser (e.g. Nd:YAG laser)
- Gas discharge laser (e.g. CO_2 laser or excimer laser)

CO_2 and Nd:YAG lasers have been used successfully for many years for a wide range of industrial applications in surface technology. With wavelengths of 10.6 μm (CO_2) and 1.06 μm (Nd:YAG), they belong to the category of infrared lasers. Excimer lasers emit ultraviolet light with a wavelength between 0.2 and 0.35 μm and are therefore particularly suitable for special applications in microtechnology.

Paint stripping with infrared laser radiation is a thermal process, i.e. the radiation energy is absorbed by the paint layer to be removed and converted into heat. This results in localized heating above the decomposition temperature of the organic paint binder. The decomposition products enter the gas phase or flow off as smoke particles.

The ultraviolet radiation of excimer lasers is strongly absorbed by most paint systems (absorption rate > 90%). Their photon energy is high enough (3–5 electron volts) to specifically break electronic bonds in the polymer molecules of the paint layer in addition to abruptly heating the top coat. The combination of this photolytic effect and the formation of a hot material vapor plasma achieves an efficient stripping result, especially with thin paint films (up to about 10 μm).

Laser paint stripping is particularly economical for small areas and thin paint film thicknesses. Currently, a paint stripping speed of more than 2 cm^2/s at a film thickness of 50 μm can already be achieved with a 100 W paint stripping laser [15]. For large surfaces and coating thicknesses of more than 100 μm, laser paint stripping is generally no longer economical.

9.5 Paint Stripping with Plasma

Low-pressure gas plasma can not only be used for the pre-treatment of plastic surfaces before bonding or painting, it can also be used for stripping processes. The process depends on the selection of gases and the structure of the base material.

Low temperature is a plasma gas discharge under reduced pressure. In this process, energy is supplied to a process gas or a gas mixture in an evacuated chamber by applying an electric field. This leads to ionization and fragmentation as well as to the excitation of the gas flow. The activated and, depending on the gas selected, very reactive or aggressive gas reacts with the exposed surfaces of the material to be treated. The evacuated gases are characterized by excellent gap mobility, which allows a very homogeneous treatment even with complicated surface geometries [16].

Since only a small percentage of the reactive gas is used directly for the surface process, only non-critical, gaseous degradation products are produced as residues in addition to these gases.

The intensity of the attack on the surface can be controlled by the choice of process gas and by changing the parameters power, pressure, gas flow, process time and temperature. Thus, the paint removal can also take place without changing the basic material properties. The typical cleaning process is achieved by burning the organic layer. These are converted into water, nitrogen oxides and carbon dioxide and desorbed. The surface temperatures during the process are between room temperature and 50 to 70 °C, so that most materials can be treated in this process. Depending on the process source and the residence time in the chamber under process conditions, the temperature can also be significantly higher, which may well be desirable due to the accelerated removal of the contamination.

By using oxidizing and reducing gases, it is possible to completely remove the thinnest layers of paint, release agents or impurities from metals, plastics, glasses and ceramics. Even pocket holes and undercuts are cleaned completely.

Since the process takes place in a vacuum, paint stripping in low-pressure gas plasma is possible exclusively in a batch process. After loading the chamber on trays or rotating baskets, pumping is performed and the process gas is fed into the chamber. The plasma is then ignited by switching on the various frequencies – in the GHz, MHz or KHz range. The process can be easily controlled via a inspection or temperature control. After paint stripping or cleaning, the chamber is ventilated and discharged.

Pigments or other inorganic components remain on the surface in the form of dust. This dust must be removed before further processing of the parts.

9.6 Inductive Paint Stripping/Vortex Paint Stripping

With inductive paint stripping, only an extremely thin layer of the surface is heated, and only to the point where the paint becomes soft and loses its adhesion. While still warm, the workpiece is pulled through a brush arrangement to remove the soft paint. The arrangement of the brushes can be adapted to the respective workpieces and are either stationary or rotating or move linearly to the workpiece surface. For complex shaped parts, a vacuum blasting unit is provided instead of the brushes.

The advantage of the process is a very fast paint stripping which is indicated with a total time of only 8 s [17]. The process can be used to remove all types of paint, and partial paint stripping is also possible.

9.7 Choice of Paint Stripping Process

Paint stripping should be fast, simple and cost-effective. Depending on which material and which coating are used, different processes and means are available. The decisive factor for the customer or paint stripper is which process delivers the desired quality at the lowest cost.

A checklist for the selection of the appropriate process, but also in solving the problem itself paint stripping or to the service provider [18], can look like this.

1. How much parts/area to be stripped?
2. What are the advantages of the different methods?
3. Where is a post-treatment of the surfaces necessary?
4. What waste products are produced?

In the case of chemical paint stripping, it is necessary to clarify [1]:

- Paint stripping effect sufficiently fast
- Attack of the base material
- occupational safety specifications
- Waste disposal ensured and affordable
- Plant technology proven and affordable

Table 9.2 Paint stripping processes and necessary after-treatments or resulting waste products [20]

	Post-treatment					
Procedure	Blasting	Blow off	High pressure water	Ultrasound cleaning	No post-treatment	Waste products
Mechanical						
Maximum water pressure					+	Cured paint particles
Wet blasting					+	Blasting medium contaminated with paint
Glass Bead Blasting	–	–	–	–	+	Blasting medium contaminated with paint
Shot blasting	–	–	–	–	+	Blasting medium contaminated with paint
Dry ice blasting	–	–	–	–	+	Fine paint particles
Thermal (pyrolysis/cold)						
Laser paint stripping					+	Ash Decomposition products
Fluidized Bed	+	+	+	+		Ash residue
Carbonization gas oven	+	+	+	+		Ash residue
Liquid nitrogen	+	+	–	–	+	Cured paint particles
Chemical						
Chemical immersion bath paint stripping			+	+		Paint sludge
Paint stripping paste			+	+		Paint sludge, paste residues

- Suitable space available
- Suitable staff available and trained
- Cost comparison: in-house paint stripping versus contract paint stripping
- Weighting of flexibility with own paint stripping

Table 9.2 gives an overview of the paint stripping processes for which post-treatment with a second process is necessary and where the waste products are produced.

To guarantee a certain quality also in paint stripping, a quality label for paint stripping was founded by paint stripping companies and paint stripping chemical manufacturers – the Qualistrip [19]. The aim of this quality label is to ensure that the paint stripping process, the agents and the result of the paint stripping provides a surface that is suitable for coating.

References

1. Antony, W.: Entlacken von beschichtetem Aluminium und Stahl, GSB Schulungskurs Vorbehandlung. Schwäbisch Gmünd (2001)
2. Visser, A., Haberland, J., Budde, F.H.: Strahlen mit Trockeneis, Teil 1. JOT 2 (1998)
3. Interne Information der Fa. ESC GmbH, Daimlerstr. 15, 72351 Geislingen
4. Werner, T.: Chemisch entlacken mit mechanischer Unterstützung. JOT 5 (1998)
5. Lonergan, J.M.: Dry-ice blasting. In: Aerospace Engineering, 3, Bd. 13, Nr. 3. Warrendale Pennsylvania (1993)
6. Bhagat, R., Krueger, B.: Cleaning energized equipment using dry ice pellets. Proceedings of the 6th International Conference on Transmission and Distribution Construction and Live-Line Maintenance, 09, IEEE, Piscataway, N.J., USA (1993)
7. Lewis, M.W.: Contamination Control and Reduction Through Dry Ice Media Blasting; Society of Vacuum Coaters, 35th Annual Technical Conference Proccedings, Baltimore, MD, USA (1992)
8. N. N.: Strahlen mit Kohlendioxid. Umweltgerechte und materialschonende Strahlmethode. Industrieanzeiger. **113**(51) (1991)
9. N. N.: Strahlmittel, das sich in Luft auflöst, Kohlendioxid als Strahlmittel. Metalloberfläche. **45**(5) (1991)
10. Stratford, S.: Perspectives on surface preparation with CO_2 blasting, Conference SSPC 90: Sights, Sounds and Visions of the future, Nashville, USA, Publ.: Steel Structures Painting Council, Pittsburgh, Penn. USA, Bd. 9205-72-0214 (1990)
11. Donath, S., Niewolik, R.: Strahlen mit CO_2-Trockeneis, Vortrag anlässlich einer VDI-Tagung, Düsseldorf (1993)
12. Visser, A., Haberland, J., Budde, F.H.: Strahlen mit Trockeneis, Teil 2. JOT. **38**(3), 44–49 (1998)
13. Momber, A.: Entlacken mit der Hochdruck-Wasserstrahltechnik. JOT. **41**(5), 18–22 (2001)
14. Lotze, R., Birkel, J., Wissenbach, K.: Entlacken mit Laserstrahlung – Neue industrielle Anwendungen. JOT. **39**(8), 44 (1999)
15. Büchter, E.: Lackabtrag mit Festkörperlaser rechnet sich. JOT. **40**(3), 110–116 (2000)
16. Schulz, J.: Entlackung empfindlicher Oberflächen im Plasma. JOT. **41**(5), 24–25 (2001)
17. Bauer, H.: Induktives Entlacken – effizient und umweltfreundlich. JOT. **41**(1), 28–31 (2001)
18. Couvreux, U.: Make or Buy – Selbst entlacken oder zum Dienstleister. JOT. **38**(11), 45 (1998)
19. N. N.: Qualitätssiegel für Entlackung. JOT. **8** (2000)
20. Allmüller, H.: Möglichst einfach und mit dem richtigen Verfahren entlacken. Besser Lackieren. **2**(2) (2002)

Environmental and Energy Aspects 10

10.1 Emissions from the Stoving Process [1]

After application, powder coatings must be cured at temperatures between 130 °C and 240 °C, depending on the binder system and the substrate to be coated. During this process, small mass losses in the form of emissions may occur in the curing oven.

In the curing of powder coatings, a distinction is made between two types of chemical crosslinking reactions:

Polyaddition (addition polymerization)

- Epoxy powder coatings
- Hybrid powder coatings (polyester/epoxy)
- Polyester powder coatings (hardener: aromatic glycidyl esters)
- Polyurethane powder coatings
- Acrylic powder coatings

Polycondensation (condensation polymerization)

- Polyester powder coatings (hardener: β-hydroxyalkylamide)

In the case of polyaddition, the crosslinking reaction does not result in any release products and the mass loss during baking is usually 0.5–1.0%. Due to the crosslinking reaction during polycondensation, additional water splitting occurs. This can result in a mass loss of 1.5–3.0%.

When using externally blocked polyurethane hardeners (e.g. isocyanate hardener blocked by ε-caprolactam), an additional cleavage of the blocking agent occurs before the actual crosslinking reaction.

J. Pietschmann, *Industrial powder coating*,
https://doi.org/10.1007/978-3-658-37592-8_10

The stoving losses during the curing of addition-curing powder coatings with hardeners free of decomposition products are largely composed of water (approx. 80%) as well as highly volatile organic compounds (approx. 20%). Pigments, fillers and polyester resins may contain water, which evaporates in the curing oven.

Examples of volatile organic substances:

- Low molecular weight resin and hardener components and residual monomers
- Additives (such as benzoin, waxes)
- Matting agent
- Levelling agent

The volatile organic substances can condense over time on colder zones of the stove or exhaust stacks.

These volatile organic substances in the exhaust air have been investigated in the past for their hazard potential for the environment and the workplace, but it was not possible to determine them precisely. Therefore, no information on MAK values can be given. Dioxins, furans, formaldehyde and other toxic components could not be detected. Therefore, a health hazard can be excluded if the plant technology is intact. In order to prevent a possible odour nuisance or irritation of the respiratory tract and mucous membranes, it is recommended to generally prevent the emission of furnace exhaust gases in factory buildings. In this way, possible contamination of the ambient air in working areas can be prevented in advance.

10.2 Life Cycle Assessment of a Powder Coating

The sustainability of a product or a process is increasingly being addressed and used as a decision criterion alongside the economic and qualitative criteria for the application of a process or product.

In order to be able to assess the environmental impact of a coated object, the overall picture must be considered. Therefore, an assessment according to the "cradle-to-grave"or "cradle-to-cradle"approach is necessary. This includes the impact of the product during manufacture, during its useful life and at the end of its life. As this data is often difficult to obtain or not available at all, most life cycle assessments (LCA) define a limit to the scope of the calculations.

In their study, Jos Verlaak and Dave Morris [2, 3] used real data from a representative group of Dutch aluminium powder coaters as the basis for their data. The powder formulations referred to powders in a shade of white (off-white) for exterior use. For the production of the resin and the powder coating, data from a respective manufacturer were used. All other data on pigments, fillers and raw materials for the resin could be calculated from databases. The scope of this LCA is the cradle-to-gate approach as shown in Fig. 10.1.

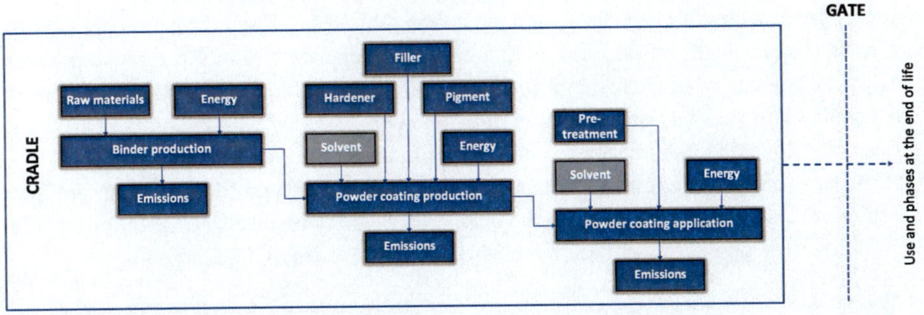

Fig. 10.1 Cradle-to-gate approach to life cycle assessment [2]

In this study of the carbon footprint were considered: The impact of pre-treatment, powder coating, curing, waste and transport.

As a result, based on the data provided, the carbon dioxide emissions caused directly and indirectly and other greenhouse gases were balanced. The reference value was 1 m2 of coated aluminium sheet, pre-treated with a yellow chromate containing Cr (VI). This is usually stated in tonnes of CO_2 equivalents, or t CO_2-eq for short.

The calculation resulted in a footprint of 2.36 kg CO_2-eq per m^2.

This figure is made up as follows:

Curing	1.5 kg CO_2-eq/m^2
Coating/lacquers	0.381 kg CO_2-eq/m^2
Pre-treatment	0.21 kg CO_2-eq/m^2
Waste	0.18 kg CO_2-eq/m^2
Transport and packaging	0.1 kg CO_2-eq/m^2

10.2.1 Application Parameters and their Influence

Energy use has the greatest influence here. 1.5 kg CO_2-eq/m^2 are attributable to energy use. Of this, 1.1 kg CO_2-eq comes from natural gas combustion and the rest from electricity consumption. It is estimated that a reduction of 40 °C in the curing temperature can reduce CO_2 emissions by 0.25 kg. Further reductions can be achieved through optimal oven design and efficient utilization.

10.2.2 Coating Material and Waste

The powder coating itself is responsible for 0.38 kg CO_2-eq/m^2. This value refers to the amount required for the coating, plus the overspray. The most significant reductions can be achieved by reducing the overspray and the coating thickness. Even if the powder coating is circulated, this is not infinite and there are losses of approx. 5% of the sprayed powder.

1 kg of powder coating, for exterior use in off-white colour has a CO_2 equivalent of 3.6 kg CO_2. This depends on the formulation, and here especially on the binders used, the pigments followed by the energy required for production.

Powder coatings for exterior use generally have a lower carbon footprint than products for interior use. This is mainly due to the bisphenol A epoxy resin. The calculations of four commonly used powder formulations resulted in the following CO_2 footprints:

Exterior white, HAA	3.5 kg CO_2-eq/kg
Exterior grey	2.7 kg CO_2-eq/kg
Exterior, clear varnish	3 kg CO_2-eq/kg
Interior white, 50/50 hybrid	4.1 kg CO_2-eq/kg
Epoxy powder coating	6.47 kg CO_2-eq/kg [4]

Options for reducing the carbon footprint of powder coatings include reducing the pigment content, especially TiO_2, the resins used and the energy used to produce the coating material.

10.2.3 Resins

Figure 10.2 shows a breakdown of the shares of a polyester resin in the $_{CO2\ footprint}$. The aromatic acids used account for the largest share, followed by the diols and the energy required for production. The use of renewable-based raw materials is one possible way to reduce the CO_2 footprint. However, the use of bio-based raw materials does not automatically mean a low CO_2 footprint. Often much more energy has to be used to convert the plant materials into the required raw materials. In addition, a conflict can arise in relation to the food cultivation or valuable soils, such as rainforests, are used for this purpose.

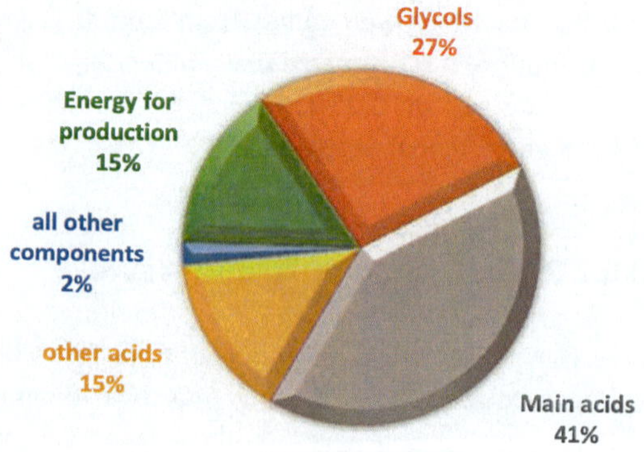

Fig. 10.2 CO_2 footprint of a polyester resin [2]

10.2.4 Pre-Treatment

The ecological footprint of all chemicals used for pretreatment is relatively low for a coating containing CrVI, at 0.2 kg CO_2-eq/m^2. However, if the human toxicity (CLM method) is considered in the assessment, the impact of the pretreatment is significantly higher.

10.2.5 The Idea of Life Cycle Assessment and its Main Features

Life cycle assessment is a comprehensive approach that systematically records, evaluates and presents as many environmental impacts as possible that are related to operational activities. In doing so, not only the impacts occurring in the company are considered, but also use/usage and disposal. In addition to defining the period under consideration, the life cycle inventory primarily records the main material flows, such as purchases of raw materials, semi-finished goods, finished goods and emissions to soil, water and air. Environmental indicators are used in the impact assessment, such as GWP, eutrophication, KEA cumulative energy input or Eco Indicator 99 for damage modelling.

Essential characteristics of a life cycle assessment method

- The recording of material and energy flows (inputs and outputs) as well as resource requirements (area, minerals, etc.) and other environmental impacts such as noise, odour, emissions
- The recording of these variables over the entire life cycle, from cradle to grave, "cradle to grave".
- The evaluation of these influences according to ecological criteria.

Life cycle assessments are thus quantitative methods for assessing the environmental impacts of human activities. They provide information on the effects of a product, a process, an operation or a measure on the environment.

If companies want to act sustainably, they need detailed information on the environmental impact of their products or processes in order to make the appropriate decisions or to be able to argue their case to stakeholders. With the help of a life cycle assessment, environmental impacts are calculated and quantified.

Today, sustainability is defined according to the Brundtland Report (Rio 1987). Sustainable development must take into account the dimensions of society, economy and environment.

But the term, or rather the action, is much older. The roots of sustainability lie in Saxony. The principle of sustainability was first formulated around 300 years ago. Hans Carl von Carlowitz, chief miner at the court of the Elector of Saxony in Freiberg (Saxony, Germany), demanded in his work "Sylvicultura oeconomica" in 1713 that only as much wood should be felled as could grow back through planned reforestation, sowing and planting, and is therefore regarded as the creator of the concept of sustainability in forestry. In forestry, the term was taken up and translated into English and French, among other languages. In the course of the twentieth century, the English sustainable became sustainable development as a result of the international conferences of the United Nations.

In 1997, the EU explicitly formulated three pillars of sustainability in its Contract of Amsterdam. According to these, sustainability encompasses not only the natural heritage, but also economic achievements and social and societal benefits, such as democratic structures and a fair distribution of income [5].

10.3 Recycling of Powder Coatings [6]

In principle, powder coatings should be used as efficiently as possible in operation and the aim must be to pass on as little as possible as "waste" for recycling. Depending on the coating system and the settings, the colours and colour changes used, the geometry of the workpieces to be coated, the "overspray"can be between 1 and 50%. Here, overspray also includes all powder coating residues from the system and fine dusts from the filter systems. In addition to these, there are further quantities from stock, incorrect batches and overstocked goods. The value of these quantities, which do not justify incineration or landfill, should not be underestimated.

Powder coating residues can be recycled and used as a modified binder in non-woven insulating materials for cars, in sound-insulating building materials or in household appliances.

The Porocom system [7] is a process for the use of waste powder in noise-insulating building materials. The powder coating is melted around granulate/grains to produce noise-insulating elements for interior and exterior use, e.g. in road construction, façade elements or also for sports or concert halls. Recycled materials such as broken glass or slag are used as granulate. The type of powder coating does not play a role in this application.

GRP laminates

Waste powders of certain types are used for fibre bonding of glass fibre reinforced plastic parts (GRP). Possible applications can be found in the manufacture of silos, kayaks and tanks. The parts can then be painted over. This application concerns rather local customers for small quantities.

Textile fibre bonding

This method is by far the best recycling process in terms of environmental friendliness and economic efficiency. Highly reactive epoxy systems are produced here on the basis of used powder coatings. In this way, environmentally harmful phenolic resins and latex can be replaced in the automotive industry and in the manufacture of carpeting.

Powder coating

One solution for certain waste powders is to reuse them as varnish. Especially in the case of overstocked goods or special colour shades, the use as a varnish can be an economical solution. The extrusion and grinding of fine dusts into new powder coating is also theoretically possible. However, mixing different manufacturers, colour shades and types in the residual powder leads to incompatibilities and thus to inferior products. This application is occasionally used by manufacturers for the reprocessing of their own products or the old powders of the customers are taken back, separated by type.

Other applications can be found as an additive in the production of polyurethane or polypropylene sheets, for pressing molded parts, in the production of nonwovens for sound reduction. These applications are often developed ad hoc and individually for individual companies that find a buyer in their niche.

One recycling option, albeit with the lowest recovery level, is energy recovery. Powder coatings can be mixed with sawdust and other high-calorific waste materials to form a fuel. The fly ash can also be incorporated into cement as an aggregate.

10.4 Energy Management in the Paint Shop

Rising energy prices, cost pressure and the desire for sustainable resource management require new solutions. Painting systems are among the largest energy consumers in the production process.

Regardless of which options are used to increase energy efficiency, the following applies: An economic concept requires a careful analysis of the system energy balance [8].

10.4.1 Analysis of Processes and Optimisation

A considerable part of the energy used in the coating process is accounted for by the heating of the pretreatment baths, the air conditioning as well as the dryers and curing ovens. Optimizing the process parameters can already result in significant savings here.

Pre-treatment: Up to 15% of the thermal connection value of the pre-treatment can be saved if the temperature of the baths is reduced by 5 °C. It should be noted that reliable degreasing, depending on the type of greasing, requires a certain temperature/action time so that the greases can be dissolved. Another approach can be to reduce the number of zones.

Powder and powder application: The use of low-temperature powder coatings also makes a positive contribution. Up to 15% of the original energy consumption can be saved if the baking temperature is lowered by 20 °C.

Changing the temperature and humidity in the powder booth enclosure can also save energy. For example, if a temperature of 23 °C and a humidity of 50% are specified, these parameters can usually be shifted to 27 °C and 65% humidity respectively without any loss of quality. This means that less energy is required to maintain constant climatic conditions.

10.4.2 Saving Energy Through System Components and Control Technology

The use of energy-saving plant components, such as efficiency-optimised IE2 electric motors, formerly EFF1, is not the question. Refrigeration systems, cooling units with direct evaporators offer significantly better efficiencies compared to chillers. Likewise, free coolers in the refrigeration system, which are used among other things to cool the powder booth enclosure, are a good solution.

Frequency converters offer advantages for electric motors that are often operated at partial load. This is particularly the case with drives for fans of supply air units. Depending on the filter condition, they compensate for different operating points and make it possible to dispense with damper valves.

A frequency converter is not advantageous for pumps that are operated at only one defined duty point. On the other hand, it proves to be advantageous with a variable flow rate or a variable pressure. If the spray pressure is to be set from the control station, a frequency converter is indispensable.

The highest savings can be achieved in terms of control technology through optimised standby operation. A pause control for pumps and fans significantly reduces energy consumption. The control system reduces their output accordingly during work breaks or the transport of empty hangers, as well as during different coating times of individual large parts.

Process parameters adapted to the mass throughput also allow effective savings: These include, for example, an optimally adjusted recirculation air volume in dryers and cooling zones whose air volume varies depending on the workpiece temperature. On the design side, weight-optimized product carriers offer a great deal of savings potential.

10.4.3 Minimising Heat Losses: Heat Recovery

In order to minimize heat losses in the dryer, A-locks are used today. The parts enter the dryer from below at a gradient angle of 30°, pass through it and leave the dryer again at entry level. The special design of the dryer makes use of the physical behaviour. Compared to conventionally designed dryers, this results in only a small amount of heat escaping when the parts enter and leave the kiln, and thus in only a small loss of energy.

In the cooling zone, the total installed cooling air volume can be drastically reduced by cascading the cooling air in two or three stages. This also leads to clear savings in energy consumption.

If all the above measures have been exhausted, heat recovery additionally reduces consumption. This can be achieved by using heat exchangers and heat pumps. For heat recovery, there are customised two-stage energy concepts with heat exchangers and heat wheels. The exhaust air of all heat emitters is used.

As an example for energy optimization, a powder coating plant with a six-zone spray pretreatment, two powder booths without enclosure, Power&Free conveyors, infrared preheating zones and a powder baking oven with cooling zone as well as water dryers with cooling zone was taken as a basis.

The analysis showed the following result:

- At 82%, thermal demand accounts for the largest share of energy costs.
- The pretreatment, the dryers with cooling zone and the powder booths each consume one third of the electrical energy.
- Dryers dominate thermal consumption with 75%.

The largest heat content is provided by the infrared gelation zone with 5000 Nm³/h exhaust air, whose temperature is 185 °C. The integrated heat recovery enables heating in the pre-treatment. For this purpose, the exhaust air is passed through a heat exchanger. There it cools down to around 100 °C and enters the heating circuit of the pre-treatment. If the pre-treatment heating circuit does not require any heat, the exhaust air can be fed to the second stage via a bypass. .

Any missing heating energy is supplemented from the district heating network in this plant. By using the waste heat from the infrared gelation zone to heat the pretreatment, 150 kW/h can be saved. The investment in the heat exchanger pays for itself after only 1.9 years.

In a second step, the 5000 Nm³/h exhaust air from the infrared zone is combined with other exhaust air streams. After heat recovery, the exhaust air has a temperature of approximately 100 °C for heating the pre-treatment.

- Exhaust air of the water dryer, 1000 Nm³/h, 120 °C
- Partial flow of exhaust air from the cooling zone, 20,000 Nm³/h, 40 to 50 °C Extraction of hall air from under the ceiling, 29,000 Nm³/h, 30 to 45 °C.

The temperature of the total air flow of 55,000 Nm³/h is approximately 55 °C. The flow is then filtered, fed via a heat wheel to a fan and transported to the outside. In counterflow to this, outside air is drawn in, which also first passes through a filter and then the heat wheel. In the process, the supply air flow, which is 47,000 Nm³/h according to the total air budget, is heated to around 20 °C before it is discharged into the hall for heating. If required, the air can be reheated via the heating coil. With an operating time of about 1800 h per year, the heating wheel enables energy savings of 470,000 kWh per year.

Numerous processes generate waste heat at a low temperature level, the recovery of which via conventional heat exchangers is uneconomical. Here, the use of heat pumps is a suitable alternative. The key factor here is the so-called Coefficient of Performance (COP) of the heat pump. This value indicates the ratio between the electrical power supplied to the compressor and the resulting heat output. Ideally, it should be between four and five. For example, a COP value of 4.5 means that a heat output of 450 kg/W is achieved with 100 kg/W of electrical energy supplied. The COP value is mainly influenced by the refrigerant used. Figure 10.3 shows an example of a concept for heat recovery from various emitters for heating hall air and pre-treatment.

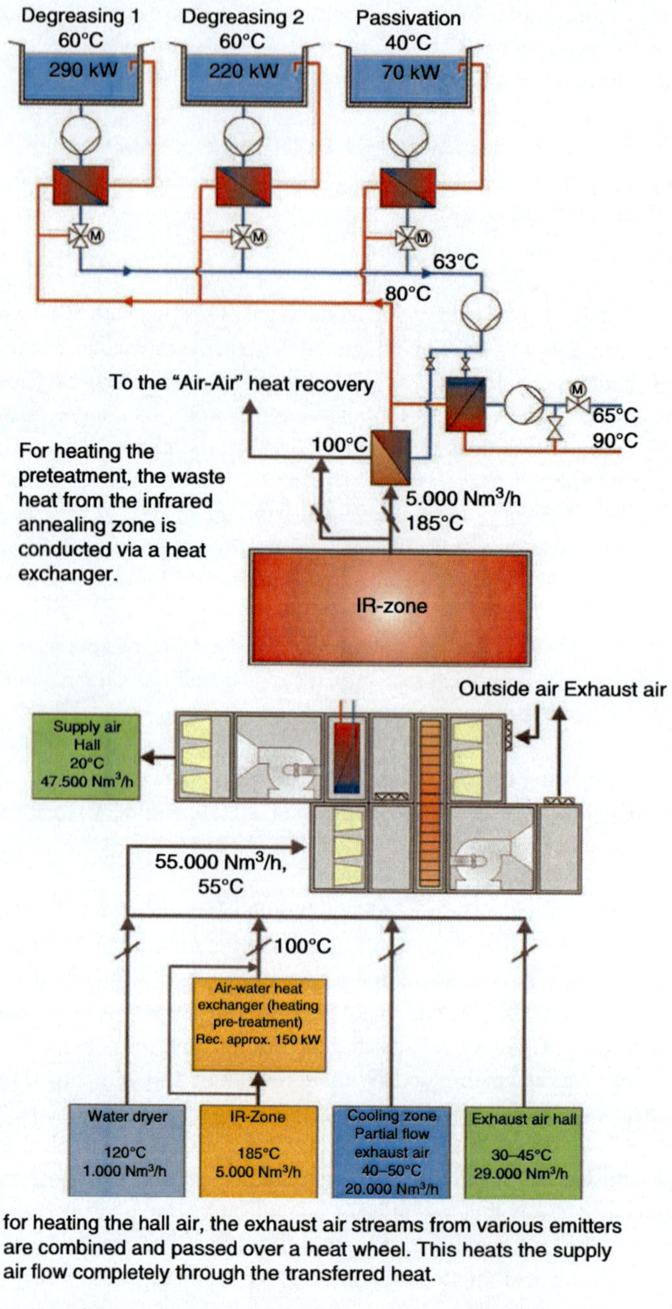

Fig. 10.3 Example of a heat recovery concept

10.4.4 Energy Comparison of a Liquid and Powder Coating System [9]

The energetic comparison of a wet and a powder painting process is carried out using the example of a painting plant for metal office furniture. There are no differences here with regard to pretreatment. In contrast to the wet painting booths, which are operated in fresh/exhaust air mode, powder painting can be carried out in recirculating air mode. This eliminates the heating energy requirement for fresh air heating. In powder coating, the coating is applied in one operation. This saves the intermediate drying and the second coat application. Despite the higher baking temperature with powder coating, there is a lower energy requirement for curing. Figure 10.4 compares the energy consumption of the two processes (Bayerisches Landesamt für Umwelt 2006).

The most important measures and recommendations for optimizing paint shops are listed in the following checklist.

1. Planning and organisational measures
   - energetic evaluation of new acquisitions on the basis of VDMA 24378
   - energetically best process in use for the respective application
   - alternative painting process possible, e.g. replacement of the atomizing water-based painting process by powder coating
   - examination of new procedures
   - optimized operation with regard to capacity utilization
   - switching off pumps and fans during colour changes or stopps
2. Optimization of the pretreatment
   - Immersion pretreatment system energetically more advantageous than spray pretreatment
   - thermally insulate heat-carrying components of the system

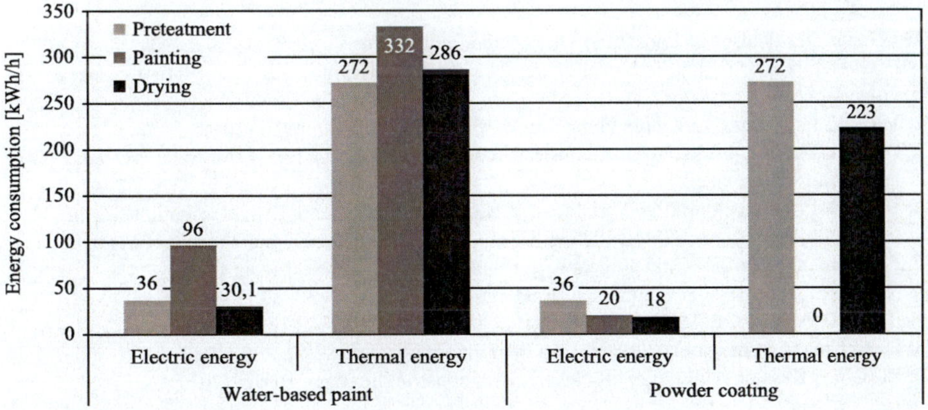

Fig. 10.4 Energy comparison of a water-based and a powder coating system [9]

  - Electrolyte heating (pre-treatment) heating: condensing boiler technology and/or a heat cascade
  - Use of low temperature cleaning agents
  - With a vapor suction the use of a condenser with heat recovery is recomented for the return of the exhaust air into the hall
  - Are damper valves still in use for injection pressure control?
  - Use of a combined heat, power and cooling system
3. Optimization of the adhesive water drying
  - Optimize the temperature level in the adhesive water dryer
  - Avoidance of adhering water residues, drops and puddles
  - A-zone or automatic doors
  - Heat recovery of the dryer for bathroom heating
  - Dehumidify circulating air to be able to lower the drying temperature
4. Optimization of the paint application
  - Reduction of the air sinking speed to the necessary minimum
  - Conversion of manual painting to automatic painting
  - Heat and moisture recovery of the exhaust air in fresh/exhaust air mode
  - Adaptation of the openings to the workpiece to reduce the air volumes in powder coating booths
5. Optimization of paint drying
  - Reduction of temperature; drying with dehumidified air
  - alternative drying and curing methods (induction, infrared curing, UV curing or microwaves)
  - Low temperature powder coating

Literature

1. Verband der deutschen Lack- und Druckfarbenindustrie e. V. Mainzer Landstraße 55, D-60329 Frankfurt; www.WirSindFarbe.de; November 2017; Emissionen aus dem Einbrennprozess von Pulverlacken
2. Verlaak, J., Morris, D.: Kleine Fußabdrücke. Farbe und Lack 06.2016 (2016)
3. DSM carbon footprint study for industrial coatings applied on a metal substrate: www.dsmpowdercoatingresin.com
4. Umwelt Produktdeklaration Verband der deutschen Lack- und Druckfarbenindustrie e. V. – Epoxid Pulverlack, nach ISO 14025 und EN 15804, Institut Bauen und Umwelt e. V. (IBU)
5. Pietschmann, J.: MAS Umwelttechnik und -management. Master Thesis (2014)
6. Nelde, A.: Altpulver – recycling statt Entsorgung. JOT. **54**(10), 30–33 (2014)
7. POROCOM BELGIUM, Knaagreepstraat 5, 8890 Moorslede, info@porocom.be
8. Hanschke, M.: Energiemanagement auf dem Prüfstand. JOT. **50**(7), 2–5 (2010)
9. Blesl, M., Kessler, A.: Energieeffizienz in der Industrie. Springer, Berlin (2017)

Further Literature on the Subject of Sustainability

Lexikon der Nachhaltigkeit, Aachner Stiftung Kathy Bays; kontakt@nachhaltigkeit.info
Dinkel, F.: Vorlesungsskript Ökobilanzen, März 2012 (2012)
DIN EN ISO 14044: Umweltmanagement, Ökobilanz Anforderungen und Anleitung, 2006-10, Beuth Verlag, Berlin
DIN EN ISO 14040: Umweltmanagement; Ökobilanz-Grundsätze und Rahmenbedingungen, 2009-11. Beuth Verlag, Berlin
Ecoinvent Datenbank.: www.ecoinvent.org. Technoparkstrasse, Zürich

Images used in Chap. 10

Eisenmann, Tübinger Str., Böblingen, D: Abb. 10.3
Blesl, M., Kessler, A.: Aus Energieeffizienz in der Industrie. Springer, Berlin (2017): Abb. 10.4